Number Systems
A Path into Rigorous Mathematics

Anthony Kay

CRC Press
Taylor & Francis Group
Boca Raton London New York

CRC Press is an imprint of the
Taylor & Francis Group, an **informa** business

A CHAPMAN & HALL BOOK

First edition published 2022
by CRC Press
6000 Broken Sound Parkway NW, Suite 300, Boca Raton, FL 33487-2742

and by CRC Press
2 Park Square, Milton Park, Abingdon, Oxon, OX14 4RN

Cover image: Postage stamp commemorating 150th birth anniversary of Richard Dedekind, whose ideas are fundamental to much of the material in this book.

ISBN: 9780367180652 (hbk)
ISBN: 9780367180614 (pbk)
ISBN: 9780429059353 (ebk)

Typeset in CMR10 font
by KnowledgeWorks Global Ltd.

Contents

Preface and Acknowledgments

For me, being a mathematician is not a profession; it is a genetic condition. One of my earliest memories, from age 3 or 4, is that having realised that the decimal system enabled me to continue counting indefinitely, I would silently count to myself while getting on with whatever I was doing; on one occasion I got to 1,112 before being interrupted, and this number has stayed in my memory ever since that day. So from an early age I appreciated mathematical concepts as abstractions; but later, during my last two years at school, a brilliant Physics teacher, Barry Jackson, showed me how exciting it can be to apply mathematics to problems in physics. This set me on a career as an applied mathematician, mainly working in fluid mechanics.

However, I never lost my love of numbers, or my appreciation of mathematical beauty for its own sake. My colleagues at Loughborough University understood this: when a new first-year module on "Numbers" was introduced to the syllabus in 2001 by Andrew Osbaldestin in his capacity as Teaching Coordinator, he decided that I was the best person to teach it. He also very kindly provided a rather vague module specification; this left me free to read around the subject and then decide for myself what I was going to teach. Chapters 3 to 7 of this book, on the rigorous theory of number systems from the Natural Numbers to the Real Numbers, contain all the material that I taught in this module, and a substantial amount of further material that could not be included in a module of 22 lectures. Chapters 8 to 10 contain material on further number systems, in particular the Complex Numbers, which I would also have included if more time had been available.

Unlike every other module that I have taught (and there have been many!), I did not feel that any of the existing books on this subject satisfied all the requirements for students: covering all the material in the module, set at the right level of mathematical sophistication, and clearly presented. So after some years, I conceived the idea of writing a book myself: I wanted to cover the theory of number systems rigorously, assuming no other mathematical knowledge apart from Naïve Set Theory; and the book should be suitable for students with no previous experience of the rigorous development of mathematical theory from axioms. So the rigour would need to be tempered with explanations that appealed to students' intuition. Just when I eventually felt ready to start writing my book, there appeared Ian Stewart and David Tall's *The Foundations of Mathematics* [Second edition], which does a brilliant job of introducing students to the rigorous theory of number systems and much else. I almost wanted to abandon my project; but Stewart and Tall have a broader canvas than I intended to cover, while I wanted to go into more detail on many aspects. I can only hope that my work approaches the standard of clarity provided by those eminent authors. Whereas Edmund Landau, whose *Foundations of Analysis* is one of the earliest books to cover the theory of number systems, describes his book as being written in "merciless telegraph style", my pursuit of clarity tends to take me to the opposite extreme. I trust that my explanations will be regarded as thorough and unambiguous, rather than simply verbose!

Although the book grew out of a module for first-year university students, it does include some more advanced material. Here is a brief guide to what you will find in each chapter. Following the introductory notes in Chapter 1, the next chapter covers all the Set Theory that readers will be assumed to know in the remainder of the book. This is Naïve Set

Theory; in a book dealing with the theory of number systems, I did not want readers to first spend time getting to grips with the axioms of Set Theory. Chapter 3, on Natural Numbers, sets the pattern for how number systems will be treated throughout the book: after defining the numbers, we develop the theory of their binary operations of addition, multiplication, and exponentiation; then their order relations, properties of bounded sets, and cardinality are covered. Chapter 4 on Integers contains the first definitions of algebraic structures (groups and rings), and also introduces methods for solving inequalities, since these will remain essentially unchanged in subsequent number systems. Chapter 5 may seem to be a digression from the main thrust of the book: we introduce primes and factorisations, representations of numbers in any base (we avoid giving base ten its usual pre-eminence), and modular arithmetic, which will all reappear later in the book; but the only new number systems defined in this chapter are the finite rings and fields of integers modulo n. Chapters 6 and 7 return to the main objective of defining number systems and examining their properties, with the theory of Rational Numbers and Real Numbers involving some notably more difficult proofs than have been required earlier in the book.

If you notice a discontinuity in style between Chapters 7 and 8, that is because the earlier chapters are based on material that I had taught for many years, whereas the later chapters consist of material that I would have liked to teach but had not previously written any detailed notes for; and also, shortly after I had started on Chapter 8, I had to take a year off writing the book due to pressure of other work. Anyway, my approach to Complex Numbers is somewhat unconventional. First, I start by treating them as "just another quadratic extension": whereas Chapter 8 is devoted to quadratic extensions of the Integers and Rationals, (and includes material that might more commonly be found in the early chapters of a book on Algebraic Number Theory), Complex Numbers are introduced as a quadratic extension of the Real Numbers. Then a large part of Chapter 9 concerns the search for how to exponentiate complex numbers: other books either assume knowledge of exponential and trigonometric functions, whose properties actually require quite sophisticated knowledge of Analysis in the Real Numbers, or else they avoid the subject of exponentiation altogether. My approach defines geometrical/trigonometrical concepts from manipulations with numbers (involving quite a lot of work with sequences, and introducing the famous numbers π and e on the way), rather than using geometry to define properties of complex numbers. After this, we look briefly at even more complicated number systems in Chapter 10. Finally, since in my lectures I always liked to draw attention to the connections between what I was teaching and other topics, my Chapter 11 very briefly discusses how the theory of number systems leads on to broader topics in mathematics. I have in recent years encouraged students to use *self-explanation* when reading proofs and derivations: this is a strategy which has been found by Lara Alcock, Mark Hodds, and Matthew Inglis to improve students' comprehension, and I would like to thank these authors for providing me with the text for their guidance notes on self-explanation which are reproduced here as Appendix A.

There are snippets of information on the historical development of the subject scattered throughout the book, but no attempt has been made to give a coherent account of the history of number systems. For readers who want to explore this further, go first to the MacTutor History of Mathematics Archive, hosted on the website of the School of Mathematics and Statistics at the University of St Andrews, from where there are links to more detailed historical articles.

There are several features (or quirks) of the book that are worthy of comment here:

- Several concepts of much wider importance in mathematics are introduced just where they are first encountered in the development of the theory of number systems. So for example, isomorphism is defined in the chapter on Integers, just where the first example of an isomorphism (between the Positive Integers and the Natural Numbers) is encountered; convergence of sequences is defined shortly after we discuss repeating representations of rational numbers in decimal or other bases, because such representations are the first example of an infinite sequence that we meet.

- Some of the notation may be regarded as quirky, or even annoying, by some readers. Where my choice of notation differs from that used by the majority of authors, I have explained the rationale behind my choice in a note in the text or in a footnote.

- The book contains Exercises within the text of each chapter and Investigations at the end of most chapters. The Exercises are for the student to fill in some of the less demanding proofs of theorems, or to gain practice with important procedures. The Investigations, many of which are based on coursework set for students when I was teaching the subject, invite the student to explore more deeply and outside the main thrust of the text. Solutions to the Exercises, and hints relating to some of the Investigations, are available online at the book's website.

- Footnotes are indicated by Roman numerals,[i] because an Arabic numeral as a superscript can be confused with a power, and the font of special symbols such as asterisk, dagger, etc. is not big enough for my requirements in some chapters.

As is usual for an undergraduate text, the book does not contain any new research results, but does involve new presentation of existing knowledge. So unlike a research monograph, I have not cited references within the text, and the bibliography only includes other books. But this does not mean that I have not consulted anything outside these books while preparing this text. For the more advanced material in later chapters I have sometimes read the original research papers; lecture notes that other academics have kindly made freely available on the Internet have sometimes helped my thinking, and my ideas have sometimes even been stimulated by posts on Mathematics StackExchange. So I would like to apologise to all those people who share their knowledge freely in this way but are not named in this book, and thank them for their generosity in the true spirit of scientific openness.

Writing this book has sometimes seemed a rather lonely endeavour, but would not have been done without the encouragement of friends, family, and colleagues. It is traditional for authors of textbooks to thank their family members for their forbearance while the author was absorbed in the task of writing; but being the stereotypically obsessive, uncommunicative, and socially awkward mathematician that I am, I doubt whether my wife and son noticed much difference in my behaviour while I was writing the book; so I would just like to thank them for their tolerance of me at all times. As well as the published and internet sources mentioned above and in the bibliography, my thinking has been stimulated by colleagues in the Department of Mathematical Sciences and the Mathematics Education Centre at Loughborough University, who have always been a pleasure to work alongside, even when more senior management has not been so conducive to a good atmosphere. Regarding the latter, I would particularly like to thank Mark Biggs, who during his brief tenure as Dean of Science at Loughborough made life so unpleasant that I was pleased to accept the offer of going part-time on a teaching-only contract, which meant that I had the time to start writing this book. Finally I would like to thank the editorial staff at CRC Press, first for taking on this project and then for their encouragement and also forbearance when deadlines slipped.

---

[i]Like this one.

Chapter 1

Introduction: The Purpose of This Book

1.1   A Very Brief Historical Context

In the ancient world, mathematical ideas were developed for practical purposes: in relation to astronomical observing, which was vital for timekeeping (days, months, and years being defined by the positions of astronomical objects relative to observers on Earth); for accountancy, to keep records of possessions and for trading; and for surveying land. However abstraction, the consideration of mathematical concepts such as number, size, and shape independently of any application, also developed notably in ancient Greece. The idea of rigorous proof, that new concepts could be firmly and indisputably ascertained by a process of logical deduction based on previous knowledge, also appeared in ancient Greece. But the subsequent development of mathematics was not a smooth process of building theory rigorously on previously established knowledge; rather, concepts were often developed in a rather intuitive manner, and only much later proved rigorously. For example, the ideas of differential and integral calculus as originated by Isaac Newton and Gottfried von Leibniz in the late 17th century and further developed by Leonhard Euler and the Bernoullis in the 18th century were adequate for all applications needed at that time but involved poorly defined notions of infinitesimal (vanishingly small) changes in quantities. It was not until the 19th century that rigorous definitions of limit, continuity, convergence, and other notions that form the subject now known to mathematicians as Analysis were developed. All these notions are founded on the properties of numbers, but was there actually a rigorous definition of numbers? No; such definitions did not appear until the second half of the 19th century, in particular in the work of Richard Dedekind who asked as the title of one of his books, *Was sind und was sollen die Zahlen?* [What are numbers and what should they be?]. It is the objective of the present book to present a rigorous theory of numbers, not only for the intrinsic interest of the topic but also as a way into rigorous mathematics for students who have not previously encountered this approach.

1.2   The Axiomatic Method

We have already referred to the idea of new concepts being based on previous knowledge, but what is that previous knowledge based on? Clearly there needs to be a starting point. All mathematical reasoning must ultimately be founded on **axioms**. Axioms are statements which are defined to be true; some axioms are statements of what seems to be obvious, others are definitions of mathematical concepts or objects. In any case, they are the "rules of the game" which everyone accepts to be true. But we cannot set axioms arbitrarily; they need to give rise to "useful" mathematics, which relates to our intuitive concepts of number and all the other mathematical notions used in applications in natural sciences and many areas of human activity; and they need to be self-consistent, i.e. they must not allow contradictory deductions to be made.

Thus the axiomatic method which we adopt proceeds from the axioms by first proving a **theorem** based only on the axioms, then proving further theorems which may be based on previous theorems as well as on the axioms, building a structure of theoretical knowledge in which everything ultimately rests on the axioms. Some theorems will be described as **lemmas**, which are those that are not of much interest in themselves but are needed in order to prove later theorems, or as **corollaries**, which are those that follow almost immediately from the previous theorem.

All axioms, theorems, and any other statements in any mathematical argument are **propositions**. A proposition is a statement that can only be true or false, so excludes self-contradictory statements (for example, "This sentence is false") and a huge variety of statements made in ordinary discourse. That is not to say that the truth or falsehood of a proposition must be known. Indeed there exist propositions in mathematics which are known to be undecidable: all that is known about such a proposition is that no process of logical deduction based on the accepted axioms is capable of proving either truth or falsehood. Below, we sometimes use a single letter P or Q to symbolise a general proposition; we also occasionally use the somewhat tautological phrase "P is true" for emphasis (it is tautological because writing, "Paris is in France" is sufficient to indicate the truth of the proposition that Paris is in France; we don't write, " 'Paris is in France' is true."). We shall also use the phrase, "P is false" for what in the standard notation of logic can be written $\neg P$ ("not P"). If a proposition concerns a class of objects, we may write $P(x)$ where x denotes a general object from the class; for example, if x denotes cities, $P(x)$ might be the proposition, "x is in France".

1.3   The Place of Number Systems within Mathematics

The Natural Numbers which we use to count discrete objects are the first mathematical concept that we encounter as children, so we naturally regard numbers as the most fundamental aspect of mathematics. However, the ancient Greeks regarded geometrical concepts (lines and shapes, with their lengths and angles) as fundamental, whereas modern mathematicians tend to regard the concept of **sets** as underlying the whole of mathematics. In particular, our axioms for number systems will be presented in the language of Set Theory. So, before starting on the axiomatic theory of number systems, we present an informal review of the required background in Set Theory in Chapter 2; a student who already has some knowledge of sets, relations and functions and their notation could probably go straight into Chapter 3 on Natural Numbers, and only refer back to Chapter 2 if and when they encounter an unfamiliar concept from Set Theory.

A single number system is not sufficient for all purposes. Natural Numbers are fine for counting objects, but not for accountancy where one may be in credit or debit. The Integers, which suffice for the latter purpose, cannot deal with dividing objects into equal parts. So we need fractions (Rational Numbers), but these are not adequate for measuring quantities that vary continuously (length, weight, etc.). The Real Numbers which suffice for such measurements are then found not to be capable of providing solutions to perfectly well-formed equations, so we then define the Complex Numbers.

Leopold Kronecker thought that *"Die ganzen Zahlen hat Gott gemacht, alles andere ist Menschenwerk"* [God made the whole numbers, all else is the work of humans[i]]. However, the approach in this book follows the philosophy of Dedekind, that *"Die Zahlen sind freie Schöpfungen des menschlichen Geistes"* [Numbers are free creations of the human mind], since for each of the five number systems mentioned above we shall write down axioms which were originally devised by human thinkers. No kind of number is supposed to simply exist without the need for such axioms to be laid down, although all number systems must demonstrate utility in describing the natural world and/or facilitating human activity.

Having defined our number systems, what do we do with them? We have mentioned some practical applications of numbers above, but what is the next step in the process of building a body of theory by logical deduction? There are various directions that one can take. Most obvious is Number Theory, which investigates the properties of the Natural Numbers and the Integers in depth. Secondly, there is Analysis, which deals with functions of Real Numbers and of Complex Numbers, laying the rigorous foundations for differential and integral calculus and then extending to consider multi-dimensional spaces. Importantly, Analysis replaces vague concepts of "infinitely large" or "infinitesimally small" with rigorous definitions based on the properties of Real Numbers; infinity is not a number, and statements like "$1 \div 0 = \infty$" have no place in mathematics! Thirdly, the study of arithmetic operations in number systems yields some of the basic concepts of Abstract Algebra, in which one investigates the structures arising from operations on sets of objects without reference to the kind of object in the set. Some of the elementary concepts in all these areas of mathematics will be discussed in this book where they arise naturally, but readers wishing to investigate any of these topics in greater depth should find a textbook devoted to the topic.

1.4 Mathematical Writing, Notation, and Terminology

Any mathematical argument, even if written mostly or entirely in symbols, should be capable of being read as grammatically correct English. Every symbol has a meaning which can be expressed as a word or words. For example, = means "equals"; and "$A = B$" can be read as a simple sentence, "A equals B". Note here that A and B are names given to some mathematical objects, so do not need to be expressed in words. We have here assumed an intuitive notion of equality: two mathematical expressions are equal if and only if they are representations of the same object. Equality can be defined more formally as an equivalence relation (see Section 2.3.1).

The symbol := means "defined to be". So "$A := B$" (where A and/or B would typically be mathematical expressions rather than single symbols) defines the expression A to be equal to the expression B. This is a stronger statement than simply saying that $A = B$; but once the definition has been made, it is certainly true that $A = B$.

The symbol $\Rightarrow$ means "implies": if P and Q are propositions, $P \Rightarrow Q$ ("P implies Q") means that Q is a logical consequence of P, and is equivalent to writing, "If P, then Q". The symbol $\Leftrightarrow$ indicates a two-way implication: $P \Leftrightarrow Q$ means that P implies Q *and* Q implies P, or equivalently that Q is true *if and only if* P is true. You must never confuse $\Rightarrow$ or $\Leftrightarrow$ with = : the symbol = appears between mathematical objects, whereas $\Rightarrow$ and $\Leftrightarrow$ can only appear between propositions. If A, B, C, D are mathematical objects, the sentence

---

[i] *Ganzen Zahlen* (literally, "whole numbers") is sometimes translated as "integers"; but would God have created anything as abhorrent as negative numbers? It is probably better to translate the phrase as "natural numbers".

$A = B \Rightarrow C = D$ does make sense: "$A = B$" is a proposition, as is "$C = D$", and the sentence asserts that if the former proposition is true, then the latter is also true. But "$A \Rightarrow B$" does not make sense. The symbol $\therefore$ ("therefore") essentially means the same as $\Rightarrow$ ("implies"), and many authors eschew the former; however, if one is reading a mathematical argument as an English sentence, there are situations where the conjunction "therefore" seems more grammatically correct than the verb "implies", and so both symbols appear in this book.

A forward slash / through any symbol is used to indicate the negation of the meaning of the symbol. For example, $\neq$ for "is not equal to", $\not\Rightarrow$ for "does not imply". However, to negate a proposition we use $\neg$; so $\neg P$, means "not P", or "P is false".

The meaning of "and" is clear, but "or" is more problematic. "P and Q" means that *both* propositions P and Q are true (in normal usage and in mathematics), whereas "P or Q" can have different meanings in normal usage. In mathematics, at least in this book, it will always mean the inclusive "or", i.e. that any of the following may be true: P but not Q; Q but not P; both P and Q.

The delimiters [ ], { } and ( ) are all often informally referred to as brackets; but strictly [ ] are **brackets**, { } are **braces** and ( ) are **parentheses**. In many situations, each of them has a particular meaning and they must not be confused; for example, see the notation described below for sets, ordered sets and equivalence classes. However, in other situations, especially in arithmetical calculations, they are used to indicate the order in which operations are to be done (with the material within a set of delimiters being evaluated before combining the result of that evaluation with objects outside the delimiters). In that case, any of the delimiters may be used, although the first choice is usually parentheses, with the other delimiters only being introduced if an expression requires multiple delimiters.

1.5 Logic and Methods of Proof

We do not require a detailed discussion of formal logic, but there are a few concepts that are important to understand.

Often a "direct proof" is possible, starting from known propositions and finishing with the proposition that is to be proved, using a sequence of steps (see Section 2.3.3 on *Transitivity and Proofs* for a justification of using a sequence of steps). However, there are also many cases where indirect methods are better.

Proof by contradiction involves starting with the negation of the proposition to be proved. A logical argument based on this negated proposition is then pursued, ultimately leading to a proposition that contradicts some known truth or contradicts the negated proposition that was assumed at the start. Since the argument has been logical, the only possible source of the contradiction is the original negated proposition. So the negated proposition cannot be true, and hence the original proposition must be true.

Sometimes a statement of the form $P \Rightarrow Q$ is most easily proved via its **contrapositive**. We first need to distinguish between contrapositive and converse.

Let P and Q be propositions, and suppose that $P \Rightarrow Q$: if P is true, then Q is true. The **converse** of $P \Rightarrow Q$ is the statement $Q \Rightarrow P$; if both $P \Rightarrow Q$ and $Q \Rightarrow P$, we have $P \Leftrightarrow Q$. Indeed, to prove that $P \Leftrightarrow Q$, we usually need to do a two-stage proof, first demonstrating an implication $P \Rightarrow Q$ and then demonstrating the converse implication, $Q \Rightarrow P$.

On the other hand, the statement $\neg Q \Rightarrow \neg P$ ("not Q implies not P", or more clearly, "if Q is false, then P is false"), is the **contrapositive** of $P \Rightarrow Q$. The contrapositive is logically equivalent to the original statement. This last sentence is itself an "if and only if" statement: it is saying that if $\neg Q \Rightarrow \neg P$, then $P \Rightarrow Q$, and also (the converse) if $P \Rightarrow Q$, then $\neg Q \Rightarrow \neg P$. To verify this, first suppose that $\neg Q \Rightarrow \neg P$, i.e. P being false is a logical consequence of Q being false; then P cannot be true if Q is false; so if P is true, then Q cannot be false, i.e. Q must be true. For the converse, suppose that $P \Rightarrow Q$: if P is true, then so is Q; thus P could not have been true if Q is found to be false, i.e. Q being false implies that P was false. The practical implication is that if it is easier to prove a contrapositive statement than the statement as originally framed, that suffices to prove the original statement.

An example to illustrate these ideas: consider the statement, "All cats are black", which can be clarified as "If an object is a cat, then that object is black". The converse is: "if an object is black, then it is a cat". Clearly saying that "All cats are black" does not eliminate the possibility of a dog being black: the original statement is not equivalent to the converse. But the contrapositive is: "if an object is not black, then it is not a cat". If you know that all cats are black, and you find a non-black object, you can be certain that it is not a cat: the original statement has implied the contrapositive.

When proving a proposition about a class of object (for example, all numbers in a number system), it is important to note that any proof must be completely general; showing the proposition to be true for some example(s) taken from the class of object is not sufficient. On the other hand, it is only necessary to find a single object in that class which does not satisfy the proposition in order to prove the proposition to be false; such an object is a **counterexample** to the proposition. Finding a single white cat proves the proposition that "All cats are black" to be false; whereas finding any number of black cats (examples satisfying the proposition) would not prove it to be true – unless you were certain that you had found every cat in existence. That would be a **proof by exhaustion**, where each member of a class of objects is verified to satisfy a proposition; not often a practical method, although sometimes a large class of objects may be split into sub-classes, and the "proof by exhaustion" involves examining whether each sub-class satisfies the proposition.

If you want to fully understand a proof, you will need to put in a certain amount of effort. It is recommended that you use the "self-explanation" strategy set down by Lara Alcock, Mark Hodds, and Matthew Inglis: their guidance notes, "How to read proofs: the self-explanation strategy" are reproduced as Appendix A at the back of this book. The strategy essentially involves ensuring that you fully understand the reasoning in each step of the proof before proceeding to the next line. There is nothing new or revolutionary about this: it is what successful mathematicians have been doing for centuries; but the great insight of Alcock, Hodds, and Inglis was to realise that it is not something that students of mathematics automatically know how to do when they first meet the rigorous development of theory by the axiomatic method; hence the need for guidance notes. To help you in your self-explanation as you read this book, a reason will often be stated for each step in a proof; the reasons will be in italics in brackets.

Chapter 2

Sets and Relations

2.1 Sets

After the emphasis placed on the Axiomatic Method in the Introduction, it may surprise you that we shall not provide an axiomatic definition of **Set**, instead relying on the naïve notion that a set is a collection of objects; the objects are called **members** or **elements** of the set. We shall add that the specification of a set must be unambiguous and must not admit self-contradictory conclusions. This requirement is not trivial, as is demonstrated most famously by Russell's paradox: if we allow a set to be a member of itself, then we can also specify a set M which consists of all sets that are *not* members of themselves. If M is a member of itself, then the specification of M implies that it is not a member of itself; if M is not a member of itself, then its specification implies that it is a member of itself. Such a self-contradiction is undesirable; systems of axioms for Set Theory have been developed, most notably the system named after Ernst Zermelo and Abraham Fraenkel, to ensure that no such paradox can arise. But the study of Zermelo-Fraenkel Set Theory is a substantial undertaking in itself, and is not necessary for our development of the theory of number systems.

There are two ways to specify a set. We may simply list its members:

$$S := \{a, b, c\}.$$

Here, the set S is defined to consist of the members a, b, c, and we have introduced the notation that braces $\{\ \}$ are always used to enclose the specification of a set. Membership of a set is denoted using the symbol $\in$, as in

$$a \in S,$$

and non-membership by the symbol $\notin$, as in

$$d \notin S.$$

The second way of specifying a set is by means of a rule, typically a property which a member of a previously defined set must have. For example, if C has been defined as the set of all countries in Europe, we may write

$$E := \{c \in C : c \text{ is in the European Union}\}$$

to specify the set of countries in the EU. Note that an object is a member of the set *if and only if* it has the property given after the colon in such a specification.

The **empty set** $\emptyset$ is defined to have no members; so if x is any object whatsoever, then $x \notin \emptyset$. It is sometimes useful to refer to the **universal set**, which consists of all objects which are capable of being members of a set; we shall use the symbol Ω for the universal set.

2.1.1   Quantifiers

The symbol $\exists$ means **there exists**, and the symbol $\forall$ means **for all**; these symbols are called **quantifiers**. If $P(x)$ is a proposition regarding members of a set X (where x stands for a general member of X), the quantifiers are typically used in sentences of the forms, "$\exists x \in X$ such that $P(x)$" and "$\forall x \in X, P(x)$". The meaning is normally clear from simply reading such a sentence using the words "there exists" or "for all" as appropriate (and reading out the proposition $P(x)$ in English); but formally, "$\exists x \in X$ such that $P(x)$" means that $\{x \in X : P(x)$ is true$\} \neq \emptyset$, and "$\forall x \in X, P(x)$" means $\{x \in X : P(x)$ is true$\} = X$.

It is opportune here to discuss **uniqueness**, although there is no quantifier symbol for "unique". We shall often make statements of the form "there exists a unique $x \in X$ such that $P(x)$". Informally, this means that there is one and only one member of X for which $P(x)$ is true.[i] More formally, it means that $\exists x \in X$ such that $P(x)$ is true *and* that if $y \in X$ with $y \neq x$, then $P(y)$ is false. The latter part of that definition may usefully be rewritten as "if $P(x)$ and $P(y)$ are both true with $x \in X$ and $y \in X$, then $x = y$"; for the usual way of proving the uniqueness of an $x \in X$ satisfying $P(x)$ is to suppose that there also exists $y \in X$ satisfying $P(y)$, and then to show that in fact x and y must be the same object.

2.1.2   Subsets and Equality

If every member of a set S is also a member of another set T, we say that S is a **subset** of T, and we write

$$S \subseteq T.$$

So the statement $S \subseteq T$ is equivalent to

$$x \in S \Rightarrow x \in T.$$

Suppose now that $S \subseteq T$ and also $T \subseteq S$, or equivalently that $x \in S \Leftrightarrow x \in T$. Every member of one of the sets is a member of the other, and we say that the sets are **equal**; we write

$$S = T.$$

Alternatively, if $S \subseteq T$ but $T \nsubseteq S$, i.e.

$$x \in S \Rightarrow x \in T \quad \text{but} \quad \exists y \in T \text{ such that } y \notin S,$$

we say that S is a **proper subset** of T, and we write

$$S \subset T.$$

Example 2.1.1. Let $S := \{a, b, c\}$, $T := \{a, b, c, d, e\}$ and $U := \{c, b, e, a, d\}$. Then $S \subset T$, $S \subset U$ and $T = U$.

The observation that $T = U$ in this example emphasises that when a set is specified by listing its members, the order of the members in the list is immaterial (given that any designation of equality is intended to imply that the equal objects should be regarded as representations of the same thing).

---

[i]Since "one" is a number and will not be defined until Chapter 3, the phrase "one and only one" should strictly not be used in the present chapter; nevertheless, it is useful here and later in this chapter to avoid clumsy sentence constructions. Similarly, we may need to use other number words, "two" etc. in this chapter to avoid clumsiness of expression. Pedantic readers may wish to rephrase sentences containing such words in terms of existence, uniqueness, etc.

2.1.3 Union, Intersection, and Complement

The **union** of sets A and B, denoted $A \cup B$, is the set which consists of all objects which are in A or in B. Formally,

$$A \cup B := \{x \in \Omega : x \in A \text{ or } x \in B\},$$

where Ω is the universal set. If A and B have been defined by listing their members, then any objects which are members of both A and B should not be written twice when listing the members of $A \cup B$. Clearly, $A \cup B = B \cup A$.

Example 2.1.2. Let $S := \{a, b, c\}$ and $V := \{c, d, e\}$. Then $S \cup V = \{a, b, c, d, e\}$.

We can consider the union of multiple sets: for example, given the sets defined in Example 2.1.1 above, we can write $S \cup T \cup U = \{a, b, c, d, e\}$.

The **intersection** of sets A and B, denoted $A \cap B$, is the set which consists of all objects which are in both A and B. Formally,

$$A \cap B := \{x \in \Omega : x \in A \text{ and } x \in B\}.$$

Clearly, $A \cap B = B \cap A$.

Example 2.1.3. Let $S := \{a, b, c\}$ and $V := \{c, d, e\}$. Then $S \cap V = \{c\}$.

We can consider the intersection of multiple sets: for example, given the sets defined in Example 2.1.1 above, we can write $S \cap T \cap U = \{a, b, c\}$.

The **complement** of B in A, denoted $A \backslash B$,[ii] consists of all objects which are in A but not in B, regardless of whether $B \subseteq A$. Formally,

$$A \backslash B := \{x \in A : x \notin B\}.$$

Example 2.1.4. Let $S := \{a, b, c\}$, $T := \{a, b, c, d, e\}$ and $W := \{e, f, g\}$. Then $T \backslash S = \{d, e\}$, $T \backslash W = \{a, b, c, d\}$, and $S \backslash W = \{a, b, c\}$.

Exercise 2.1.5. *Consider the following sets of Greek letters:*
$A := \{\alpha, \beta, \gamma, \delta\}, B := \{\gamma, \delta, \alpha\}, C := \{\gamma, \delta, \epsilon\}, D := \{\gamma, \delta\}, E := \{\epsilon\}$.
Write out the following sets as lists of members (or as $\emptyset$ if appropriate):
$A \cup B, \quad A \cap B, \quad A \backslash B, \quad B \backslash A, \quad A \backslash C, \quad C \backslash A, \quad B \cap C, \quad D \cup E,$
$C \cup D \cup E, \quad C \cap D \cap E, \quad E \backslash D$.
Which of the sets are equal?
Which of the sets are proper subsets of each other?

Exercise 2.1.6. *Which of the following are true for all non-empty sets A, B?*
 (i) $A \cap B \subseteq A \cup B$
 (ii) $A \cap B \subset A \cup B$
 (iii) $A \backslash B \subset A \cup B$
 (iv) $A \backslash B \neq A \cap B$
 (v) $A \backslash B \neq A \cup B$
Do any of your answers change if we allow A or B to be empty sets?
Give a counterexample in any case where you have said that a statement is false.

A useful theorem on complements is the following:

Theorem 2.1.7. *If $A \subseteq B$, with both $A \subseteq S$ and $B \subseteq S$, then $S \backslash B \subseteq S \backslash A$.*

Proof. "$A \subseteq B$" means: "If $x \in A$, then $x \in B$". The contrapositive of this is: "If $x \notin B$, then $x \notin A$", which is equivalent to "$S \backslash B \subseteq S \backslash A$", where S is some set which all objects under consideration are members of. $\qquad \square$

[ii]Some authors write $A - B$ rather than $A \backslash B$; but we shall see later that the symbol "$-$" has multiple meanings in relation to numbers, and we do not want to also use it in relation to sets!

2.1.4   Ordered Sets

We have stated that the order in which the members of a set are listed is immaterial, but we do sometimes need to define the order of members in a set. We use parentheses to indicate such an ordered set: for example, (a, b) is an **ordered pair** consisting of the objects a and b in that order, so that $(a, b) \neq (b, a)$ in contrast to the sets we have considered previously, for which $\{a, b\} = \{b, a\}$. An ordered pair may be defined formally by

$$(a, b) := \{\, \{a\}, \{a, b\} \,\},$$

an ordered triple by

$$(a, b, c) := \{\, \{a\}, \{a, b\}, \{a, b, c\} \,\},$$

and more general ordered sets by an obvious extension of this method of listing subsets. However, the intuitive notion of listing members of a set in a specified order will be sufficient for our purposes.

Given sets A and B, we may obtain ordered pairs by selecting a member of A followed by a member of B; the set of all ordered pairs (a, b) obtained in this way from given sets A and B is called the **Cartesian product** of A and B, denoted $A \times B$.[iii] So

$$A \times B := \{(a, b) : a \in A, b \in B\}.$$

The extension to more general ordered sets is obvious, with the Cartesian product $A \times B \times \cdots \times Z$ consisting of all ordered sets $(a, b, \ldots, z)$ with $a \in A, b \in B, \ldots, z \in Z$.

2.2   Relations between Sets

2.2.1   Relations in General

Given sets A and B, a **relation** $R(A, B)$ may be defined simply as a subset of $A \times B$. Given $a \in A$ and $b \in B$, we may say that a is related to b by R if and only if the ordered pair (a, b) is a member of the subset $R(A, B)$. We refer to $R(A, B)$ as a relation **between** A and B.

In contrast to this rather stark definition, the way that useful relations in mathematics are usually defined is by means of a rule: a is related to b if and only if a and b satisfy some requirement expressed in words and/or mathematical symbols. We may also adopt some symbol such as $\sim$ to indicate a relation, i.e. $a \sim b$ when a is related to b; this notation is used in Section 2.3.

Example 2.2.1. Given the sets of geographical entities,

$$G := \{\text{ Paris, Lyon, Berlin, Frankfurt, Hamburg, Zürich }\},$$
$$H := \{\text{ France, Germany, Switzerland }\},$$

the relation,

"is in the country called" (or more simply, "is in")

is defined by the meaning of the English words, and specifies a subset $R(G, H)$ of $G \times H$. For example, Hamburg is in Germany, but Lyon is not in Switzerland; so in terms of ordered pairs, (Hamburg, Germany) $\in R(G, H)$ but (Lyon, Switzerland) $\notin R(G, H)$.

[iii]The symbol "$\times$" will later be used in its more familiar setting between two numbers; it is important to recognise that its meaning there is completely different from the meaning introduced here in connection with sets.

2.2.2 Functions

A relation $R(A, B)$ is a **function** if, for *each* $a \in A$, there exists a *unique* $b \in B$ such that $(a, b) \in R(A, B)$. Again, this is a rather stark definition and is not the usual way we think about functions. We generally refer to a function **from** A to B and we use the notation $f : A \rightarrow B$, where f denotes a rule that relates each member of A to a unique member of B; the set A is called the **domain** of f, denoted dom f, and B is its **codomain** when the sets are related by f in this way. Given $a \in A$ and the unique $b \in B$ which is related to it by the function f, we write $f(a) = b$ ("f of a equals b") or $f : a \mapsto b$ ("f maps a to b"), and we say that b is the **image** of a under f. Note the difference between the symbols $\mapsto$ and $\rightarrow$; the latter is used between sets, the former between the members of those sets. The set of images of all members of A under f is called the **range** of f, so the range of f is a subset of its codomain. We may denote the range of f as $f(A)$, which is really a shorthand for $\{ f(a) : a \in A \}$.

Example 2.2.2. The relation given in Example 2.2.1 is a function: each of the cities in the domain set G is in one and only one country in the codomain set H. However, if we add Madrid to the list of cities in G to get the set

$$G_* := \{ \text{ Paris, Lyon, Berlin, Frankfurt, Hamburg, Zürich, Madrid } \},$$

the relation, "is in the country called", would not be a function from G_* to H: there does not exist a country in H that Madrid is in. Alternatively, if we add Basel to G to get the set

$$G_\sharp := \{ \text{ Paris, Lyon, Berlin, Frankfurt, Hamburg, Zürich, Basel } \},$$

we would again not have a function: each of Switzerland, Germany, and France contains part of the urban area of Basel, i.e. we can say "Basel is in x", where x is any of the countries in H: there is not a *unique* member of H with Basel in it.[iv]

Given $f : A \rightarrow B$ with range $f(A)$, there may be another function g which maps each $b \in f(A)$ to a unique object c in a further set C. So we have $g(b) = c$, where $f(a) = b$, and simple substitution allows us to write $g(f(a)) = c$. We have defined a new function, denoted $g \circ f : A \rightarrow C$, called the **composition** of f and g, which maps each $a \in A$ to a unique $c \in C$. Note that the domain of g must contain every $b \in f(A)$, but may contain other objects, i.e. we require $f(A) \subseteq \text{dom} \, g$.

Example 2.2.3. If we let $f : G \rightarrow H$ be the function, "is in the country called", as in Example 2.2.2, and then define g to be the function, "is in the continent called", then the composition $g \circ f$ would be a function mapping each of the cities in G to the continent Europe. The domain of g would need to be defined to include all the countries in H (but not any country such as Russia which is in more than one continent).

A function from A to B is a **surjection** if, for each $b \in B$, there exists a member of A which is mapped to b by f. So if f is a surjection, its range is the same as its codomain.

Example 2.2.4. The relation in Example 2.2.1 is a surjection from G to H: there is a city (or many cities) from the set G in each of the countries in the set H. However, if we consider the set

$$K := \{ \text{ France, Germany, Switzerland, Italy } \},$$

we would still have a function, but not a surjection, from G to K: there is no member of G in Italy.

[iv]As well as extending across national borders, Basel is also notable as the birthplace of great mathematicians: Leonhard Euler and the Bernoulli family.

A function from A to B is an **injection** if, for each $b \in f(A)$, there is a *unique* member of A that is mapped to b by f (recall that $f(A) \subseteq B$). So if $a_* \neq a_\sharp$, then $f(a_*) \neq f(a_\sharp)$ when f is an injection. The contrapositive of the last statement gives an alternative way of identifying f as an injection: if $f(a_*) = f(a_\sharp)$, then $a_* = a_\sharp$.

Example 2.2.5. The relation in Example 2.2.1 is not an injection from G to H: both Paris and Lyon are in France, so the uniqueness requirement is not satisfied. However, if we consider the set

$$J := \{ \text{ Paris, Berlin, Zürich } \},$$

then the relation, "is in the country called" is an injection from J to H. Indeed, it is also an injection from J to the set K defined in Example 2.2.4. From the list of countries in K, the range of the function is the set

$$\{ \text{ France, Germany, Switzerland } \},$$

and each of these countries has a unique member of J in it.

Observe that a surjection must have every member of its codomain mapped *to*, while a function must have every member of its domain mapped *from*; and an injection can only have a unique member of the domain mapping *to* any member of the codomain, while a function can only have a unique member of the codomain mapped *from* any member of the domain. So the requirements in the definitions of "surjection" and "injection" together are identical to those in the definition of "function", but with the roles of "domain" and "codomain" exchanged. Thus, if we have a function which is both a surjection and an injection, the domain and codomain satisfy identical requirements. Such a function is called a **bijection**: it maps each member of the domain to a unique member of the codomain, and each member of the codomain is mapped from a unique member of the domain. Clearly we can reverse the roles of domain and codomain: given a bijection $f : A \to B$, we can define its **inverse function** $f^{-1} : B \to A$ (which is also a bijection) by[v]

$$f^{-1} : b \mapsto a \Leftrightarrow f : a \mapsto b.$$

Example 2.2.6. The injection from J to H in Example 2.2.5 is in fact a bijection: each city in the set J is in one and only one country in H, and each country in H contains a unique city in J. The inverse of the function "is in" is the function "contains"; for example, Berlin is in Germany, Germany contains Berlin.

Note that an injection f from A to B is the same thing as a bijection from A to a *subset* of B; specifically, to $f(A) \subseteq B$. Each member of $f(A)$ is the image of a unique member of A under the injection f, so f is a bijection to $f(A)$. In Example 2.2.5, the relation "is in the country called" is an injection from J to K, and is a bijection from J to a subset of K, specifically to $H \subset K$.

One simple but important example of a bijection is the identity function id_A on a set A:

Definition 2.2.7. For any set A the function $\mathrm{id}_A : A \to A$ is defined by

$$\mathrm{id}_A : a \mapsto a \ \forall a \in A.$$

---

[v]The notation using the superscript -1 to indicate an inverse function is dreadful, with its scope for confusion with -1 power, i.e. reciprocal (to be defined formally in Chapter 6). Nevertheless, the notation has become so well established that the present author feels unable to refuse to use it. Readers will simply need to remember that whenever the superscript -1 is attached to the symbol for a function, it means inverse function.

So the identity function simply maps each member of A to itself. The inverse of the identity function is obviously also the identity function. Formally, we can define **equality** using the identity function:

Definition 2.2.8. For $a, b \in A$,

$$a = b \text{ if and only if } \mathrm{id}_A : a \mapsto b.$$

Exercise 2.2.9. *Consider again the sets defined in Exercise 2.1.5 above:*

$$A := \{\alpha, \beta, \gamma, \delta\}, B := \{\gamma, \delta, \alpha\}, C := \{\gamma, \delta, \epsilon\}, D := \{\gamma, \delta\}, E := \{\epsilon\}.$$

Below is a list of relations between pairs of these sets, with each relation specified formally as a set of ordered pairs. In each case, identify the relation as either: a bijection; an injection but not a surjection; a surjection but not an injection; a function which is neither an injection nor a surjection; or a relation which is not a function.

(i) $R(D, E)$: $\{(\gamma, \epsilon), (\delta, \epsilon)\}$
(ii) $R(E, D)$: $\{(\epsilon, \gamma), (\epsilon, \delta)\}$
(iii) $R(C, D)$: $\{(\gamma, \gamma), (\epsilon, \delta), (\delta, \gamma)\}$
(iv) $R(C, D)$: $\{(\gamma, \gamma), (\epsilon, \gamma), (\delta, \gamma)\}$
(v) $R(D, A)$: $\{(\gamma, \alpha), (\delta, \delta)\}$
(vi) $R(B, C)$: $\{(\gamma, \epsilon), (\alpha, \gamma), (\gamma, \delta\}$
(vii) $R(B, C)$: $\{(\gamma, \epsilon), (\alpha, \gamma), (\delta, \delta\}$
(viii) $R(B, C)$: $\{(\gamma, \epsilon), (\alpha, \gamma), (\delta, \epsilon)\}$

2.3 Relations on a Set

If a relation R is between a set A and that same set, we have $R(A, A)$ which is described as a relation **on** A. Several types of relation on a set are important for the development of the theory of number systems.

2.3.1 Equivalence Relations

Definition 2.3.1. A relation $\sim$ on a set A is an **equivalence relation** if it has the following three properties for all $a, b, c \in A$:

- It is **reflexive**: $a \sim a$;

- It is **symmetric**: if $a \sim b$, then $b \sim a$;

- It is **transitive**: if $a \sim b$ and $b \sim c$, then $a \sim c$.

It is clear that both our intuitive notion of equality and Definition 2.2.8 satisfy the requirements of an equivalence relation: for any objects a, b, c, we have $a = a$; if $a = b$, then $b = a$; and if $a = b$ and $b = c$, then $a = c$. But equality is not the only kind of equivalence relation.

Example 2.3.2. The relation, "is in the same country as", is an equivalence relation on the set of cities G defined in Example 2.2.1. To verify this:

- Every city is in the same country as itself (reflexive);

- If city a is in the same country as city b, then b is in the same country as a (symmetric);

- If city a is in the same country as city b and city b is in the same country as city c, then a is in the same country as c (transitive).

However, Berlin is not the same as Frankfurt, although they are in the same country: this is an equivalence relation which cannot be described as equality.

Consider now the following subsets of G:

$$G_F := \{ \text{ Paris, Lyon } \}; G_D := \{ \text{ Berlin, Frankfurt, Hamburg } \}; G_S := \{ \text{ Zürich } \}.$$

Within each subset, every member of the subset is equivalent to (i.e. in the same country as) every other member; but no member of any subset is also a member of a different subset. So every member of G is a member of a unique subset from the list $\{G_F, G_D, G_S\}$. These observations illustrate the following definition and theorem.

Definition 2.3.3. A **partition** P of a set X is a set of non-empty subsets of X, such that every member of X is a member of a unique subset in P.

Example 2.3.4. The set $\{G_F, G_D, G_S\}$ is a partition of the set G.

Theorem 2.3.5. *Every equivalence relation on a set A generates a partition of A. The converse is also true: every partition of A yields an equivalence relation on A.*

Proof. The converse is easy to see: given a partition of A into subsets, the relation, "is in the same subset as", is an equivalence relation on A (in a similar way to "is in the same country as" in Example 2.3.2).

Now suppose that there is an equivalence relation $\sim$ on A. Take any $a \in A$ and let $[a]$ be a subset of A consisting of all members of A which are related to a by $\sim$: for each $x \in A$ we have $x \in [a]$ if and only if $a \sim x$. Reflexivity ($a \sim a$) guarantees that each $a \in A$ is a member of such a subset.

It only remains to verify that a is not a member of any other subset, i.e. that the subset it belongs to is unique. Let $[b]$ be another subset, defined in the same way as $[a]$ but with respect to another member $b \in A$. Suppose $a \in [b]$ as well as $a \in [a]$. Now, $a \in [b]$ means $b \sim a$. Let x_a be any member of $[a]$, so $a \sim x_a$; then transitivity yields $b \sim x_a \ \forall x_a \in [a]$, so that all members of $[a]$ are also members of $[b]$; so $[a] \subseteq [b]$. By symmetry, $a \sim b$ if $b \sim a$, and reversing the roles of a and b in the previous sentence shows that $[b] \subseteq [a]$. Hence $[a] = [b]$: the different subsets that a was supposed to belong to are in fact the same, i.e. a is a member of a *unique* subset. $\square$

Each subset in the partition of A generated by an equivalence relation is called an **equivalence class**; it consists of members of A which are all equivalent to each other, i.e. it is a set of the form $[a]$ defined in the above proof. Indeed, the notation $[a]$ is used to denote the equivalence class containing the object a. If the equivalence relation is equality, i.e. if equivalent objects are considered to be representations of the same object, then each equivalence class contains a single member of A. More generally, an equivalence relation may be completely specified by listing the members of each equivalence class.

Equivalence relations and bijections are the two most important concepts from this chapter for the development of the theory of number systems. The following theorem connects these two concepts.

Theorem 2.3.6. *Let T be a set of sets, $T = \{A, B, C, \ldots\}$. Then the relation, "has a bijection to", is an equivalence relation on T.*

Proof. Reflexivity: the identity function provides a bijection from each set to itself.

Symmetry: if there is a bijection $f : A \to B$, then there is an inverse bijection $f^{-1} : B \to A$.

Transitivity: if $f : A \to B$ and $g : B \to C$ are bijections, then so is the composition $g \circ f : A \to C$: for each $a \in A$ there is a unique $c \in C$ such that $c = g(f(a))$, and for each $c \in C$ there is a unique $a \in A$ such that $a = f^{-1}(g^{-1}(c))$. □

2.3.2 Order Relations

We could define a relation, "is further north than", on our set of cities G. It would be transitive: if city a is further north than city b and city b is further north than city c, then city a is further north than city c. However, it is not reflexive (a city cannot be further north than itself) and is not symmetric: if a is further north than b, then b is definitely not further north than a. Indeed, the last sentence defines the relation to have **strict antisymmetry**:

Definition 2.3.7. A relation $\sim$ on a set A is **strictly antisymmetric** if

$$a \not\sim a \quad \text{and} \quad a \sim b \Rightarrow b \not\sim a \quad \forall a, b \in A.$$

This still allows the possibility that neither $a \sim b$ nor $b \sim a$ is true; in our example, this would apply to cities a and b at the same latitude.

Now, the symmetric relation $a = b$ cannot be true at the same time as a strictly antisymmetric relation $a \sim b$ or $b \sim a$; but with the relation, "is further north than", $a = b$ is not even true when neither $a \sim b$ nor $b \sim a$ (for different cities at the same latitude). A relation for which it *is* true that $a = b$ if and only if neither $a \sim b$ nor $b \sim a$, in addition to the requirements of strict antisymmetry, is called a **trichotomy**. More concisely:

Definition 2.3.8. A relation $\sim$ on a set A is **trichotomous** if, given any $a, b \in A$, one and only one of the following is true:

$$a \sim b \quad \text{or} \quad b \sim a \quad \text{or} \quad a = b.$$

Definition 2.3.9. A relation that is transitive and trichotomous is called a **strict order**.

It is also useful to define **weak antisymmetry**.

Definition 2.3.10. A relation $\simeq$ on A is weakly antisymmetric if it is reflexive ($a \simeq a$) and also

$$a \simeq b \text{ and } b \simeq a \Rightarrow a = b.$$

Thus it is possible to have both $a \simeq b$ and $b \simeq a$, but only if a and b are actually the same object. This may seem very different from our definition of strict antisymmetry; however, if $\sim$ is a strictly antisymmetric relation and we then define a relation $\simeq$ as

$$a \simeq b \quad \text{if} \quad a \sim b \quad \text{or} \quad a = b,$$

the relation $\simeq$ is weakly antisymmetric. To see this, first note that the option that $a = b$ makes $\simeq$ reflexive; secondly, the strict antisymmetry of $\sim$ makes it impossible for $a \simeq b$ and $b \simeq a$ without $a = b$, thus fulfilling the definition of weak antisymmetry. Note that we can reverse the process of defining a weakly antisymmetric relation from a strictly antisymmetric one: given a weakly antisymmetric relation $\simeq$, a relation $\sim$ defined by

$$a \sim b \quad \text{if} \quad a \simeq b \quad \text{and} \quad a \neq b$$

is strictly antisymmetric.

We can then use weak antisymmetry to define a weak version of trichotomy and a weak order analogously to the way strict antisymmetry is used to define trichotomy and strict order.

Definition 2.3.11. A relation $\simeq$ on a set A is **weakly trichotomous** if, for any $a, b \in A$, at least one of the following is true:

$$a \simeq b \quad \text{or} \quad b \simeq a,$$

with both being true only if $a = b$.

Definition 2.3.12. A relation that is transitive and weakly trichotomous is called a **weak order**.

So the relation, "is at least as far north as", is a weak order on a set of cities.

Exercise 2.3.13. *For each of the following relations on a set of people, state whether it is: reflexive; symmetric; transitive; strictly antisymmetric; weakly antisymmetric. Hence classify each of the relations as: an equivalence relation; a strict order; a weak order; or none of these.*
[Note: the relations below are given as phrases, such that the names of two people who may or may not be related can be appended at the beginning or end. The word "related" is used here in the mathematical sense!]
   (i) is older than
   (ii) was born in the same year as
   (iii) was born in the same or an earlier year than
   (iv) is a brother of
   (v) is a sibling (i.e. brother or sister) of
   (vi) is a first cousin of
   (vii) is an uncle or aunt of
   (viii) is an ancestor of
   (ix) has lived in the same country as
   (x) has lived in a different country than

2.3.3   Transitivity and Proofs

We frequently want to prove equality (or sometimes other equivalence relations) or an order (strong or weak) between quantities. All these relations are transitive. To prove that $a \sim z$ (where "a" and "z" stand for some expressions), a sequence of steps,

$$a \sim b$$
$$\sim c$$
$$\vdots$$
$$\sim z,$$

will suffice if $\sim$ is a transitive relation.

When proving an equality $a = z$ in this way, it is sometimes easier to do two separate calculations, firstly showing that $a = m$ and secondly that $z = m$, where "m" is a third expression. This applies in particular when a and z are both complicated expressions: it is easier to simplify than to complicate a mathematical expression, so the proof may be possible by simplifying both a and z to the same simpler expression m. [On the other hand, something that is not legitimate (but which the author has seen students do very often!) is to start with the equality to be proved, $a = z$, and then in subsequent lines of the "proof" to simultaneously simplify both sides of the equality until one eventually arrives at $m = m$.

This is backwards logic: starting with what one wants to verify, and finishing with what one previously knows to be true.]

In the case where $\sim$ is a strict order and $\simeq$ is the corresponding weak order (i.e. $a \simeq b$ when either $a \sim b$ or $a = b$), a sequence of steps involving at least one instance of $\sim$ amongst a sequence of $\simeq$ relations is sufficient to prove the strict order, $a \sim z$. This is because every step satisfies the weak order, so $a \simeq z$, but any instance of the strong order eliminates the possibility that $a = z$. Note that the definition of weak antisymmetry provides a useful technique for proving equalities: there are many circumstances where the best way to prove that $a = b$ involves the use of a weak order $\simeq$, by proving that both $a \simeq b$ and $b \simeq a$.

Greater care needs to be taken when attempting to prove inequalities when no order relation is available. The relation $\neq$ ("not equal to") is not transitive: if $a \neq b$ and $b \neq c$, it is still possible that $a = c$. However, if we have a sequence of equalities but with just a single inequality,

$$a = b$$
$$\vdots$$
$$= l$$
$$\neq m$$
$$= n$$
$$\vdots$$
$$= z,$$

we do know that $a \neq z$: since equality is an equivalence relation, the quantities from a to l are all in one equivalence class, and the quantities from m to z are all in a different equivalence class. Obviously a similar argument applies to proving negation of more general equivalence relations.

Another kind of proof is where we need to show an implication, that $P \Rightarrow Q$ where P and Q are propositions. The relation $\Rightarrow$ is transitive; so a sequence of propositions, starting with P and ending with Q, and with each proposition implying the next, will yield a valid proof.

2.4 Binary Operations and Algebraic Structures

A **binary operation** on a set A is simply a function whose domain is $A \times A$ and whose codomain is A. In other words, we take any two members of A in a specified order, and the binary operation yields a further member of the same set. The property that the "output" of the operation is a member of the same set as the "inputs" is referred to as being **closed**. We shall frequently define a rule for operating with two members of a set A and immediately check that it is indeed closed so that it satisfies the definition of a binary operation.

Rather than the usual function notation, $f : A \times A \to A$ and $f(r, s) = t$ (where $r, s, t \in A$), binary operations are usually denoted using a symbol placed between the two inputs, in the form $r * s = t$. Indeed, when discussing abstract properties of binary operations on a general set A we shall use the symbols $*$ or $\vee$ to indicate the operations; but you may have realised already that the familiar arithmetic operations that we do with numbers, using symbols like $+$ and $\times$, apparently satisfy the definition of binary operations (although we shall see that we need to be careful about which operations are closed on which number systems). However, numbers are just one class of mathematical object on which binary operations may be defined; geometrical symmetries, matrices, polynomials, and various other classes of object have operations defined on them which have similar properties to arithmetic operations on numbers. A set of objects with one or more binary operations satisfying some list of properties is called an **algebraic structure**, and the study of these structures without reference to the type of object in the set is called **abstract algebra**. Algebraic structures are classified according to the list of properties satisfied by the operation(s), with the various classes given names like "group", "ring", and "field". We shall develop very little theory in abstract algebra, but as we discuss the properties of arithmetic operations on number systems we shall point out where the definitions of various classes of algebraic structure are satisfied. Discussion of the "properties" has deliberately been avoided here, because the author feels that the best way to introduce them is by actually deriving them in the concrete context of arithmetic on numbers.

Chapter 3

Natural Numbers, $\mathbb{N}$

3.1 Peano's Axioms

People have used the Natural Numbers to count objects for many millennia, and it is perhaps because they seemed to intrinsically exist that it took until the late 19th century for an axiomatic definition of them to appear. The axioms we present below are essentially those published by Giuseppe Peano in 1889, based on ideas in Dedekind's *Was sind und was sollen die Zahlen?* which appeared the previous year. The five axioms define the set $\mathbb{N}$ as follows.

Axiom 3.1.1. $\mathbb{N}$ *has a member, denoted* 1.

This axiom defines $\mathbb{N}$ to be non-empty.

Axiom 3.1.2. *There is a function* $S : \mathbb{N} \to \mathbb{N}$, *called the **Successor Function**.*

The domain and codomain of S are both $\mathbb{N}$, i.e. $\forall n \in \mathbb{N}, \exists S(n) \in \mathbb{N}$. So far, the only member of $\mathbb{N}$ that we know of is 1, so we apply the function S to 1 and introduce a symbol, $2 := S(1)$. We now have a new member of $\mathbb{N}$, and we can apply S to that and introduce a further symbol, $3 := S(2)$. We could proceed further, introducing more symbols for more members of $\mathbb{N}$ – but wait! How do we know that $S(1)$ is not equal to 1 itself, or that $S(3)$ is not 2, i.e. going back to a previously introduced member of $\mathbb{N}$? The next two axioms ensure that our idea of introducing a *new* member of $\mathbb{N}$ each time we apply the Successor Function is valid.

Axiom 3.1.3. $1 \neq S(n) \ \forall n \in \mathbb{N}$.

The number 1 is not the successor of any natural number; the successor process never brings us back to where we started.

Axiom 3.1.4. S *is an injection.*

From the definition of "injection", this means that if $m \neq n$, then $S(m) \neq S(n)$ (or, as contrapositive, if $S(m) = S(n)$, then $m = n$): no two natural numbers can have the same successor. This prevents the successor process from going back to a number that has previously been defined as a successor.

Axiom 3.1.5 (The Principle of Induction). *Suppose* $A \subseteq \mathbb{N}$. *If: (a)* $1 \in A$, *and (b)* $n \in A \Rightarrow S(n) \in A$, *then* $A = \mathbb{N}$.

Here A is a set defined by some property, for example, an equation to be satisfied. We may want to prove a theorem that *every* Natural Number has this property. The axiom states that two conditions are sufficient for this: (a) the number 1 must have the property, and (b) a *consequence* of a general natural number n having the property is that $S(n)$

must also have the property. Note that we do not need to show that any particular natural number apart from 1 has the property.

We can liken Induction to having a ladder which extends upwards from the ground, maybe extending out of sight. We want to know if we are able to reach any rung of the ladder. To do this, it is sufficient, (a) to be able to step onto the first rung, and (b) when already on the ladder, to be able to move from one rung to the next.

Induction is often presented as a useful technique for proving that a theorem is true for all natural numbers – which it certainly is – but it is far more than that. It is the foundation on which the Natural Number system is built (as we shall see when developing the theory of Natural Numbers); and it is specific to this one number system (do not attempt to prove theorems for any other number system using induction!).

Some terminology associated with induction is as follows. The verification that 1 satisfies the theorem is called the **anchor** or **basis** of the induction; the supposition that some general n satisfies the theorem is called the **inductive hypothesis**; and the process of showing that $S(n)$ satisfies the theorem (using the inductive hypothesis as a premise) is called the **inductive step**. Note that the inductive hypothesis will typically take exactly the same form as the theorem to be proved (because the hypothesis is the supposition that the theorem is true for some $n \in \mathbb{N}$), and the inductive step then involves verifying a similar proposition but with $S(n)$ substituted for n throughout. Of course, the variable on which the induction is done may be symbolised by any letter, not necessarily "n". When writing proofs by induction below, we shall write "(a)" and "(b)" as above as pointers to indicate respectively the anchor and the inductive step.

It is useful at this stage to draw attention to an immediate consequence of the axioms:

Corollary 3.1.6. *If $n \in \mathbb{N}\backslash\{1\}$, $\exists m \in \mathbb{N}$ such that $n = S(m)$.*

This says that every natural number apart from 1 is the successor of another natural number. A formal proof of this corollary would seem pedantic: simply let A be the set of numbers which are either 1 or the successor of a natural number, and the corollary follows immediately from the Principle of Induction.

There is one aspect of the above axioms that is controversial. We have used the symbol 1 ("one") for the distinguished member of $\mathbb{N}$ which is not the successor of another member and which is used when anchoring an induction; but many mathematicians prefer the symbol 0 ("zero"). And at this stage it is simply a different choice of symbols; only when arithmetic operations are defined (in Sections 3.2–3.4) do differences between the properties of 1 and 0 emerge. We shall define 0 as an Integer in the next chapter, after which you may like to return to this chapter and rewrite any of the definitions, theorems, and proofs in Sections 3.2–3.4 that would be different with 0 as the basis member of $\mathbb{N}$ and having the same properties as it does in the Integers (or you can do this on a first reading of these sections, if you have an intuitive understanding of zero and don't want to wait pedantically until it is formally defined). In fact, Peano did use 0 as the basis member of his axioms, although Dedekind had chosen 1. But most importantly, whereas definitions of arithmetic operations in $\mathbb{N}$ are affected by the choice of 0 or 1, induction is not. The Principle of Induction only requires there to be a basis member, and indeed proofs by induction can be anchored at any Integer, being valid for all numbers generated by the Successor Function from that basis member.

In order to finally settle the controversy over whether 0 should be included in $\mathbb{N}$, it has been suggested that the following rules should be adopted.

Rule 0 0 is a natural number.

Rule 1 All valid rules shall be identified by natural numbers.

[Think about it!][i]

3.2   Addition of Natural Numbers

The Principle of Induction is used not only to prove theorems but also to define operations in $\mathbb{N}$; hence our claim that it is the foundation of the Natural Number system. We now define the operation of addition inductively.

Definition 3.2.1. $\forall n, m \in \mathbb{N}$:
  (a) $n + 1 := S(n)$;
  (b) $n + S(m) := S(n + m)$.

By an "inductive definition", we mean that we have first defined the operation (adding) with the number 1, and then defined the operation with $S(m)$ on the supposition that we already know how to operate with m. [Note that with two symbols (n and m) representing natural numbers, we can do induction on either of them, but not both at the same time. Here we have done induction on m.] We can see immediately that our definition of addition satisfies the requirements of a binary operation: since $n \in \mathbb{N}$ and $m \in \mathbb{N}$ the domain is $\mathbb{N} \times \mathbb{N}$; while on the right-hand sides of both parts of the definition, Axiom 3.1.2 gives that $S(n) \in \mathbb{N}$ and, from the inductive hypothesis that $n + m \in \mathbb{N}$, that $S(n + m) \in \mathbb{N}$; so the codomain is $\mathbb{N}$.

Note that part (a) of the definition gives rise to the common practice of verifying a property for $n + 1$, rather than $S(n)$, in the inductive step of proofs by induction. However, the use of $S(n)$ yields greater clarity in proofs, and so this will be the practice throughout this book.

We should check that our formal definition of addition accords with the intuitive understanding of addition that we have had since childhood. Firstly, part (a) of the definition says that adding 1 yields the "next" number. Next, having defined $2 := S(1)$ and $3 := S(2)$ (see following Axiom 3.1.2), we can add 2 using part (b) of Definition 3.2.1:

$$n + 2 = n + S(1) = S(n + 1),$$

i.e. $n + 2$ is the next number after $n + 1$. Similarly,

$$n + 3 = n + S(2) = S(n + 2),$$

so $n + 3$ is the next number after $n + 2$; and these results and the obvious continuation of the process are certainly what we expect intuitively. However, it is more important to verify certain general properties of addition that we have become so familiar with that we take them for granted; firstly, the rule that when we add two numbers, it doesn't matter which way round we write the numbers either side of the $+$ sign. This is not at all obvious from Definition 3.2.1, and we shall see that it takes considerable effort to prove it from that definition; we shall require two lemmas before proving the desired theorem.

Lemma 3.2.2. $1 + n = n + 1 \ \forall n \in \mathbb{N}$.

Proof. (a) For $n = 1$:

$$1 + n = 1 + 1$$
$$= n + 1.$$

---
[i]Thanks to Joshua Ross for drawing my attention to these rules.

(b) Given the inductive hypothesis, $1 + n = n + 1$, we must show that $1 + S(n) = S(n) + 1$. Now,

$$
\begin{aligned}
1 + S(n) &= S(1 + n) &&[\textit{Def. 3.2.1, (b)}]\\
&= S(n + 1) &&[\textit{Inductive hypothesis}]\\
&= (n + 1) + 1 &&[\textit{Def. 3.2.1, (a)}]\\
&= S(n) + 1. &&[\textit{Def. 3.2.1, (a)}] \qquad \square
\end{aligned}
$$

Lemma 3.2.3. $(m + 1) + n = (m + n) + 1 \ \forall m, n \in \mathbb{N}$.

Proof. By induction on n:
(a) For $n = 1$:

$$
\begin{aligned}
(m + 1) + n &= (m + 1) + 1\\
&= (m + n) + 1.
\end{aligned}
$$

(b) Given the inductive hypothesis, $(m + 1) + n = (m + n) + 1$, we must show that $(m + 1) + S(n) = (m + S(n)) + 1$. Now,

$$
\begin{aligned}
(m + 1) + S(n) &= S((m + 1) + n) &&[\textit{Def. 3.2.1, (b)}]\\
&= S((m + n) + 1) &&[\textit{Inductive hypothesis}]\\
&= S(S(m + n)) &&[\textit{Def. 3.2.1, (a)}]\\
&= S(m + S(n)) &&[\textit{Def. 3.2.1, (b)}]\\
&= (m + S(n)) + 1. &&[\textit{Def. 3.2.1, (a)}] \qquad \square
\end{aligned}
$$

Theorem 3.2.4. $m + n = n + m \ \forall m, n \in \mathbb{N}$.

Proof. By induction on m:
(a) For $m = 1$:

$$
\begin{aligned}
m + n &= 1 + n\\
&= n + 1 &&[\textit{Lemma 3.2.2}]\\
&= n + m.
\end{aligned}
$$

(b) Given the inductive hypothesis, $m + n = n + m$, we must show that $S(m) + n = n + S(m)$. Now,

$$
\begin{aligned}
S(m) + n &= (m + 1) + n &&[\textit{Def. 3.2.1, (a)}]\\
&= (m + n) + 1 &&[\textit{Lemma 3.2.3}]\\
&= (n + m) + 1 &&[\textit{Inductive hypothesis}]\\
&= S(n + m) &&[\textit{Def. 3.2.1, (a)}]\\
&= n + S(m). &&[\textit{Def. 3.2.1, (b)}] \qquad \square
\end{aligned}
$$

We have shown that the same result is obtained regardless of which number is placed first and which is second in the addition of natural numbers. This property of a binary operation on a set of mathematical objects, that the result does not depend on the order in which the objects are placed, is called **commutativity**. So an abstract definition is:

Definition 3.2.5. A binary operation $*$ on a set A is said to be **commutative** if

$$r * s = s * r \ \forall r, s \in A.$$

We have shown in Theorem 3.2.4 that addition is commutative on $\mathbb{N}$. Many binary operations on many other kinds of mathematical objects are commutative, but there also exist very many non-commutative binary operations.

Our next theorem concerns the order in which two successive additions are done.

Theorem 3.2.6.
$$(m + n) + p = m + (n + p) \ \forall m, n, p \in \mathbb{N}.$$

Proof. By induction on p: the choice of the right-hand variable is because in the definition of addition the induction is on the right-hand variable. Proofs by induction on m or n would be possible, but would involve the use of the commutative property, which is not required in the proof presented below.

(a) For $p = 1$:

$$
\begin{aligned}
(m + n) + p &= (m + n) + 1 & \\
&= S(m + n) & [\textit{Def. 3.2.1, (a)}] \\
&= m + S(n) & [\textit{Def. 3.2.1, (b)}] \\
&= m + (n + 1) & [\textit{Def. 3.2.1, (a)}] \\
&= m + (n + p).
\end{aligned}
$$

(b) Given the inductive hypothesis, $(m + n) + p = m + (n + p)$, we must show that $(m + n) + S(p) = m + (n + S(p))$. Now,

$$
\begin{aligned}
(m + n) + S(p) &= S((m + n) + p) & [\textit{Def. 3.2.1, (b)}] \\
&= S(m + (n + p)) & [\textit{Inductive hypothesis}] \\
&= m + S(n + p) & [\textit{Def. 3.2.1, (b)}] \\
&= m + (n + S(p)). & [\textit{Def. 3.2.1, (b)}] \qquad \square
\end{aligned}
$$

Here we have three objects combined using two successive instances of the same binary operation; and the theorem says that the same result is obtained regardless of which of those binary operations is performed first. This property of a binary operation is called **associativity**. So an abstract definition is:

Definition 3.2.7. A binary operation $*$ on a set A is said to be **associative** if

$$(r * s) * t = r * (s * t) \ \forall r, s, t \in A.$$

We have shown in Theorem 3.2.6 that addition is associative on $\mathbb{N}$. Associativity is frequently found as a property of binary operations in many areas of mathematics; indeed, it is more common than commutativity, but still not entirely universal.

The associative property of addition on $\mathbb{N}$ means that if we are adding many natural numbers, there is no need to include any brackets to indicate the order of doing operations, and the commutative property means that we can rearrange the numbers in any order we like. So for example, if m, n, p, q are natural numbers, $(m + n) + (p + q)$ can be rearranged as $(n + (q + p)) + m$, or as $q + p + n + m$ where the absence of brackets means that we can do the additions in any order we like.

We shall see that commutativity and associativity of addition are common to all our number systems. However, the following theorem is very particular to the natural numbers, and indeed to our choice of 1 (rather than 0) as the basis number.

Theorem 3.2.8. *Given any $m \in \mathbb{N}$, $\nexists n \in \mathbb{N}$ such that $n + m = m$.*

Note that the commutative property means that if this theorem is true, then it is also true that $\nexists n \in \mathbb{N}$ such that $m + n = m$. Thus, if we add two natural numbers, the result is *not* equal to either of the numbers being added.

Proof. By induction on m:
(a) For $m = 1$:

$$n + m = n + 1$$
$$= S(n) \qquad\qquad [Def.\ 3.2.1,\ (a)]$$
$$\neq 1 \qquad\qquad [Axiom\ 3.1.3]$$
$$= m.$$

(b) Given the inductive hypothesis, $n + m \neq m$, we must show that $n + S(m) \neq S(m)$. Now,

$$n + S(m) = S(n + m) \qquad\qquad [Def.\ 3.2.1,\ (b)]$$
$$\neq S(m). \qquad\qquad [Inductive\ hypothesis\ and\ Axiom\ 3.1.4]. \qquad \square$$

Note how in both the anchor and the inductive step a single inequality among a sequence of equalities yields a valid verification of inequality (see Section 2.3.3 on "Transitivity and Proofs"). Also note that in the final step the inductive hypothesis implies that $n + m \neq m$, and Axiom 3.1.4 then implies that $S(n + m) \neq S(m)$.

If there did exist a natural number n which, when added to any other $m \in \mathbb{N}$, gave a result equal to that m, the number n would be an **identity element** for addition. The abstract definition is:

Definition 3.2.9. Let $*$ be a binary operation on a set A. If $\exists i_* \in A$ such that

$$i_* * a = a \quad \text{and} \quad a * i_* = a \quad \forall a \in A,$$

then i_* is an **identity element** for the operation $*$ on A.

So if the identity element is combined with any member of A, that member is left unchanged. To prove that a member of A is an identity element for $*$, it needs to be combined with a general member of A both on the left and the right of the binary operation, unless $*$ is already known to be commutative in which case either left or right is sufficient. Theorem 3.2.8 says that there is no identity element for addition in $\mathbb{N}$.

The next theorem may appear to be introducing the concept of subtraction; however, that will be formally defined at the end of this chapter.

Theorem 3.2.10. $\forall m, n, p \in \mathbb{N}$: *if $n + m = p + m$, then $n = p$.*

This is the **Cancellation Law for addition** (we are cancelling the addition of m, not subtracting m). Unlike previous theorems and lemmas in this section, which have been equalities or an inequality, we are here needing to prove an implication. This is done by a sequence of implications (see Section 2.3.3 on "Transitivity and Proofs").

Proof. By induction on m.
(a) For $m = 1$:

$$n + m = p + m \Rightarrow n + 1 = p + 1$$
$$\Rightarrow S(n) = S(p) \qquad\qquad [Def.\ 3.2.1,\ (a)]$$
$$\Rightarrow n = p. \qquad\qquad [Axiom\ 3.1.4]$$

(b) Given the inductive hypothesis, $n + m = p + m \Rightarrow n = p$, we must show that $n + S(m) = p + S(m) \Rightarrow n = p$. Now,

$$n + S(m) = p + S(m) \Rightarrow S(n + m) = S(p + m) \qquad [\textit{Def. 3.2.1, (b)}]$$
$$\Rightarrow n + m = p + m \qquad [\textit{Axiom 3.1.4}]$$
$$\Rightarrow n = p. \qquad [\textit{Inductive hypothesis}] \quad \square$$

Exercise 3.2.11. *Using only Peano's Axioms, Corollary 3.1.6 and the definition of addition in $\mathbb{N}$, prove that there do not exist $m, n \in \mathbb{N}$ such that $m + n = 1$.*

3.3 Multiplication of Natural Numbers

As with addition, we provide an inductive definition of this operation, first defining multiplication by 1 and then defining multiplication by $S(m)$ on the supposition that we know how to multiply by m.

Definition 3.3.1. $\forall n, m \in \mathbb{N}$:
   (a) $n \times 1 := n$;
   (b) $n \times S(m) := (n \times m) + n$.

This satisfies the definition of a binary operation: the domain is clearly $\mathbb{N} \times \mathbb{N}$ and, on the right-hand sides of parts (a) and (b), respectively, $n \in \mathbb{N}$ and $(n \times m) + n \in \mathbb{N}$ because $n \times m \in \mathbb{N}$ (by inductive hypothesis) and addition is closed on $\mathbb{N}$. It is also easy to see that the definition accords with the intuitive notion that multiplying n by m is the same as adding m lots of n together: part (b) says that if you want to multiply n by the next number after m, you simply need to add another n.

Notes on Notation. *The familiar "times" symbol $\times$ is used here, but it also has other uses in mathematics; in particular, it is used for Cartesian product of sets. There should be no confusion, since in one case there are numbers (or symbols representing general numbers) on either side of the $\times$ sign, while in the other case there are symbols representing sets. An alternative symbol to $\times$ for multiplication is the dot, $\cdot$; indeed, most mathematicians prefer the dot and deprecate the use of $\times$ for multiplication; but dots also have other uses in mathematics, as well as being easily overlooked when hand-written, so we shall use the "times" symbol.*

It is well-established practice to omit any symbol between quantities being multiplied, i.e. to write nm for $n \times m$; we will follow this practice, only writing the $\times$ symbol where it is necessary to emphasise the operation. Another well-established convention is that where an expression involves both multiplication(s) and addition(s), the multiplication(s) are done before the addition(s) unless indicated otherwise by brackets.

We are familiar with the idea that like addition, multiplication of numbers is both commutative and associative; but as with addition, these properties need to be rigorously verified since they are not obvious from Definition 3.3.1. However, the proofs by induction are somewhat similar to those for the commutative and associative properties of addition, so we shall leave them as an exercise for the reader, as well as the proofs of the two lemmas that are required before the commutativity theorem.

Lemma 3.3.2. $1 \times n = n \times 1 \ \forall n \in \mathbb{N}$.

Proof. See Exercise 3.3.6.   □

This lemma, together with part (a) of Definition 3.3.1, yields that

$$n \times 1 = n \quad \text{and} \quad 1 \times n = n,$$

i.e. the number 1 satisfies Definition 3.2.9 as being an identity element for multiplication in the Natural Numbers. We can now ask whether there are any more identity elements for multiplication in $\mathbb{N}$, or is 1 the only natural number with this property? We could prove a theorem specific to multiplication in $\mathbb{N}$, but it is more useful to have a general theorem which applies to all binary operations in all number systems, and indeed in sets of other kinds of mathematical object.

Theorem 3.3.3. *If an identity element exists for a binary operation $*$ on a set A, it is unique.*

Proof. Suppose that i_1 and i_2 are both identity elements for $*$ on A. Then

$$\begin{aligned} i_1 &= i_1 * i_2 && [i_2 \text{ is an identity element}] \\ &= i_2. && [i_1 \text{ is an identity element}] \end{aligned}$$

We have shown that the supposed two identity elements are in fact the same.   □

This abstract theorem may be invoked whenever we find an identity element for any operation in any number system; indeed, it means that we can always refer to *the* identity element rather than *an* identity element for the given operation in the given number system. Returning to our consideration of the number 1 in $\mathbb{N}$, we now see that 1 is the unique identity element for multiplication in $\mathbb{N}$.

Lemma 3.3.4. $(m+1)n = mn + n \ \forall m, n \in \mathbb{N}$.

Theorem 3.3.5. $m \times n = n \times m \ \forall m, n \in \mathbb{N}$.

Exercise 3.3.6. *Prove that multiplication is commutative on $\mathbb{N}$. This requires you to first prove Lemmas 3.3.2 and 3.3.4, which will both be used when proving the commutativity in Theorem 3.3.5. [Hint: to prove Lemma 3.3.4, you should do the induction on n, the right-hand quantity in the multiplications, since the definition of multiplication is inductive on the right-hand variable.]*

The last theorem states that multiplication is commutative on $\mathbb{N}$. Before going on to the associative property, we prove a theorem which combines multiplication with addition.

Theorem 3.3.7. $(m+n)p = mp + np \ \forall m, n, p \in \mathbb{N}$.

Proof. By induction on p, since the induction is on the right-hand variable in Definition 3.3.1, and p is the right-hand variable here.
(a) For $p = 1$:

$$\begin{aligned} (m+n)p &= (m+n) \times 1 \\ &= m+n && [Def.\ 3.3.1,\ (a)] \\ &= m \times 1 + n \times 1 && [Def.\ 3.3.1,\ (a)] \\ &= mp + np. \end{aligned}$$

(b) Given the inductive hypothesis, $(m + n)p = mp + np$, we must show that $(m + n)S(p) = mS(p) + nS(p)$. Now,

$$
\begin{aligned}
(m + n)S(p) &= (m + n)p + (m + n) &&[\textit{Def. 3.3.1, (b)}]\\
&= (mp + np) + (m + n) &&[\textit{Inductive hypothesis}]\\
&= (mp + m) + (np + n) &&[\textit{Addition is commutative and associative}]\\
&= mS(p) + nS(p). &&[\textit{Def. 3.3.1, (b)}] \ \square
\end{aligned}
$$

Corollary 3.3.8. $p(m + n) = pm + pn \ \forall m, n, p \in \mathbb{N}.$

This follows from Theorem 3.3.7 because multiplication is commutative on $\mathbb{N}$.

Together, Theorem 3.3.7 and Corollary 3.3.8 constitute a **distributive law**: we say that **multiplication distributes over addition** in $\mathbb{N}$. The abstract definition of a distributive law is as follows.

Definition 3.3.9. Let $*$ and $\vee$ be binary operations on a set A. The operation $\vee$ **distributes over** $*$ if both

$$
r \vee (s * t) = (r \vee s) * (r \vee t)
$$

and

$$
(s * t) \vee r = (s \vee r) * (t \vee r)
$$

for all $r, s, t \in A$.

To prove distributivity, only one of the equalities in the definition needs to be verified if the operation $\vee$ is already known to be commutative on A.

We conclude this section with an associativity theorem for multiplication on $\mathbb{N}$, for which the proof will require use of the distributive law.

Theorem 3.3.10. $m \times (n \times p) = (m \times n) \times p \ \forall m, n, p \in \mathbb{N}.$

Exercise 3.3.11. *Prove that multiplication is associative on* $\mathbb{N}$, *i.e. Theorem 3.3.10. [Which variable should you do induction on, in order to avoid having to use the commutative theorem?]*

As with addition, the combination of associative and commutative properties means that we can rearrange a sequence of successive multiplications and the numbers being multiplied in any order we want, and brackets are not required. Of course, combinations of multiplication with addition can only be rearranged in ways specified by the distributive law given by Theorem 3.3.7 and Corollary 3.3.8.

3.4 Exponentiation (Powers) of Natural Numbers

Before giving the definition, a few comments about notation. Binary operations are generally represented by a symbol, such as $+$ or $\times$, but in the case of multiplication we have a shorthand which omits the symbol: nm stands for $n \times m$. The generally accepted notation for exponentiation is n^m, which involves no symbol for the operation and has a similar status to "nm" for multiplication; but it is useful to have a symbol, and since the symbol $\wedge$ has become generally accepted in computing we shall write $n \wedge m$ rather than n^m when we want to emphasise the operation of exponentiation. In particular, we use this symbol in the inductive definition:

Definition 3.4.1. $\forall n, m \in \mathbb{N}$:

(a) $n \wedge 1 := n$;

(b) $n \wedge S(m) := n \times (n \wedge m)$

Checking that this does define a binary operation proceeds similarly to the check for multiplication; in particular, on the right-hand side of part (b), $n \wedge m \in \mathbb{N}$ by inductive hypothesis and then $n \times (n \wedge m) \in \mathbb{N}$ because multiplication is closed on $\mathbb{N}$. The parentheses in the latter expression may be omitted if we adopt the usual convention that where an expression involves both multiplication(s) and exponentiation(s), the exponentiation(s) are done before the multiplication(s) unless indicated otherwise. The definition does accord with the familiar idea that taking the m'th power of n means multiplying n by itself m times: part (b) says that to get the next power of n after the m'th, you have to multiply by another n. Looking at the similarity between the definitions of multiplication and exponentiation, you may see that it should be possible to define further binary operations in a similar way, with each definition having an inductive step involving the previous operation. This idea is pursued in Investigation 1 at the end of this chapter.

There are familiar laws for combining powers, and these are stated in the following theorems. The proofs for most of them have been left as exercises for the reader, since they are proofs by induction of a similar type to those already seen.

Theorem 3.4.2. $1^n = 1 \ \forall n \in \mathbb{N}$.

Exercise 3.4.3. *Prove Theorem 3.4.2*

Note that exponentiation is *not* commutative: comparing part (a) of Definition 3.4.1 with Theorem 3.4.2, we see that $1^n \neq n^1$ unless $n = 1$. [In fact, there is only one pair of Natural Numbers n, m for which $m^n = n^m$ with $m \neq n$, namely 2 and 4. It is interesting to try to prove that there are no other such cases, although you may wish to wait until prime factorisations have been discussed in Chapter 5 before attempting this.] Furthermore, the number 1 is *not* an identity element for exponentiation, since that would require $1^n = n$ as well as $n^1 = n$.

Theorem 3.4.4. $n^m \times n^p = n^{m+p} \ \forall n, m, p \in \mathbb{N}$.

Proof. By induction on p, since the induction is on the power in Definition 3.4.1, and on the right-hand variable in the definition of addition.

(a) For $p = 1$:

$$
\begin{aligned}
n^m \times n^p &= n^m \times n^1 \\
&= n^m \times n && [\textit{Def. 3.4.1, (a)}] \\
&= n \times n^m && [\textit{Multiplication is commutative}] \\
&= n^{S(m)} && [\textit{Def. 3.4.1, (b)}] \\
&= n^{m+1} && [\textit{Def. 3.2.1, (a)}] \\
&= n^{m+p}.
\end{aligned}
$$

(b) Given the inductive hypothesis, $n^m \times n^p = n^{m+p}$, we must show that $n^m \times n^{S(p)} = n^{m+S(p)}$. Now,

$$
\begin{aligned}
n^m \times n^{S(p)} &= n^m \times (n \times n^p) && [\textit{Def. 3.4.1, (b)}] \\
&= n \times (n^m \times n^p) && [\textit{Multiplication is commutative and associative}] \\
&= n \times n^{m+p} && [\textit{Inductive hypothesis}] \\
&= n^{S(m+p)} && [\textit{Def. 3.4.1, (b)}] \\
&= n^{m+S(p)}. && [\textit{Def. 3.2.1, (b)}] \quad \square
\end{aligned}
$$

Theorem 3.4.5. $(np)^m = n^m \times p^m \; \forall n, p, m \in \mathbb{N}$.

Exercise 3.4.6. *Prove Theorem 3.4.5*

If we write Theorem 3.4.5 as

$$(n \times p) \wedge m = (n \wedge m) \times (p \wedge m),$$

we can see that it has the same form as the second equation of Definition 3.3.9 (with $\times$ in place of $*$ and $\wedge$ in place of $\vee$). So Theorem 3.4.5 could be regarded as a "one-sided distributive law": exponentiation distributes over multiplication only from the right. To have a true distributive law (i.e. "two-sided"), we would also need $n^{pm} = n^p \times n^m$, and it is easy to find a counterexample to show that this is not true (e.g. with $n = 2, p = 1, m = 1$). The next theorem shows what n^{pm} does equal.

Theorem 3.4.7. $n^{pm} = (n^p)^m \; \forall n, p, m \in \mathbb{N}$.

Exercise 3.4.8. *Prove Theorem 3.4.7*

Note that exponentiation is not associative: $(n^p)^m \neq n^{(p^m)}$ in general. Again, it is easy to find a numerical counterexample to verify this.

The four theorems on exponentiation in this section are often referred to collectively as the **Laws of Indices**.

Addition, multiplication, and exponentiation are arithmetic operations. Later, we shall define inverses of these operations, but for now, we turn to another property of many number systems: order.

3.5 Order in the Natural Numbers

We are familiar with the idea that one number may be **greater than**, or **less than**, another number; but how should we provide precise definitions of these relations between numbers in the framework of our definition of $\mathbb{N}$? One intuitive idea is that a natural number n is greater than another natural number m if n was generated after m by the Successor process. Framing this idea in a precise mathematical form is awkward but not impossible: we could define a "p'th Successor function" S^p inductively by

$$S^1(n) := S(n)$$
$$S^{S(p)}(n) := S(S^p(n)),$$

and then say that $n > m$ if $\exists p \in \mathbb{N}$ such that $n = S^p(m)$. This definition is not only awkward to write down, it would also be awkward to use in proving theorems; and it turns out that there is a much simpler, but equivalent, way to define "greater than" and "less than" in $\mathbb{N}$:

Definition 3.5.1.

    (i) $n > m$ if $\exists p \in \mathbb{N}$ such that $m + p = n$;

    (ii) $n < m$ if $\exists q \in \mathbb{N}$ such that $m = n + q$.

[From the definition of addition you can see that adding p is equivalent to finding the "p'th successor" as defined above.]

It is clear from Definition 3.5.1 that $m > n$ if and only if $n < m$. So in fact it is only necessary to define one of these relations, but it is useful to be able to write down both.

We now state two theorems which are specific to the Natural Number system, followed by theorems that the relations $>$ and $<$ are trichotomous and transitive, and so qualify as *strict orders*, as defined in Section 2.3.2; which is why we can refer to them as **order relations**.

Theorem 3.5.2.
$$S(m) > m \ \forall m \in \mathbb{N}.$$

Proof. $S(m) = m + 1$, so Definition 3.5.1(i) is satisfied with $n = S(m)$ and $p = 1$. □

Theorem 3.5.3.
$$n > 1 \ \forall n \in \mathbb{N}\backslash\{1\}.$$

Proof. Corollary 3.1.6 states that if $n \in \mathbb{N}\backslash\{1\}$, then $n = S(m)$ for some $m \in \mathbb{N}$. But then $n = m + 1 = 1 + m$, which means that $n > 1$ according to Definition 3.5.1(i). □

Theorem 3.5.4 (Trichotomy). *Given any $m, n \in \mathbb{N}$, one and only one of the following is true: (i) $n = m$, or (ii) $n > m$, or (iii) $n < m$.*

This property appears different to Definition 2.3.8 of trichotomy, which is in terms of equality and one other relation denoted $\sim$; but by taking either $>$ or $<$ to play the role of $\sim$ and noting that $m < n$ if and only if $n > m$, we see that it is the same property.

Proof of Theorem 3.5.4. First we prove that *not more than one* of the relations may be true between any given $m, n \in \mathbb{N}$.

From Theorem 3.2.8, if either $m + p = n$ or $n + q = m$, then $m \neq n$. So according to Definition 3.5.1, we cannot have $n = m$ if either $n > m$ or $n < m$. It remains to show that we cannot have both $n > m$ and $n < m$. If both were true, then

$$\begin{aligned} n &= m + p & [\textit{Def. of } n > m] \\ &= (n + q) + p & [\textit{Def. of } n < m] \\ &= n + (q + p); & [\textit{Addition is associative}] \end{aligned}$$

but $(q + p) \in \mathbb{N}$, so our result $n = n + (q + p)$ contradicts Theorem 3.2.8. Hence it cannot be true that both $n > m$ and $n < m$.

Now we prove that one of the three relations between n and m must be true, by induction on m.

(a) For $m = 1$:

If $n = 1$, then $n = m$. But if $n \neq 1$, then $n > 1$ [*Theorem 3.5.3*], i.e. $n > m$. So one of the three relations holds in all cases when $m = 1$.

(b) The inductive hypothesis is: one of the relations, $n = m$, $n > m$, or $n < m$ is true. We need to show that one of the relations, $n = S(m)$, $n > S(m)$ or $n < S(m)$ is true. Consider each of the cases in the inductive hypothesis.

(i) If $n = m$, then $n + 1 = S(m)$, so $n < S(m)$. [*Def. 3.5.1 (ii)*]

(ii) If $n > m$, then $n = m + p$ for some $p \in \mathbb{N}$. Then either:

$p = 1$, so $n = m + 1$, $\therefore n = S(m)$;

or:

$p \neq 1$, so $p = S(k) = k + 1$ for some $k \in \mathbb{N}$,

$$\therefore n = m + (k + 1)$$
$$= (m + 1) + k$$
$$\therefore n = S(m) + k,$$
$$\text{i.e. } n > S(m).$$

(iii) If $n < m$, then $n + q = m$ for some $q \in \mathbb{N}$. Then

$$(n + q) + 1 = S(m)$$
$$\therefore n + (q + 1) = S(m)$$
$$\therefore n < S(m). \qquad [\textit{Def. 3.5.1 (ii), noting that } (q + 1) \in \mathbb{N}]$$

Using just our definition of the $>$ and $<$ relations and our definition and theorems about addition, we have shown that in each of the three cases in the inductive hypothesis, one of the three relations is true between n and $S(m)$. $\qquad \square$

Theorem 3.5.5 (Transitivity).

> If $n > m$ and $m > k$, then $n > k$.
>
> If $n < m$ and $m < k$, then $n < k$.

Exercise 3.5.6. *Prove Theorem 3.5.5, using only Definition 3.5.1 and properties of addition.*

We have now verified that the relations $>$ and $<$ are strict orders. From the discussion in Section 2.3.2, we can then define corresponding weak orders, $\geq$ ("greater than or equal") and $\leq$ ("less than or equal"), where $n \leq m$ if and only if $m \geq n$. These relations have the useful property that if both $n \geq m$ and $n \leq m$, then $n = m$: there are many situations where the easiest way to show equality between two expressions is to show that both the $\geq$ and $\leq$ relations hold between them.

We now consider the effect on order relations of doing arithmetic operations on the numbers involved in the relation.

Theorem 3.5.7.
$$n > m \Leftrightarrow n + k > m + k \quad \forall n, m, k \in \mathbb{N},$$

i.e. order is preserved by addition and by cancellation of addition.

Proof. It is necessary to prove the implication both ways. We show below that $n > m \Rightarrow n + k > m + k$; the converse, that $n + k > m + k \Rightarrow n > m$, is left as an exercise for the reader; it will require use of the Cancellation Law, Theorem 3.2.10. Now,

$$n > m \Rightarrow n = m + p \quad \text{for some } p \in \mathbb{N}$$
$$\Rightarrow n + k = (m + p) + k$$
$$\Rightarrow n + k = (m + k) + p$$
$$\Rightarrow n + k > m + k. \qquad \square$$

Exercise 3.5.8. *Prove that $n + k > m + k \Rightarrow n > m$ for $n, m, k \in \mathbb{N}$.*

Before considering the effect of other arithmetic operations on order relations, we consider a property of $\mathbb{N}$ which may seem obvious, but is useful in proving various theorems, and is very different from what is found in some later number systems.

Theorem 3.5.9.

$$\nexists k \in \mathbb{N} \ \text{such that} \ n < k < S(n), \ \text{for any} \ n \in \mathbb{N}.$$

An informal statement of this theorem is that there is a "gap" between any natural number n and its successor, with no other natural number within that gap.

Proof. By induction on n:
(a) We need to show that $\nexists k \in \mathbb{N}$ such that $1 < k < S(1)$. We prove this by contradiction, supposing that such a k does exist. It satisfies two order relations.

First, $1 < k$, so $k \neq 1$ [*Trichotomy*], so $k = S(j) = j + 1$ for some $j \in \mathbb{N}$.

Next, $k < S(1)$. Since $k = j + 1$, this yields $j + 1 < 1 + 1$. Cancelling the addition of 1 according to Theorem 3.5.7, we have $j < 1$. But Theorem 3.5.3 only allows $j = 1$ or $j > 1$, so by trichotomy we cannot have $j < 1$. This contradiction implies that no k exists such that $1 < k < S(1)$.
(b) Given the inductive hypothesis that $\nexists k \in \mathbb{N}$ such that $n < k < S(n)$, we need to show that $\nexists l \in \mathbb{N}$ such that $S(n) < l < S(S(n))$. Supposing that such l does exist, a contradiction can be obtained by a procedure similar to that in the anchor step.

Exercise 3.5.10. *Complete the inductive proof of Theorem 3.5.9.*   □

Corollary 3.5.11.

$$(i) \ n < m \Leftrightarrow S(n) \leq m. \qquad (ii) \ n > m \Leftrightarrow n \geq S(m).$$
$$(iii) \ n < S(m) \Leftrightarrow n \leq m. \qquad (iv) \ S(n) > m \Leftrightarrow n \geq m.$$

Proof of (i). To verify that $n < m \Rightarrow S(n) \leq m$, suppose that it is *not* true that $S(n) \leq m$. Then by trichotomy, $S(n) > m$. To have both $n < m$ and $S(n) > m$ would imply $n < m < S(n)$, contradicting Theorem 3.5.9. So we must have $S(n) \leq m$.

Conversely, to verify that $S(n) \leq m \Rightarrow n < m$, suppose that it is *not* true that $n < m$. Then by trichotomy, $n \geq m$. To have both $S(n) \leq m$ and $n \geq m$ would imply $S(n) \leq n$ (by transitivity), contradicting Theorem 3.5.2. So we must have $n < m$.

Parts (ii), (iii), and (iv) can be verified similarly.   □

We now consider the effect of multiplication on an order relation.

Theorem 3.5.12.

$$\forall n, m, k \in \mathbb{N} : \quad \text{If} \ n > m, \quad \text{then} \ nk > mk.$$

This says that order is preserved by multiplication. But beware: this is a theorem that only applies in $\mathbb{N}$, not in any of our subsequent number systems.

Exercise 3.5.13. *Prove Theorem 3.5.12. You will need the distributive law as well as the definition of the relation $>$.*

We proved a cancellation law for addition in equalities while discussing addition in Section 3.2, and we now have Theorem 3.5.7 which includes cancellation of addition in order relations. However, no cancellation law for multiplication in equalities was stated when multiplication was originally discussed, and Theorem 3.5.12 does not include cancellation in order relations. We shall now see that trichotomy provides the best way of proving cancellation theorems for multiplication in both equalities and order relations.

Theorem 3.5.14. *For $n, m, k \in \mathbb{N}$:*
(i) If $nk = mk$, then $n = m$;
(ii) If $nk > mk$, then $n > m$.

Proof of (i). Theorem 3.5.12 states that $n > m \Rightarrow nk > mk$ and it immediately follows that $n < m \Rightarrow nk < mk$ (from the fact that $n > m$ if and only if $m < n$). Now, by trichotomy,

$$n > m \text{ or } n < m \Leftrightarrow n \neq m :$$

"not equals" is equivalent to "greater than or less than". So we have: $n \neq m \Rightarrow nk \neq mk$. The contrapositive (i.e. logical equivalent) of the last statement is: $nk = mk \Rightarrow n = m$.

Part (ii) can be proved by the same process of first considering the two "unwanted" relations under trichotomy. $\qquad\square$

Because exponentiation is not commutative, we need two theorems on the effect of exponentiation on order relations, one for a relation between the bases and the other for a relation between the powers. The same warning applies to these theorems as to Theorem 3.5.12: they only apply in $\mathbb{N}$. The theorems involve two-way implications: proofs that exponentiation preserves order require the use of the definition of exponentiation as well as the definition of order, and proofs that the cancellation of exponentiation also preserves order can then be framed using trichotomy as for Theorem 3.5.14. The proofs are left as exercises for the reader.

Theorem 3.5.15. *For $n, m, k \in \mathbb{N}$, $n > m \Leftrightarrow n^k > m^k$.*

Exercise 3.5.16. *(i) Prove that if $n > m$, then $n^k > m^k$ for $m, n, k \in \mathbb{N}$. [Hint: use induction on k, and recall Theorem 3.5.12 on multiplication with order relations.]*
(ii) Prove that if $n^k > m^k$, then $n > m$. [Hint: look at the proof of Theorem 3.5.14.]

Theorem 3.5.17. *For $n, m, k \in \mathbb{N}$ with $k > 1$: $n > m \Leftrightarrow k^n > k^m$.*

Note that if $k = 1$, then $k^n = k^m = 1 \; \forall n, m \in \mathbb{N}$.

Exercise 3.5.18. *Prove Theorem 3.5.17. [Hint: it may be useful to prove a lemma such as $k^q > 1$ for all $k > 1$ and $q \in \mathbb{N}$. You may use any of the theorems on exponentiation and any of the theorems on order up to Theorem 3.5.14.]*

3.6 Bounded Sets in $\mathbb{N}$

Trichotomy means that order can be determined between any two numbers. We can then extend this ordering to sets of as many numbers as we like, and we have an intuitive idea of what is meant by the "greatest member" and "least member" of such a set of numbers. In this section, we will define such ideas rigorously, which is important because the four main ordered number systems in this book (Natural Numbers, Integers, Rational Numbers, and Real Numbers) each have distinctive properties in relation to the existence of greatest and least members of sets of their numbers.

Before looking at the distinctive properties of $\mathbb{N}$, we need to define some concepts and prove some theorems which apply in any number system with order relations $<$ and $>$ satisfying trichotomy and transitivity. We shall use the notation $\mathbb{S}$ to denote a general ordered number system, A for a subset of $\mathbb{S}$, and give some simple examples using subsets of $\mathbb{N}$.

Definition 3.6.1. (i) A set $A \subseteq \mathbb{S}$ is **bounded above** if $\exists k \in \mathbb{S}$ such that $a \leq k \; \forall a \in A$. Any such k is called an **upper bound** for A.

(ii) A set $A \subseteq \mathbb{S}$ is **bounded below** if $\exists l \in \mathbb{S}$ such that $a \geq l \; \forall a \in A$. Any such l is called a **lower bound** for A.

Example 3.6.2. Let $A = \{3, 6, 7\} \subseteq \mathbb{N}$. Then:
9 is an upper bound for A: considering each member of A, we have

$$3 \leq 9, \; 6 \leq 9, \; 7 \leq 9.$$

2 is a lower bound for A: considering each member of A, we have

$$3 \geq 2, \; 6 \geq 2, \; 7 \geq 2.$$

Definition 3.6.3. (i) Suppose $g \in A$ and g is an upper bound for A; then g is a **greatest member** or **maximum** of A; we write $g = \max A$.
   (ii) Suppose $h \in A$ and h is a lower bound for A; then h is a **least member** or **minimum** of A; we write $h = \min A$.

Note the two conditions required for a "greatest member" of A: membership of the set A, and being an upper bound; and similarly for a "least member".

Example 3.6.4. Let $A = \{3, 6, 7\} \subseteq \mathbb{N}$. Then:
   7 is a greatest member of A: $7 \in A$ and $a \leq 7 \; \forall a \in A$ (so 7 is an upper bound).
   3 is a least member for A: $3 \in A$ and $a \geq 3 \; \forall a \in A$ (so 3 is a lower bound).

The use of the indefinite article ("*a* greatest member", "*a* least member") in the above definition and example may have seemed rather odd: why not *the* greatest or least member? The two conditions that constitute the definitions of "greatest member" and "least member" do not immediately imply that there can only be one number satisfying the conditions; this uniqueness needs to be proved.

Theorem 3.6.5. *If* $\max A$ *exists, it is unique. If* $\min A$ *exists, it is unique.*

We shall prove the case of $\max A$, and you might like to write down the similar proof for $\min A$. As usual with uniqueness proofs, we start by supposing non-uniqueness and show that the two supposed greatest members are in fact the same. To do this, we use the observation (see following Exercise 3.5.6) that if both the relations $\geq$ and $\leq$ apply between two numbers, those numbers must be equal.

Proof of Theorem 3.6.5. Suppose that g_1 and g_2 are both greatest members of A. Then $g_1 \in A$ and g_2 is an upper bound for A, so $g_1 \leq g_2$ [*Definition 3.6.1*]. Also, $g_2 \in A$ and g_1 is an upper bound for A, so $g_2 \leq g_1$. We now have $g_1 \leq g_2$ and $g_1 \geq g_2$, so that $g_1 = g_2$: the two supposed greatest members are in fact the same. $\qquad\square$

If a set A is bounded above, then there is a non-empty set of upper bounds for A; in this book, we shall adopt the notation that U_A is the set of upper bounds for a set A. Similarly, if A is bounded below, we shall denote by L_A the set of lower bounds for A. It is understood that U_A and L_A (if they exist) are subsets of the same number system as A.

Definition 3.6.6. (i) Suppose the set $A \subseteq \mathbb{S}$ is bounded above. If U_A has a least member, then $\min U_A$ is called the **least upper bound** or **supremum** of A, denoted $\sup A$.
   (ii) Suppose the set $A \subseteq \mathbb{S}$ is bounded below. If L_A has a greatest member, then $\max L_A$ is called the **greatest lower bound** or **infimum** of A, denoted $\inf A$.

Note that if a set has a supremum or infimum, that supremum or infimum is *unique*. This follows from the definitions of supremum and infimum as least and greatest members of sets of upper and lower bounds, with Theorem 3.6.5 telling us that least and greatest members are unique.

Example 3.6.7. Let $A = \{3, 6, 7\} \subseteq \mathbb{N}$. We shall temporarily abandon any semblance of rigour, by noting:

(i) $U_A = \{7, 8, 9, \ldots\}$, where we do not specify which numbers are included in "$\ldots$", but simply observe that it is intuitively clear that $\min U_A = 7$, so that by definition of supremum, $\sup A = 7$;

(ii) $L_A = \{1, 2, 3\}$, where we do not verify rigorously that we have a complete list of members of L_A (try it yourself!), but simply observe that $\max L_A = 3$, so that by definition of infimum, $\inf A = 3$.

Comparing the results in this example with those found in Example 3.6.4, we see that $\sup A = \max A$ and $\inf A = \min A$. An obvious question is, "Could we have known that the supremum would be the same as the maximum, and the infimum the same as the minimum, without having to calculate the supremum and infimum?" The following theorem answers that question; indeed, if we had wanted to calculate $\sup A$ and $\inf A$ rigorously in Example 3.6.7, we would essentially have needed to follow the argument in the proof of this theorem.

Theorem 3.6.8. *(i) If* $\max A$ *exists, then* $\sup A$ *exists and* $\sup A = \max A$.
(ii) If $\min A$ *exists, then* $\inf A$ *exists and* $\inf A = \min A$.

Proof of (i). Let $g = \max A$; so $g \in U_A$ and $g \in A$ [*Def. 3.6.3 (i)*]. Let u be an upper bound for A. Since $g \in A$, we have $g \leq u$, by Definition 3.6.1 for upper bound; indeed $g \leq u \; \forall u \in U_A$, which means that g is a lower bound for U_A. We have shown that $g \in U_A$ and g is a lower bound for U_A, so $g = \min U_A = \sup A$ [*Def. 3.6.6 (i)*].

A similar proof applies to part (ii) of the theorem. $\qquad \square$

If the supremum and maximum were identical in all cases, and similarly for the infimum and minimum, then there would be no point in defining supremum and infimum separately from maximum and minimum. But the theorem is a one-way implication: if a maximum or minimum exists, then a supremum or infimum will be identical to the respective maximum or minimum; but if a supremum or infimum exists, the theorem does not guarantee the existence of a maximum or minimum. Theorems to be proved below imply that in $\mathbb{N}$ the supremum and maximum are actually identical in all cases, and similarly for infimum and minimum. The same will later be seen to be true in the Integers, but in the Rational and Real Numbers, we shall see that sets may have a supremum and/or infimum without the respective maximum and/or minimum existing. Having established the concepts of least and greatest member, we now prove theorems on these concepts which are specific to the Natural Numbers.

Theorem 3.6.9 (The Well-Ordering Principle). *Every non-empty subset of* $\mathbb{N}$ *has a least member.*

Like the Principle of Induction, the Well-Ordering Principle is distinctive to the Natural Numbers, and not found in any other number system. In fact, it can be used as an axiom for $\mathbb{N}$, instead of the Principle of Induction: given the first four of Peano's Axioms, in which the properties of the Successor function are established, we can state that there is a strict order $<$ (as defined in Section 2.3.2 in purely set-theoretic terms) with $n < S(n) \; \forall n \in \mathbb{N}$; this allows us to define "least member", so we can then pose the Well-Ordering Principle as the fifth axiom. One can then prove the Principle of Induction from the Well-Ordering Principle; but our approach is to begin with Induction as an axiom and then prove Well-Ordering. The fact that each principle can be proved from the other means that there is complete logical equivalence between them, so the choice of which principle to use as an axiom makes no difference to the properties of $\mathbb{N}$.

Before launching into the proof, we outline the method. Recall that a least member of a set A is both a lower bound for A and a member of A. It is easy to show that any non-empty

$A \subseteq \mathbb{N}$ is bounded below, so the set L_A is non-empty; we shall then show that there exists one lower bound $l \in L_A$ such that $S(l) \notin L_A$ (i.e. the successor of l is *not* a lower bound), and that this l is a member of A; so l is the required least member.

Proof of Theorem 3.6.9. Given $A \subseteq \mathbb{N}$, we have $a \geq 1 \ \forall a \in A$ [*Theorem 3.5.3*]. So 1 is a lower bound for A: $1 \in L_A$, so the set L_A is non-empty.

If $a \in A$, then $S(a) \notin L_A$, since $S(a) > a$ whereas a lower bound would have to be $\leq a$. But $S(a) \in \mathbb{N}$; so $L_A \neq \mathbb{N}$.

We already have $1 \in L_A$; if we had $S(l) \in L_A$ for every $l \in L_A$, induction would give $L_A = \mathbb{N}$, contradicting our finding that $L_A \neq \mathbb{N}$. So there must exist $l \in L_A$ with $S(l) \notin L_A$. Because l is a lower bound for A, $a \geq l \ \forall a \in A$.

We now prove by contradiction that $l \in A$: we suppose that $l \notin A$, which means that $a \neq l \ \forall a \in A$. Combining this with $a \geq l$ (see previous paragraph), we have $a > l \ \forall a \in A$. Corollary 3.5.11 (ii) then gives $a \geq S(l) \ \forall a \in A$. But this means that $S(l)$ is a lower bound for A, contradicting our earlier definition of l as a number for which $S(l) \notin L_A$. This contradiction means that our supposition that $l \notin A$ was false. So we have $l \in A$ which, combined with $l \in L_A$, means that l is the least member of A. $\qquad \square$

Next, notice how the following theorem on greatest member differs from the Well-Ordering Principle:

Theorem 3.6.10. *If a non-empty subset of $\mathbb{N}$ is bounded above, then it has a greatest member.*

Whereas *every* non-empty subset of $\mathbb{N}$ has a least member, having a greatest member is conditional on being bounded above. [Of course, we could write "if and only if" in this theorem, since the definition of greatest member includes being bounded above.] To prove this theorem one can use some of the ideas in the proof of the Well-Ordering Principle, but care is required; it is not simply a case of replacing "<" with ">", etc.

Exercise 3.6.11. *Prove Theorem 3.6.10.*

3.7   Cardinality, Finite and Infinite Sets

In Section 1.3, we stated that Set Theory is considered to be the foundation for all of mathematics, in particular the theory of number systems; but certain important concepts in Set Theory are defined in terms of Natural Numbers, and these will be explored here. In particular we shall define **finite** and **infinite** sets; many of the theorems about finite sets will seem so obvious as not to need proving, but it is important that we do provide rigorous proofs, especially since the properties of infinite sets are often far from obvious.

An important tool for our definitions and proofs is the **counting set**, C_n, defined for each $n \in \mathbb{N}$. The definition is inductive:

Definition 3.7.1. (a) $C_1 := \{1\}$;
(b) $C_{S(n)} := C_n \cup \{S(n)\}$.

Thus we have $C_2 = \{1, 2\}$, $C_3 = \{1, 2, 3\}$, etc. The following theorem may be regarded as giving an alternative definition of these counting sets:

Theorem 3.7.2. *For each $n \in \mathbb{N}$:*

$$C_n = \{k \in \mathbb{N} : k \leq n\}.$$

Note that this immediately implies that n is the greatest member of C_n.

Proof. By induction on n.

(a) For $n = 1$: from Theorem 3.5.3 and trichotomy, $k \leq 1$ only if $k = 1$. So

$$\{k \in \mathbb{N} : k \leq 1\} = \{1\} = C_1.$$

(b) The inductive hypothesis is that $C_n = \{k \in \mathbb{N} : k \leq n\}$, i.e. that $k \in C_n$ if $k \leq n$ and that $k \notin C_n$ if $k > n$. We need to show that $k \in C_{S(n)}$ if $k \leq S(n)$ and that $k \notin C_{S(n)}$ if $k > S(n)$, where Definition 3.7.1 says that $k \in C_{S(n)}$ if and only if either $k \in C_n$ or $k = S(n)$.

Now, $k \leq S(n)$ means that either: $k = S(n)$, so $k \in C_{S(n)}$ immediately, or: $k < S(n)$, in which case $k \leq n$ [*Corollary 3.5.11 (iii)*] so $k \in C_n$ [*Inductive Hypothesis*] and hence $k \in C_{S(n)}$ [*Definition 3.7.1*].

On the other hand, if $k > S(n)$, then $k \neq S(n)$, and also $k \notin C_n$ because $S(n) > n$ and transitivity then yields $k > n$; but the conditions, $k \neq S(n)$ and $k \notin C_n$, are together equivalent to $k \notin C_{S(n)}$ [*Definition 3.7.1*]. $\square$

The development of the theory below makes much use of Theorem 2.3.6, that the existence of a bijection between sets is an equivalence relation, together with the following theorem about counting sets:

Theorem 3.7.3. *There does not exist a bijection between C_n and C_m if $n \neq m$.*

Proof. First note that $n \neq m$ means $n > m$ or $n < m$, but we only need to consider the case $n > m$. This is because if $n < m$, then $m > n$, and the reasoning below can be repeated with the notations n and m swapped. So we proceed by induction on n, proving that there is no bijection between C_n and C_m if $n > m$.

(a) If $n = 1$, there does not exist $m \in \mathbb{N}$ such that $n > m$: the set C_m cannot exist, and C_n certainly cannot have a bijection to a non-existent set.

(b) The inductive hypothesis is that there does not exist a bijection between C_n and C_m, for any $m < n$. We need to show that there does not exist a bijection between $C_{S(n)}$ and C_m, for any $m < S(n)$. Our proof proceeds by assuming that there *is* a bijection $f : C_{S(n)} \to C_m$, and showing that this leads to a contradiction.

Our supposed bijection maps each member of $C_{S(n)}$ to one and only one member of C_m. In particular, it maps $S(n) \in C_{S(n)}$ to some $k \in C_m$. Now define a function $\tilde{f} : C_n \to C_m \backslash \{k\}$ by $\tilde{f}(j) = f(j)$ for each $j \in C_n$; so $\tilde{f}$ does exactly the same as f, except that for $\tilde{f}$ we have removed $S(n)$ from the domain and the corresponding number k from the codomain. We still have the one-to-one correspondence between the remaining members of the domain and codomain: $\tilde{f}$ is a bijection if and only if f is a bijection. Consider two possibilities.

(i) If $m = 1$, then we can only have $k = 1$, so $C_m \backslash \{k\} = \emptyset$. But C_n is non-empty (since $n \geq 1$), and there cannot be a bijection $\tilde{f}$ from a non-empty set to the empty set. Hence a bijection f from $C_{S(n)}$ to C_1 cannot exist, for any $n \in \mathbb{N}$.

(ii) For $m > 1$, we have $m = S(p)$ for some $p \in \mathbb{N}$, and $n > p$ (since $n > m = S(p) > p$). First consider the case where $k = m$. Then $C_m \backslash \{k\} = C_{S(p)} \backslash \{S(p)\} = C_p$, so $\tilde{f}$ is a bijection from C_n to C_p where $p < n$: this contradicts the inductive hypothesis. On the other hand, if $k \neq m$, we can define a bijection $g : C_m \backslash \{k\} \to C_p$ by $g(j) = j$ for $j \neq m$ and $g(m) = k$: see Figure 3.1. Now, bijections are transitive [*Theorem 2.3.6*], so the composition $g \circ \tilde{f}$ is a bijection from C_n to C_p, again contradicting the inductive hypothesis. $\square$

We can now define our central concepts:

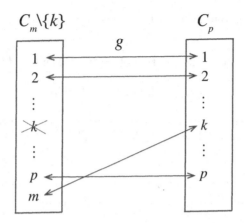

FIGURE 3.1: The bijection g between $C_m \backslash \{k\}$ and C_p, where $m = S(p)$. Each number in $C_m \backslash \{k\}$ is mapped to the same number in C_p, except that because k is absent from $C_m \backslash \{k\}$ and m is absent from C_p, we map m from the former to k in the latter.

Definition 3.7.4. If there exists a bijection between a set A and the set C_n for some $n \in \mathbb{N}$, then A is **finite** and the **cardinality** of A is n; we write $|A| = n$. If no such bijection exists for any $n \in \mathbb{N}$, then the set A is **infinite**.

Informally, the cardinality of A is the number of objects in A; we **count** the members of A by bijecting[ii] them to successive natural numbers $1, 2, 3, \ldots$, i.e. to the members of a counting set C_n.

Example 3.7.5. Consider a set of Greek letters, $A = \{\beta, \epsilon, \omega\}$. We can find a bijection $f : A \to C_3$ where $C_3 = \{1, 2, 3\}$:

$$f(\beta) = 1, \ f(\epsilon) = 2, \ f(\omega) = 3.$$

So $|A| = 3$.

We intuitively expect that counting the members of a set will give an unambiguous result:

Corollary 3.7.6. *The cardinality of a finite set is unique.*

Proof. In order for a set A to have two different cardinalities n and m, it would need bijections to both C_n and C_m. Transitivity of bijections (Theorem 2.3.6) would then imply that C_n has a bijection to C_m, contradicting Theorem 3.7.3. □

3.7.1   Some Useful Notations

When we make a bijection $f : A \to C_n$ between a set A and a counting set C_n, we can use the inverse bijection to label the members of A, by denoting $f^{-1}(k) = a_k$ for each $k \in C_n$. So the members of A are labelled as $a_1, a_2, \ldots, a_n$. This labelling is useful in the following inductive definitions of the sum and product of finitely many numbers (where those numbers may be taken from any number system in which addition and multiplication are defined).

[ii]The verb, "biject" does not appear in any dictionary that the author has seen; but it ought to exist, and so will be used in this book.

Definition 3.7.7. Given numbers $a_1, a_2, \ldots$, we define $\sum_{k=1}^{n} a_k$ by

(a) $\displaystyle\sum_{k=1}^{1} a_k := a_1;$

(b) $\displaystyle\sum_{k=1}^{S(n)} a_k := \sum_{k=1}^{n} a_k + a_{S(n)}.$

Definition 3.7.8. Given numbers $a_1, a_2, \ldots$, we define $\prod_{k=1}^{n} a_k$ by

(a) $\displaystyle\prod_{k=1}^{1} a_k := a_1;$

(b) $\displaystyle\prod_{k=1}^{S(n)} a_k := \prod_{k=1}^{n} a_k \times a_{S(n)}.$

An important example of a product is the factorial, $n!$, which is the product of the first n natural numbers:

Definition 3.7.9. For each $n \in \mathbb{N}$,

$$n! := \prod_{k=1}^{n} k.$$

3.7.2   Finite Sets, Their Subsets and Injections

Theorem 2.3.6 says that having a bijection is an equivalence relation; and an equivalence relation on a set T generates a partition of T. But what is T in this case? Theorem 2.3.6 simply refers to "a set of sets"; here we can take T to be the set of all finite sets. All finite sets with a given cardinality n have bijections to C_n and hence to each other; so for each $n \in \mathbb{N}$ there is an equivalence class consisting of all sets with cardinality n. Informally, we think of these sets which have bijections to each other as being of the same size.

But what kinds of relations can exist between finite sets of different sizes? If there is an *injection* from set A to set B, so no member of B is the image of more than one member of A, then it is intuitively obvious that A cannot be bigger than B: in terms of cardinalities, we expect that $|A| \leq |B|$. And if A is a proper subset of B, we think of A as being smaller than B: we expect that $|A| < |B|$. These are the principal ideas in this subsection; although they seem obvious, establishing them rigorously requires some effort. We first need two lemmas about subsets of counting sets, since cardinalities of finite sets are defined by reference to counting sets; the proofs of these lemmas require some reasoning similar to that in the proof of Theorem 3.7.3, but after that the proofs of the main theorems are more straightforward.

Lemma 3.7.10. *A non-empty subset of a counting set C_n has cardinality no greater than n; i.e. if $D \subseteq C_n$ for some $n \in \mathbb{N}$, then $|D| \leq n$.*

Proof. By induction:
(a) The only non-empty subset of C_1 is C_1 itself, and $1 \leq 1$.
(b) The inductive hypothesis is that if $D \subseteq C_n$, then $|D| \leq n$. Now consider D_S, a subset of $C_{S(n)}$, where $C_{S(n)} := C_n \cup \{S(n)\}$ [*Definition 3.7.1*]: we need to show that $|D_S| \leq S(n)$.

If $S(n) \notin D_S$, then $D_S \subseteq C_n$, so by the inductive hypothesis, $|D_S| \leq n < S(n)$.

On the other hand, if $S(n) \in D_S$, then $D_S = D \cup \{S(n)\}$ where $D \subseteq C_n$. There is a bijection $f : D \to C_m$ for some $m \leq n$ [*Inductive hypothesis*]. We show that $|D_S| = S(m)$ by defining a bijection $f_S : D_S \to C_{S(m)}$:

$$f_S(d) := f(d) \text{ for } d \in D,$$
$$f_S(S(n)) := S(m).$$

But $S(m) \leq S(n)$ if $m \leq n$, so $|D_S| \leq S(n)$ as required.   □

Lemma 3.7.11. *A non-empty* proper *subset of a counting set C_n has cardinality* less *than n; i.e. if $D \subset C_n$ for some $n \in \mathbb{N}$, then $|D| < n$.*

Proof. There does not exist a non-empty proper subset of C_1, so $n \neq 1$ and hence $n = S(p)$ for some $p \in \mathbb{N}$. If $D \subset C_n$, then $\exists k \in C_n$ with $k \notin D$. If $k = n$ then $D \subseteq C_p$, while if $k \neq n$, there is a bijection $g : D \to D_p \subseteq C_p$ by $g(j) = j$ for $j \neq n$ and $g(n) = k$, with D_p defined as the range of this bijection: this is similar to the bijection illustrated in Figure 3.1. By Lemma 3.7.10, a subset of C_p has a further bijection to C_m for some $m \leq p < S(p) = n$. Bijections are transitive, so $|D| = m < n$.   □

It is now easy to prove the theorem for proper subsets of general finite sets:

Theorem 3.7.12. *If $A \subset B$ and B is finite, then A is finite and $|A| < |B|$.*

Proof. If B is finite, there is a bijection $f : B \to C_n$ for some $n \in \mathbb{N}$. If $A \subset B$, this bijection will map A to a proper subset of C_n; according to Lemma 3.7.11, this subset has cardinality m for some $m < n$, and so $|A| = m < n = |B|$.   □

Note that the converse of this theorem is not true: if $|A| < |B|$, there is no need for A to be a subset of B; the two sets may consist of completely different objects. However, the contrapositive of the theorem yields a useful statement about infinite sets:

Corollary 3.7.13. *If $A \subset B$ and A is infinite, then B is infinite.*

We now consider how the existence of an injection between finite sets determines ordering between their cardinalities.

Theorem 3.7.14. *If there exists an injection $f : A \to B$ where A and B are finite sets, then $|A| \leq |B|$.*

Proof. By definition of *range*, any function $f : A \to B$ is surjective onto its range $f(A)$. So if f is injective from A to B, then it is bijective from A to $f(A)$, so $|A| = |f(A)|$. The range of a function is a subset of its codomain, $f(A) \subseteq B$, so from Theorem 3.7.12 (together with the case of equal sets having equal cardinalities), $|f(A)| \leq |B|$. Hence $|A| \leq |B|$.   □

The contrapositive of Theorem 3.7.14 is a principle, first stated by Peter Dirichlet, which is remarkably useful in many applications.

Corollary 3.7.15 (The Pigeonhole Principle). *There does not exist an injection from A to B if $|A| > |B|$, for finite sets A and B.*

An informal statement of this principle is that if n objects are put into m pigeonholes, where $n > m$, there must be a pigeonhole containing more than one object. To see that this is indeed a restatement of the theorem, let A be the set of objects and let B be the set of pigeonholes; then let $n = |A|$ and $m = |B|$ and recognise that an injection from A to B would require each object in A to be mapped to a *distinct* pigeonhole in B. Once the first m objects in A have been mapped to distinct pigeonholes, all pigeonholes contain an object but there remain further object(s) (because $|A| > |B|$), which can only be mapped to pigeonholes that already have an object.

Example 3.7.16. At the time of writing, the author is teaching a module on number systems to a class of 110 students. He already knows that they will not all get different marks in the exam. This is because the possible marks are natural numbers from 1 to 100, together with 0 which was briefly mentioned at the end of Section 3.1; the pigeonholes are the 101 possible marks, and the objects to be placed in the pigeonholes are the 110 students. Since $110 > 101$, there must be a mark obtained by more than one student. In more formal language, it is not possible to make an injective mapping from the set of students to the set of possible marks, because the cardinality of the student set is greater than that of the mark set.

Exercise 3.7.17. *A square tableau with n columns and n rows has one of the numbers, $1, 2, 3$ entered in each of the $n \times n$ positions. The sums of the numbers in each row, in each column, and in each of the two diagonals are calculated. Is it possible for each of these sums to be different? [Hint: How many rows, columns, and diagonals are there? How many possible values are there for the sum of n numbers each chosen from $\{1, 2, 3\}$? You should prove the lemma that if $n = m + p$, then the cardinality of the set $\{k \in \mathbb{N} : m \le k \le n\}$ is $p + 1$.]*

An important consequence of the Pigeonhole Principle is the following:

Theorem 3.7.18. *If finite sets A, B have $|A| = |B|$, and a function $f : A \to B$ is an injection, then it is a bijection.*

The theorem says that when two finite sets have the same cardinality, a function between them cannot be injective without also being surjective.

Proof. If $f : A \to B$ is not a surjection, there exists an element of B not in $f(A)$, the range of f: so $f(A) \subset B$, and hence $|f(A)| < |B| = |A|$. Now, if f is an injection, having $|f(A)| < |A|$ would contradict Corollary 3.7.15. So f cannot be an injection without also being a surjection.

3.7.3   Finiteness and Boundedness of Sets

The first theorem here applies not just to $\mathbb{N}$, but to any of the subsequent number systems that have an order relation satisfying the requirements of trichotomy and transitivity.

Theorem 3.7.19. *Every finite subset of an ordered number system has a greatest member and a least member.*

Proof. Given $A \subseteq \mathbb{S}$, where $\mathbb{S}$ is an ordered number system, we use induction on the cardinality of A. If $|A| = n$, we can label the members of A as $a_1, a_2, \ldots, a_n$.
(a) For $n = 1$, so that $A = \{a_1\}$, it is clear that the single member a_1 satisfies the definitions of both greatest member and least member (Definition 3.6.3).
(b) The inductive hypothesis is that if a finite subset of $\mathbb{S}$ has cardinality n, then it has a greatest member and a least member. We now consider a set $A_{S(n)} \subseteq \mathbb{S}$ with cardinality $S(n)$. We can write $A_{S(n)} = A_n \cup \{a_{S(n)}\}$ where A_n is a set with cardinality n. By the inductive hypothesis, A_n has a greatest member g and a least member l.
    If $a_{S(n)} < l$, then $a_{S(n)}$ satisfies the requirements to be the least member of $A_{S(n)}$, since $a_{S(n)} < l \le a_k$ for each $a_k \in A_n$; otherwise (by trichotomy), $a_{S(n)} \ge l$, and $l = \min A_{S(n)}$. Similarly, if $a_{S(n)} > g$, then $a_{S(n)}$ satisfies the requirements to be the greatest member of $A_{S(n)}$; otherwise, $a_{S(n)} \le g$, and $g = \max A_{S(n)}$. So in all cases $A_{S(n)}$ has a greatest member and a least member. □

Whereas the last theorem applies to any ordered number system, the next applies specifically to $\mathbb{N}$.

Theorem 3.7.20. *A non-empty subset of $\mathbb{N}$ is finite if and only if it is bounded above.*

Proof. Theorem 3.7.19 immediately tells us that finiteness implies being bounded above (because a greatest member is an upper bound). Conversely, let u be an upper bound for a set $A \subseteq \mathbb{N}$. Then $A \subseteq C_u$, where C_u is the counting set for the natural number u; so by Lemma 3.7.10, A is finite with $|A| \leq u$.   □

3.7.4   Infinite Sets

We have so far only defined cardinality for finite sets, using bijections to counting sets; but the idea of using bijections to specify cardinality is completely general, applying to infinite as well as finite sets:

Definition 3.7.21. If there exists a bijection between sets A and B, then $|A| = |B|$.

Because having a bijection is an equivalence relation, there is no possibility of ambiguity in the cardinality of any set.

The ordering of cardinalities for finite sets was specified by injections according to Theorem 3.7.14; we had to *prove* this theorem, because the cardinalities of finite sets are natural numbers for which order has already been defined. But for infinite sets we can use injections to *define* the ordering of cardinalities, in particular, the symbol $\leq$ when used between cardinalities of infinite sets:[iii]

Definition 3.7.22. If there exists an injection $f : A \to B$ where A, B are infinite sets, then $|A| \leq |B|$.

We can identify a set as being infinite if it does not have a bijection to any counting set (see Definition 3.7.4); but an alternative definition of an infinite set is provided by the following theorem:

Theorem 3.7.23. *A set is infinite if and only if it has a bijection to a proper subset of itself.*[iv]

Proof. We first show that having a bijection to a proper subset implies that a set is infinite. From Theorem 3.7.12, a finite set and its proper subset have different cardinalities; but finite sets having bijections to each other must have the same cardinality. Thus a set with a bijection to a proper subset of itself cannot be finite.

Before proving the converse, we consider an important example of a set with a bijection to a proper subset of itself. The Successor function is a bijection from $\mathbb{N}$ to its proper subset $\mathbb{N}\backslash\{1\}$, since each natural number has a unique successor, and each natural number apart from 1 is the successor of a unique natural number. So $\mathbb{N}$ is infinite. Furthermore, for any $M \in \mathbb{N}$, the function $f : n \mapsto n + M$ is a bijection from $\mathbb{N}$ to $\{m \in \mathbb{N} : m > M\}$: if $m > M$, there is a unique $n \in \mathbb{N}$ such that $m = n + M$. This is important because it means that we

---

[iii]We have not yet actually specified the cardinality of any particular infinite set (although we will do so at the end of this section); so for now the cardinality of an infinite set is a rather abstract concept, but we can still define relations between such cardinalities.

[iv]This idea was introduced by Georg Cantor in the 1870s and formalised by Dedekind, but the concept can be found in earlier writings, at least in theology which shares with mathematics a concern about the infinite. The famous hymn, *Amazing Grace*, written by John Newton in the 1790s, includes a verse which refers to the prospect of spending eternity in heaven:

> When we've been there ten thousand years,
> Bright shining as the sun,
> We've no less days to sing God's praise
> Than when we'd first begun.

So the set of days spent in heaven has a bijection ("We've no less days") to its proper subset in which the days in the first ten thousand years have been removed.

can use induction to prove a theorem for all Natural Numbers greater than some given M, i.e. for all $m \in \mathbb{N}$ with $m \geq M + 1$, by anchoring it at $m = M + 1$ and then performing the inductive step in the usual way. Our bijection implies that the theorem has been anchored at $n = 1$, as required by the Principle of Induction, so it is proved for all $n \in \mathbb{N}$ and hence for all $m \in \mathbb{N}$ with $m > M$.

Exercise 3.7.24.

(i) Find a natural number $M > 1$ such that $2^{M+1} > (M + 1)^2$. Then prove by induction that $2^m > m^2$ for all $m \geq M + 1$

(ii) It is actually true that $2^1 > 1^2$. What goes wrong if you try to use this as an anchor for a proof that $2^n > n^2$ for all $n \in \mathbb{N}$?

Having identified $\mathbb{N}$ as an infinite set, we now note that $\mathbb{N}$ can be considered the "smallest" infinite set, in the following sense:

Theorem 3.7.25. *If a set has a bijection to a subset of $\mathbb{N}$, it is either finite or has a bijection to $\mathbb{N}$.*

This means that any set that is infinite but appears smaller than $\mathbb{N}$ (by having a bijection to a proper subset of $\mathbb{N}$) is in fact the same size as $\mathbb{N}$, because it has a bijection to $\mathbb{N}$ itself.

Proof. Let A be a set with a bijection to some subset of $\mathbb{N}$. Count the members of A: this means bijecting them to successive natural numbers, $1, 2, 3, \ldots$. There are two possible outcomes.

Either: for each n, when a member of A has been counted as the n'th, there exists a further member of A yet to be counted as the $S(n)$'th. By the Principle of Induction, this means that every member of $\mathbb{N}$ will be bijected in the counting process.

Or: for some $n \in \mathbb{N}$, when the n'th member of A has been counted, there are no further members of A. Then A is finite, with $|A| = n$.

These are the only possibilities: since A has a bijection to a subset of $\mathbb{N}$, it will certainly not require more numbers than there are in $\mathbb{N}$ to count the members of A. $\qquad\square$

This motivates the following definition:

Definition 3.7.26. A set is **countably infinite** if it has a bijection to $\mathbb{N}$. A set is **countable** if it is either finite or countably infinite.[v]

From Theorem 3.7.25, this definition means that a set is countable if it has a bijection to a subset of $\mathbb{N}$, i.e. an injection to $\mathbb{N}$.

We can now complete the proof of Theorem 3.7.23: we need to find a bijection to a proper subset for an infinite set A. Since we now know that $\mathbb{N}$ is the smallest infinite set, A must either have a bijection to $\mathbb{N}$ (if A is countable), or else have a proper subset with a bijection to $\mathbb{N}$. So $\exists A_N \subseteq A$ such that the members of A_N can be labelled with natural numbers, $A_N = \{a_1, a_2, \ldots\}$. The Successor Function gives us a bijection $f : A_N \to A_N \backslash \{a_1\}$ by $f(a_n) = a_{S(n)}$ $\forall n \in \mathbb{N}$. There may still be further members of A not in A_N: for each $a \in A \backslash A_N$, simply let $f(a) = a$. We now have f as a bijection from A to $A \backslash \{a_1\}$. $\qquad\square$

The idea of having a bijection to a proper subset was illustrated by David Hilbert, who conceived of a hotel with countably infinitely many rooms; thus the rooms are numbered with the Natural Numbers. One evening every room is occupied, but then another guest arrives and asks for a room. "No problem", says the manager, who simply asks all the guests

---

[v]Some authors use "countable" to mean what we have called "countably infinite". But "countable" means "able to be counted", and a finite set can certainly be counted; so it does not make sense to restrict "countable" to infinite sets.

already in the hotel to move to the next room: the guest from room 1 moves to room 2, the guest from room 2 moves to room 3, etc., and the new guest moves into room 1. The hotel manager has simply used the Successor function to biject the set of rooms to a proper subset of itself, leaving a vacant room. Hilbert's hotel stands in contrast to Dirichlet's pigeonholes: when the latter are all filled, we cannot make room for any further objects because there are only finitely many pigeonholes.

Exercise 3.7.27. *Suppose that Hilbert's Hotel is full, and then a coach arrives with countably infinitely many passengers all wanting a room. How can they all be accommodated?*

Since all *countably* infinite sets have a bijection to the same set $\mathbb{N}$ (by definition), they all have the same cardinality, which is denoted by the symbol $\aleph_0$ ("aleph-zero", in which aleph is a Hebrew letter). So according to Theorem 3.7.25 we can think of $\aleph_0$ as the smallest infinite cardinal number, and using Definition 3.7.22 we can write $\aleph_0 \leq |A|$ for every infinite set A: an injection from $\mathbb{N}$ to A would be obtained by finding $A_N \subseteq A$ with a bijection to $\mathbb{N}$ (as above), and then the inverse of this bijection maps $\mathbb{N}$ injectively to A.   □

3.8   Subtraction: The Inverse of Addition

We have covered all the essential theory of Natural Numbers; so why do we need to go any further? Why are we not content with just this one number system? In Section 1.3 we pointed out that various practical applications require different number systems; but there is also a purely mathematical answer to that question: we want to *invert* the binary operations of arithmetic. The inverse of a binary operation is not the same as an inverse function, but is defined as follows.

Definition 3.8.1. If $*$ is a binary operation on a set A, its inverse, denoted $\overline{*}$, is defined by

$$t \,\overline{*}\, s = r \quad \text{if and only if} \quad r * s = t,$$

where $r, s, t \in A$.[vi]

Here, we choose two members of A, denoted t and s, for the respective roles of output and second input of the binary operation $*$, and we seek a third member of A, here denoted r, that will fulfil the role of first input so that $r * s = t$. If such an r exists, we say that $t \,\overline{*}\, s = r$: we can do the inverse of $*$ with t and s. But in general, even when $*$ is closed on A, there is no guarantee that we will be able to find $r \in A$ such that $r * s = t$ for all choices of $s, t \in A$: the inverse, $\overline{*}$, may not be closed on A and so may not satisfy the definition of a binary operation.

One may ask, why not choose s to be the first input of $*$ and seek r as a second input, rather than the way s and r are placed in Definition 3.8.1? In other words, why not the alternative definition,

$$t \,\widetilde{*}\, s = r \quad \text{if and only if} \quad s * r = t,$$

in which $\widetilde{*}$ symbolises this "second inverse"? If the original operation $*$ is commutative, then $s * r = r * s$ so the two definitions are equivalent, and we can just use the form in Definition 3.8.1. However, for a non-commutative operation, both types of inverse may be of interest, and later we shall indeed discuss both inverses of exponentiation.

[vi] Inverses of specific binary operations have their own notations, but there is no standard notation to invert a general binary operation: the overbar is used in this book for that purpose.

Turning now to the specific case of addition in $\mathbb{N}$, to invert this operation we need to find $r \in \mathbb{N}$ such that $r + s = t$ where $s, t \in \mathbb{N}$ are given. This inverse of addition is called **subtraction**, and is symbolised with the familiar minus sign, $-$:

Definition 3.8.2.

$$t - s = r \quad \text{if and only if} \quad r + s = t, \quad \text{where } r, s, t \in \mathbb{N}.$$

The Cancellation Law for addition (Theorem 3.2.10) assures us that subtraction is **well-defined** on $\mathbb{N}$: given $s, t \in \mathbb{N}$, there can only be a unique $r \in \mathbb{N}$ satisfying $r + s = t$ (if such r exists at all). We can also cancel a subtraction: if $t - s = u - s$, let r be the common result of the subtraction; then $t - s = r \Rightarrow r + s = t$ and $u - s = r \Rightarrow r + s = u$, so that $t = u$. Furthermore, following Theorem 3.5.7 which states that the order relations $<$ and $>$ are preserved by addition and cancellation of addition, we can use similar arguments to those for equalities to show that order relations are preserved by subtraction and cancellation of subtraction. To summarise, we have the following theorem.

Theorem 3.8.3.

    (i) $t - s = u - s$ *if and only if* $t = u$;

    (ii) $t - s < u - s$ *if and only if* $t < u$.

This all applies if and only if the subtractions are possible; but is it always true that $t - s$ exists in $\mathbb{N}$ for any $t, s \in \mathbb{N}$? In other words, is subtraction *closed* on $\mathbb{N}$? Definition 3.5.1 of the order relations in $\mathbb{N}$ tells us that $\exists r \in \mathbb{N}$ such that $r + s = t$ if and only if $s < t$. So, by trichotomy, $\nexists r \in \mathbb{N}$ such that $r + s = t$ if $s = t$ or $s > t$. In particular, if $s < t$ so that $t - s \in \mathbb{N}$, then $t > s$ so that $s - t \notin \mathbb{N}$: if we can subtract s from t, then we cannot subtract t from s when we are in the Natural Number system. This motivates the definition of a new number system, in which subtraction is closed. This will be done in the next chapter.

Exercise 3.8.4. *Using only the definition of subtraction (Definition 3.8.2) and the distributive law (Theorem 3.3.7 with Corollary 3.3.8), prove that*

$$k(p - n) = kp - kn,$$

where $k, n, p \in \mathbb{N}$ with $p > n$; i.e. that multiplication distributes over subtraction in $\mathbb{N}$. [Hint: it may be useful to let $p - n = m$.]

Investigations

1. Recalling how exponentiation is defined (Definition 3.4.1), we can define further binary operations in a similar way, using a notation derived from the $\wedge$ notation for exponentiation. First, the binary operation $\wedge\wedge$ is defined by

   (a) $n \wedge\wedge 1 := n$

   (b) $n \wedge\wedge S(m) := n \wedge (n \wedge\wedge m)$.

   Then the binary operation $\wedge\wedge\wedge$ is defined by

   (a) $n \wedge\wedge\wedge 1 := n$

   (b) $n \wedge\wedge\wedge S(m) := n \wedge\wedge (n \wedge\wedge\wedge m)$.

Note that $\curlywedge\!\!\curlywedge$ is a symbol representing a single operation; it does *not* represent repeated exponentiations; and similarly for $\wedge\wedge\wedge$.

Clearly we could continue defining further operations in a similar way, but to avoid writing long rows of $\wedge$ symbols for such operations, we introduce another notation, writing $n \to m \to c$ instead of $n \wedge\wedge \cdots \wedge\wedge m$, where c is the number of $\wedge$ symbols; for instance, $n \wedge\wedge\wedge\wedge m$ can be written $n \to m \to 4$.

(i) Evaluate $3 \wedge\wedge 3$ and $2 \wedge\wedge\wedge 3$. These are fairly long calculations using the above inductive definitions, and for the final numerical evaluations you will probably need a calculator or computer software.

(ii) Show by means of numerical counterexamples that $\wedge\wedge$ is not commutative and that it is not associative.

(iii) Write down the inductive definition of the binary operation $\wedge\wedge\wedge$. [Don't use the $\to$ notation at this stage.]

(iv) Evaluate $2 \to 2 \to 4$. Form a hypothesis about the value of $2 \to 2 \to c$ for all $c \in \mathbb{N}$, and prove your hypothesis. [Hint: try to write down a single definition of $m \to n \to c$ which is inductive on c and encompasses the definitions of $\wedge\wedge$, $\wedge\wedge\wedge$, etc.]

2. Choose any $m \in \mathbb{N}$ and consider the following subsets of $\mathbb{N}$:

$$M_m := \{n \in \mathbb{N} : n = mk, k \in \mathbb{N}\} \qquad E_m := \{n \in \mathbb{N} : n = m^k, k \in \mathbb{N}\}.$$

Informally, M_m is the set of all natural number multiples of m, and E_m is the set of all natural number powers of m.

(i) Is M_m or E_m finite for any $m \in \mathbb{N}$? For cases where the sets are infinite, how do you know that they are infinite?

(ii) Is (a) addition, (b) multiplication, or (c) exponentiation closed on M_m or E_m for any (or all) $m \in \mathbb{N}$? Explain your answers.

3. Below is a "theorem" that is obviously false, together with a supposed proof by induction, apparently similar to the proof of Theorem 3.7.19. What is the fallacy in the proof?

Theorem

The members of a finite subset of any number system are all equal to each other.

Proof

We use induction on the cardinality of a set $A \subseteq \mathbb{S}$, where $\mathbb{S}$ is a number system.

(a) For $n = 1$: if a set has a single member a_1, then a_1 is equal to itself.

(b) The inductive hypothesis is that in every set of cardinality n, the members are all equal to each other. We now consider a set $A_{S(n)} \subseteq \mathbb{S}$ with cardinality $S(n)$. We can denote one member of $A_{S(n)}$ as $a_{S(n)}$, and write $A_{S(n)} = A_n \cup \{a_{S(n)}\}$ where $A_n = \{a_1, a_2, \ldots, a_n\}$ is a set of cardinality n. The set $A_{S(n)} \backslash \{a_1\} = \{a_2, a_3, \ldots, a_n, a_{S(n)}\}$ also has cardinality n.

According to the inductive hypothesis, the members of A_n are all equal to each other, and the members of $A_{S(n)} \backslash \{a_1\}$ are all equal to each other; so $a_1 = a_2 = \ldots = a_n$ and $a_2 = \ldots = a_n = a_{S(n)}$. Hence $a_1 = a_2 = \ldots = a_n = a_{S(n)}$: the members of $A_{S(n)}$ are all equal to each other, which completes the induction.

4. Any natural number can, in principle, be specified in a finite number of words in the English language. For instance, here are three specifications of the same number:

"Two hundred and seventeen" [4 words]

"Thirty-one times seven" [3 words, counting a hyphenated word as a single word]

"Six cubed plus one" [4 words]

Allowing the more informal "Thirty-one sevens", we see that this particular number can be specified in two English words, but certainly not fewer than two. Larger numbers will typically require a larger number of words to specify them; for example, although "one billion" can be specified in two words, most numbers of a similar magnitude will require many more words, e.g. "nine hundred and fifty-eight million, seven hundred and nineteen thousand, two hundred and sixty-seven" [14 words].

Now consider the number specified as follows:

"The least natural number which cannot be specified in fewer than fifteen English words"

Explain everything you can find out about this number. What conclusions can you draw on the process of specifying a natural number in words?

Chapter 4

Integers, $\mathbb{Z}$

4.1 Definition of the Integers

Given $r \in \mathbb{N}$, we can add any $s \in \mathbb{N}$ to obtain $t = r + s$, so that $r = t - s$. So there is a countably infinite set of ordered pairs $(t, s) \in \mathbb{N} \times \mathbb{N}$ such that $r = t - s$; one such ordered pair for each $s \in \mathbb{N}$. We can regard all such ordered pairs as equivalent representations of r. Let us introduce the notation $\equiv$ for this concept of "equivalence": two ordered pairs are equivalent, $(t_1, s_1) \equiv (t_2, s_2)$, if they represent the same r, i.e. if $t_1 - s_1 = t_2 - s_2 = r$. But then

$$t_1 = r + s_1 \quad \text{and} \quad t_2 = r + s_2,$$

so

$$t_1 + s_2 = (r + s_1) + s_2 \quad \text{and} \quad t_2 + s_1 = (r + s_2) + s_1. \tag{4.1}$$

The right-hand sides of the two equations in (4.1) are equal according to the commutative and associative laws for addition in $\mathbb{N}$, so that

$$t_1 + s_2 = t_2 + s_1. \tag{4.2}$$

Whereas our original notion of equivalence, $t_1 - s_1 = t_2 - s_2$, was only valid in $\mathbb{N}$ when $s_1 < t_1$ and $s_2 < t_2$, the criterion (4.2) can be applied to any ordered pairs of Natural Numbers. So we define the relation $\equiv$ on $\mathbb{N} \times \mathbb{N}$ as follows:

Definition 4.1.1.

$$(t_1, s_1) \equiv (t_2, s_2) \quad \text{if and only if} \quad t_1 + s_2 = t_2 + s_1.$$

Example 4.1.2.

$$(9, 5) \equiv (6, 2) \quad \text{because} \quad 9 + 2 = 6 + 5,$$

where we have taken $t_1 = 9, s_1 = 5, t_2 = 6, s_2 = 2$ in Definition 4.1.1. Referring to the informal discussion above that Definition, we could say that both ordered pairs $(9, 5)$ and $(6, 2)$ are representations of the natural number $r = 4$. However, Definition 4.1.1 also yields that

$$(5, 9) \equiv (2, 6) \quad \text{because} \quad 5 + 6 = 2 + 9,$$

where these ordered pairs do not represent any natural number.

It is worth noting here that two ordered pairs are equivalent according to Definition 4.1.1 if one of them can be obtained by adding, or cancelling the addition of, the same quantity to/from both members of the other pair: $(t, s) \equiv (t+p, s+p)$ because $t + (s+p) = (t+p) + s$. This is often the quickest way to check equivalence between ordered pairs.

We have been using the word "equivalent", but have not so far verified that Definition 4.1.1 satisfies the requirements of an equivalence relation.

Exercise 4.1.3. *Show that the relation $\equiv$ in Definition 4.1.1 is reflexive, symmetric, and transitive (so that it qualifies as an equivalence relation).*

We can now define an Integer:-

Definition 4.1.4. An **Integer** is an equivalence class of ordered pairs of natural numbers under the equivalence relation $\equiv$ in Definition 4.1.1.

Thus equivalent ordered pairs are representations of the same, i.e. *equal*, integers. So in calculations below, we shall use the symbol $\equiv$ at any step where we are using Definition 4.1.1 to compare ordered pairs of natural numbers, but the symbol $=$ when stating the result of the calculation as a property of integers.

We started this chapter by noting how each natural number r generates an equivalence class of ordered pairs, for which we now introduce the (temporary) notation ζ_r:[i]

Definition 4.1.5. For each $r \in \mathbb{N}$,

$$\zeta_r := \{(t,s) \in \mathbb{N} \times \mathbb{N} : t = r + s\} = \{(r+s, s) \in \mathbb{N} \times \mathbb{N} : s \in \mathbb{N}\}.$$

For example, the ordered pairs $(9,5)$ and $(6,2)$ in Example 4.1.2 are members of the equivalence class ζ_4. Equivalence classes, i.e. integers, of the form ζ_r are called **Positive Integers**, and the set of all positive integers is denoted $\mathbb{Z}_+$:

$$\mathbb{Z}_+ := \{\zeta_r : r \in \mathbb{N}\},$$

where there is one and only one positive integer ζ_r corresponding to each natural number r; in other words, there is a bijection between $\mathbb{N}$ and $\mathbb{Z}_+$.

Now, every ordered pair (t,s) in an equivalence class of the form ζ_r has $t > s$. If we now consider ordered pairs (t,s) in which $t < s$, $\exists q \in \mathbb{N}$ such that $s = q + t$. Given any $q \in \mathbb{N}$ we can then define the following set:

Definition 4.1.6.

$$\eta_q := \{(t,s) \in \mathbb{N} \times \mathbb{N} : s = q + t\} = \{(t, q+t) \in \mathbb{N} \times \mathbb{N} : t \in \mathbb{N}\}.$$

For example, the ordered pairs $(5,9)$ and $(2,6)$ in Example 4.1.2 are members of the set η_4. Like ζ_r, η_q is an equivalence class under the equivalence relation $\equiv$ in Definition 4.1.1. To verify this, note first that if $(t_1, s_1) \in \eta_q$ and $(t_2, s_2) \in \eta_q$, then

$$s_1 = q + t_1 \quad \text{and} \quad s_2 = q + t_2;$$

then, adding s_2 to both sides of the first equation and adding s_1 to both sides of the second equation, we obtain $q + t_1 + s_2 = s_1 + s_2 = q + t_2 + s_1$, from which q can be cancelled to obtain $t_1 + s_2 = t_2 + s_1$ as required by Definition 4.1.1. Hence η_q satisfies the definition of an Integer: such an integer is called a **Negative Integer**, and the set of all negative integers is denoted $\mathbb{Z}_-$:

$$\mathbb{Z}_- := \{\eta_q : q \in \mathbb{N}\}.$$

By trichotomy, there is only one further possibility to be considered for ordered pairs $(t,s) \in \mathbb{N} \times \mathbb{N}$, namely where $t = s$. All ordered pairs in this class are of the form (s,s) and are equivalent to each other according to Definition 4.1.1: $(s_1, s_1) \equiv (s_2, s_2)$ because $s_1 + s_2 = s_2 + s_1$. Thus we have a single equivalence class, i.e. a single integer, which we call **Zero**:

---

[i]ζ is the Greek letter "zeta"; and η (introduced below) is the Greek letter "eta".

Definition 4.1.7.

$$0 := \{(t,s) \in \mathbb{N} \times \mathbb{N} : t = s\} = \{(s,s) \in \mathbb{N} \times \mathbb{N} : s \in \mathbb{N}\}.$$

We have now exhausted all possibilities for ordered pairs $(t,s) \in \mathbb{N} \times \mathbb{N}$, and found that they are all members of equivalence classes of one of the forms, ζ_r, η_q, or 0. So the set of all integers is $\mathbb{Z} = \mathbb{Z}_+ \cup \mathbb{Z}_- \cup \{0\}$, where the use of a form of the letter Z derives from the German word *Zahlen*, meaning "Numbers".

4.2 Arithmetic on $\mathbb{Z}$

Each time we define a new number system, we shall define binary operations called "addition" and "multiplication" in that number system. However, when the objects in different sets are defined differently, then any binary operations in the different sets must also be defined differently. Our definitions of addition and multiplication in $\mathbb{N}$ use Induction, which only applies in $\mathbb{N}$; any binary operation in $\mathbb{Z}$ must recognise the definition of Integers as equivalence classes in $\mathbb{N} \times \mathbb{N}$. So when we define any such operation in $\mathbb{Z}$ we are not entitled to use the same names or symbols as for operations in $\mathbb{N}$ – or at least, not until we have verified that the operations in $\mathbb{Z}$ perform essentially the same function as those in $\mathbb{N}$, which will be done where we establish *isomorphism* (defined below).

With an integer being an equivalence class of ordered pairs, we use an arbitrarily chosen member of that equivalence class to represent the integer, and so define operations which combine two such ordered pairs, $(t,s), (p,n) \in \mathbb{N} \times \mathbb{N}$:-

Definition 4.2.1.
$$(t,s) \oplus (p,n) := (t + p, s + n).$$

Definition 4.2.2.
$$(t,s) \otimes (p,n) := (tp + sn, tn + sp).$$

Here $\oplus$ and $\otimes$ denote new operations, to be considered as integer versions of addition and multiplication, while on the right-hand sides of the definitions there are additions and multiplications of natural numbers within each member of each ordered pair. The operations $\oplus$ and $\otimes$ are clearly closed on $\mathbb{N} \times \mathbb{N}$: the inputs on the left-hand side are ordered pairs of natural numbers, and so are the outputs on the right-hand side for any $t, s, p, n \in \mathbb{N}$. However, we are not wanting to perform operations on ordered pairs in $\mathbb{N} \times \mathbb{N}$, but on integers which are *equivalence classes* of ordered pairs in $\mathbb{N} \times \mathbb{N}$. Different members of the same equivalence class are supposed to be representations of the same integer. Suppose that $(t_1, s_1) \equiv (t_2, s_2)$ and $(p_1, n_1) \equiv (p_2, n_2)$ according to Definition 4.1.1, so that the pairs with subscript 2 represent the same integers as the respective pairs with subscript 1. We should expect the results of doing binary operations $\oplus$ and $\otimes$ on the subscript-2 pairs to be equivalent (according to Definition 4.1.1) to the results of doing the binary operations on the subscript-1 pairs. Otherwise, the results of the binary operations would depend on how we write down the integers being operated on, which would clearly be nonsensical. The binary operations are said to be **well-defined** if the outputs don't depend on which representations of the numbers we use as inputs; checking that $\oplus$ and $\otimes$ are well-defined is simple though a little tedious:

Exercise 4.2.3. *Show that addition is a well-defined operation on the Integers. To do this,* *write down the results of the additions,*

$$(t_1, s_1) \oplus (p_1, n_1) \quad and \quad (t_2, s_2) \oplus (p_2, n_2),$$

and then show that these results are equivalent according to Definition 4.1.1 if

$$(t_1, s_1) \equiv (t_2, s_2) \quad and \quad (p_1, n_1) \equiv (p_2, n_2).$$

Definition 4.2.1 for $\oplus$ is straightforward, whereas Definition 4.2.2 is more complicated (but note that to obtain the first member of the ordered pair on the right, you add the product of first members to the product of second members of the left-hand ordered pairs, while the second member on the right is obtained by adding the two "cross-products" (first member of one pair times second member of other pair)). However, we can motivate this definition as follows. Recall that each positive integer corresponds to a unique natural number, in the sense given in Definition 4.1.5: if $t > s$ and (t, s) is a member of the equivalence class ζ_r, then $t - s = r$ for some $r \in \mathbb{N}$. So if we want to apply the operation $\otimes$ to two positive integers ζ_r and ζ_m, where $r = t - s$ and $m = p - n$, consider multiplying the natural numbers r and m:

$$r \times m = (t - s)(p - n)$$
$$= (tp + sn) - (tn + sp)$$

in which the last expression is a natural number corresponding to the ordered pair on the right of Definition 4.2.2. This is an intuitive justification of Definition 4.2.2, using familiar properties of subtraction that we have not yet formally derived; but a clearer and more rigorous way of comparing $\oplus$ and $\otimes$ with addition and multiplication in $\mathbb{N}$ is to note that corresponding to natural numbers r and m we have ordered pairs $(r + s, s)$ and $(m + s, s)$ (where we can use the same symbol s in both cases because our operations are well-defined). Then

$$(r + s, s) \oplus (m + s, s) = (r + s + m + s, s + s) \qquad [Def.\ 4.2.1]$$
$$\equiv ((r + m) + s, s) \qquad [Def.\ 4.1.1]$$

and

$$(r + s, s) \otimes (m + s, s) = ((r + s)(m + s) + s^2, (r + s)s + s(m + s)) \qquad [Def.\ 4.2.2]$$
$$= (rm + sm + rs + s^2 + s^2, rs + s^2 + sm + s^2)$$
$$\equiv (rm + s, s), \qquad [Def.\ 4.1.1]$$

where the last line follows according to Definition 4.1.1 because

$$(rm + sm + rs + s^2 + s^2) + s = (rm + s) + (rs + s^2 + sm + s^2).$$

The results of these calculations show that when we perform $\oplus$ or $\otimes$ on positive integers corresponding to natural numbers r and m, the results are the positive integers corresponding to natural numbers $r + m$ and $r \times m$. This is an example of **isomorphism**:

Definition 4.2.4. A set A with a binary operation $*$ is **isomorphic** to a set B with a binary operation $\circledast$ if:
  (i) There is a bijection $f : A \to B$;
  (ii) $f(a_1 * a_2) = f(a_1) \circledast f(a_2) \ \forall a_1, a_2 \in A$.

Here, a_1, a_2 are members of A and $f(a_1), f(a_2)$ are the corresponding members of B; for example, there is a bijection $f : \mathbb{N} \to \mathbb{Z}_+$ in which $f(r) = \zeta_r$ where the positive integer ζ_r may be represented by $(r+s, s)$. Part (ii) of the definition says that doing the operation $*$ on members of A gives a result which corresponds to the result of doing $\circledast$ on the corresponding members of B. In particular, we have shown that there are two isomorphisms between $\mathbb{N}$ and $\mathbb{Z}_+$: the results of the operation $+$ in $\mathbb{N}$ correspond to the results of $\oplus$ in $\mathbb{Z}_+$, and the results of $\times$ in $\mathbb{N}$ correspond to the results of $\otimes$ in $\mathbb{Z}_+$. Since $\oplus$ and $\otimes$ in $\mathbb{Z}_+$ operate in the same way as $+$ and $\times$ in $\mathbb{N}$, we can now cease using separate symbols and start using $+$ and $\times$ in $\mathbb{Z}$ (not just in $\mathbb{Z}_+$). Furthermore, if the numbers in $\mathbb{Z}_+$ play the same roles (with respect to the arithmetic operations) as the corresponding numbers in $\mathbb{N}$, there is no need for separate notations, and we can now use "r" for a positive integer rather than "ζ_r". For negative integers, there is a bijection $g : \mathbb{N} \to \mathbb{Z}_-$ in which $g(q) = \eta_q$ where η_q may be represented by the ordered pair $(t, q + t)$, and we shall henceforth use the familiar notation "$-q$" for the negative integer corresponding to the natural number q, rather than persisting with "η_q". Indeed, there is an isomorphism between $\mathbb{N}$ with the operation $+$ and $\mathbb{Z}_-$ with the operation $\oplus$, so that in the familiar notation $(-q) + (-p) = -(q+p)$; but there is no isomorphism between $\mathbb{N}$ with $\times$ and $\mathbb{Z}_-$ with $\otimes$.

Exercise 4.2.5. *Show that there is an isomorphism between $\mathbb{Z}_-$ with $\oplus$ and $\mathbb{N}$ with $+$, by evaluating $(t, q + t) \oplus (t, n + t)$ according to Definition 4.2.1 and using the equivalence relation (Definition 4.1.1).*

Also evaluate $(t, q + t) \otimes (t, n + t)$ and explain why there is no isomorphism between $\mathbb{Z}_-$ with $\otimes$ and $\mathbb{N}$ with $\times$.

When defining new number systems, we will always want their arithmetic operations to retain the same properties as corresponding operations in the previous number systems, as well as possibly having new useful properties. The next theorem states that addition and multiplication in $\mathbb{Z}$ do have the same properties as in $\mathbb{N}$, while subsequent theorems in this section will reveal new properties, including that we have achieved our aim of devising a number system in which subtraction is closed.

Theorem 4.2.6. *(i) Addition is commutative in $\mathbb{Z}$.*
(ii) Addition is associative in $\mathbb{Z}$.
(iii) Multiplication is commutative in $\mathbb{Z}$.
(iv) Multiplication is associative in $\mathbb{Z}$.
(v) The integer 1 is the unique identity element for multiplication in $\mathbb{Z}$.
(vi) Multiplication distributes over addition in $\mathbb{Z}$.

The isomorphisms guarantee that these theorems are true for *positive* integers since they are true for natural numbers; but to show that they are true for all integers, we need to return to the definitions of the operations in $\mathbb{Z}$ in terms of ordered pairs of natural numbers. To illustrate this procedure, we shall prove just one of the properties in Theorem 4.2.6, leaving proofs of the others as exercises for the reader.

Proof of Theorem 4.2.6 (iii). Let (t, s) and (p, n) be arbitrary ordered pairs of natural numbers. Then

$$(t, s) \times (p, n) = (tp + sn, tn + sp) \qquad [Def.\ 4.2.2]$$
$$= (pt + ns, ps + nt) \quad [Multiplication\ and\ addition\ are\ commutative\ in\ \mathbb{N}]$$
$$= (p, n) \times (t, s). \qquad [Def.\ 4.2.2] \ \square$$

Note that in the second line of the proof we used the known theorems that multiplication and addition are commutative *in the Natural Numbers* to do manipulations within each member of an ordered pair, on the way to proving that our new operation (albeit now notated with the same symbol ×, which is permissible due to the isomorphism) is commutative *in the Integers*.

Exercise 4.2.7. *Prove the remaining parts of Theorem 4.2.6, using the definitions of the operations* ⊕ *and* ⊗*. [For part (v), note that the ordered-pair form of the integer* 1 *is* $(1 + s, s)$*; once you have shown that* $(1 + s, s)$ *satisfies the definition of an identity element for multiplication, how do you know that it is* unique*, without any further calculation?]*

The next theorem is a property of $\mathbb{Z}$ which is in contrast to Theorem 3.2.8 for $\mathbb{N}$.

Theorem 4.2.8. *The number* 0 *is the unique identity element for addition in* $\mathbb{Z}$*.*

Proof. The number 0 is represented by the ordered pair (s, s), while an arbitrary integer x may be represented by (p, n). Then

$$(p, n) + (s, s) = (p + s, n + s) \qquad\qquad [\textit{Def. 4.2.1}]$$
$$\equiv (p, n), \qquad\qquad\qquad [\textit{Def. 4.1.1}]$$

i.e. $x + 0 = x$. Since addition is commutative on $\mathbb{Z}$ [*Theorem 4.2.6 (i)*], we also have $0 + x = x$, so 0 fulfils Definition 3.2.9 as an identity element for + on $\mathbb{Z}$. Theorem 3.3.3 then guarantees its uniqueness in this role. □

The objective of defining the Integers was to have a number system in which the *inverse operation* of addition is closed. In order to verify that this has been achieved, we need to define the related concept of *inverse element*.

Definition 4.2.9. Suppose a binary operation $*$ on a set A has an identity element i_*. Given $a \in A$, if there exists $\bar{a} \in A$ such that

$$a * \bar{a} = i_* \quad \text{and} \quad \bar{a} * a = i_*,$$

then $\bar{a}$ is an **inverse element** for a under the operation $*$.

Observe that, according to this definition, if $\bar{a}$ is an inverse element of a, then a is an inverse element of $\bar{a}$ also.

Before applying this abstract definition to addition in $\mathbb{Z}$, we provide a uniqueness theorem for inverse elements, also in abstract terms.

Theorem 4.2.10. *If* $a \in A$ *has an inverse element under an associative binary operation* $*$*, then that inverse element is unique.*

Proof. As usual with uniqueness proofs, we begin by assuming *non-uniqueness*: suppose that $a \in A$ has two inverses, $\overline{a_1}$ and $\overline{a_2}$, under $*$. Then

$$\overline{a_1} = \overline{a_1} * i_* \qquad\qquad [i_* \text{ is identity element}]$$
$$= \overline{a_1} * (a * \overline{a_2}) \qquad\qquad [\overline{a_2} \text{ is inverse of } a]$$
$$= (\overline{a_1} * a) * \overline{a_2} \qquad\qquad [* \text{ is associative}]$$
$$= i_* * \overline{a_2} \qquad\qquad\qquad [\overline{a_1} \text{ is inverse of } a]$$
$$= \overline{a_2}.$$

Hence $\overline{a_1}$ and $\overline{a_2}$ are actually the same. □

The application to addition in $\mathbb{Z}$ is this:

Theorem 4.2.11. *Every integer has a unique inverse element under addition. In ordered-pair representation, (n, p) is the additive inverse of (p, n).*

Proof. Consider an arbitrary integer, represented by the ordered pair (p, n). We have

$$(p, n) + (n, p) = (p + n, n + p) \qquad\qquad [\textit{Def. 4.2.1}]$$
$$\equiv (s, s), \qquad\qquad\qquad\qquad [\textit{Def. 4.1.1}]$$

where (s, s) is a representation of the identity element for addition in $\mathbb{Z}$. Commutativity of addition in $\mathbb{Z}$ then yields that $(n, p) + (p, n) \equiv (s, s)$ also. So under addition, the integer represented by (n, p) is an inverse of the integer represented by (p, n), and is the *unique* inverse, by Theorem 4.2.10. $\qquad\qquad\qquad\qquad\qquad\qquad\qquad\qquad\qquad\qquad\qquad\qquad$ □

Now, if $p > n$ then $p - n = x$ for some $x \in \mathbb{N}$ and (p, n) represents the positive integer which we initially notated as ζ_x but now write simply as x because of the isomorphisms. Furthermore, in this case (n, p) represents the negative integer which we write as $-x$. If $p < n$ the roles of the positive and negative integers are simply interchanged. Either way, we have that positive and negative integers form additive inverse pairs:

$$x + (-x) = 0 \quad \text{and} \quad (-x) + x = 0. \qquad\qquad (4.3)$$

In the case where $p = n$ we have the equation $0 + 0 = 0$. Note that (4.3) can be taken as an equation between any integer x and its additive inverse $-x$, regardless of whether x is positive or negative.

This is an appropriate point at which to introduce a definition and notation which is applied here in $\mathbb{Z}$, but also works in any number system in which there are additive inverse pairs of positive and negative numbers:

Definition 4.2.12. The **modulus** or **absolute value** of an integer x is denoted $|x|$ and defined as

$$|x| := \begin{cases} x & \text{if } x \in \mathbb{Z}_+ \\ -x & \text{if } x \in \mathbb{Z}_- \\ 0 & \text{if } x = 0. \end{cases}$$

So $|x|$ is always positive, unless it is zero; if x is already positive or zero, its modulus is simply equal to x, but if x is negative we take its additive inverse element (which is positive) as its modulus.[ii]

We now make the connection between the concepts of inverse element and inverse operation, first in abstract terms and then with application to addition in $\mathbb{Z}$.

Theorem 4.2.13. *If an inverse element $\bar{a}$ exists for some $a \in A$ under an associative binary operation $*$, then*

$$b \,\bar{*}\, a = b * \bar{a} \,\forall\, b \in A,$$

where $\bar{*}$ is the inverse operation of $*$ (see Definition 3.8.1).

[ii]The notation for modulus is the same as that for cardinality, introduced in Section 3.7. But they are completely different concepts, and there should not be any confusion, since cardinality applies to sets, whereas modulus applies to numbers.

Proof. Let $c = b * \overline{a}$ (where $c \in A$ because $\overline{a}$ exists in A and $*$ is closed on A). Then

$$
\begin{aligned}
c * a &= (b * \overline{a}) * a \\
&= b * (\overline{a} * a) && [* \text{ is associative}] \\
&= b * i_* \\
&= b, && [i_* \text{ is identity element}] \\
\therefore c &= b \,\overline{*}\, a && [Def.\ 3.8.1] \\
\text{i.e. } b * \overline{a} &= b \,\overline{*}\, a. && \square
\end{aligned}
$$

Corollary 4.2.14. *If every element of a set A has an inverse element under a binary operation $*$, then the inverse operation $\overline{*}$ is closed on A.*

Theorem 4.2.13 shows how to do the inverse operation with any two elements of A when it is closed on A. In particular, it follows from Theorem 4.2.11 and Corollary 4.2.14 that:

Corollary 4.2.15. *Subtraction is closed on $\mathbb{Z}$. For all $r, x \in \mathbb{Z}$,*

$$ r - x = r + (-x) $$

or, for integers with ordered-pair representations (t, s) and (p, n),

$$ (t, s) - (p, n) = (t, s) + (n, p). $$

Exercise 4.2.16. *The ordered-pair representation of -2 is $(s, 2 + s)$. Prove that*

$$ (-2) \times x = (-x) - x $$

by writing x as an arbitrary ordered pair of natural numbers, (p, n), and using the definition of multiplication, Theorem 4.2.11 and Corollary 4.2.15.

Note on notation. *We have used the symbol $-$ (the "minus sign") in three different ways in $\mathbb{Z}$. If x is a natural number, with x also denoting the corresponding positive integer, then the corresponding negative integer is denoted $-x$. But the symbol $-$ is also used to indicate an inverse element under addition: if x is any integer (positive, negative or zero), we write $-x$ for the integer such that $x + (-x) = (-x) + x = 0$. These two uses coincide if x is positive: for example, 3 is a positive integer, -3 is a negative integer and is also the additive inverse of 3. The third use of the symbol $-$ is for subtraction. These three uses of the minus sign will also appear in subsequent number systems.*

Example 4.2.17. In the equation

$$ 5 - (-3) = -(-8) $$

(which is true!), there are four minus signs: the first indicates subtraction, the second and fourth indicate negative integers, and the third indicates additive inverse.

We have established that subtraction is a binary operation on $\mathbb{Z}$, but it is certainly not commutative or associative. However, there are many useful properties, in particular relating subtraction or additive inverses to addition and multiplication. These could be proved using ordered-pair representations; however, a better approach is to understand that Theorems 4.2.6, 4.2.8, and 4.2.11 together establish the **algebraic structure** of $\mathbb{Z}$, and to deduce all further properties from the structure in these theorems. The main advantage of this approach is that we shall be defining further number systems with the same algebraic structure, so any properties established once on the basis of that structure will then be valid for all number systems possessing the structure. However, we first need to consider one arithmetic operation which was defined in $\mathbb{N}$ but has not yet been mentioned in the context of $\mathbb{Z}$.

It is not possible to define exponentiation in such a way that the operation is closed in $\mathbb{Z}$ and such that $\mathbb{Z}_+$ and $\mathbb{N}$ with their respective exponentiation operations are isomorphic. However, it is useful to define positive integer powers of integers, i.e. x^n where $x \in \mathbb{Z}$ and $n \in \mathbb{Z}_+$. The isomorphisms that we already have between $\mathbb{Z}_+$ and $\mathbb{N}$ mean that we can treat positive integers as being the same as natural numbers, so we shall simply say that the inductive definition of exponentiation (Definition 3.4.1) applies: the induction is on the power n which is a positive integer so is treated like a natural number, while the definition involves multiplication which we have defined for all integers x. Defining x^n in this way clearly ensures that exponentiation in $\mathbb{Z}_+$ is isomorphic to exponentiation in $\mathbb{N}$, and that the Laws of Indices established by induction in Section 3.4 still apply; but the operation is not closed in $\mathbb{Z}$: we have not defined x^n where $n \in \mathbb{Z}_-$.

4.3 Algebraic Structure of $\mathbb{Z}$

4.3.1 An Abelian Group

We have established that the addition operation has what may be regarded as a "full set" of properties in $\mathbb{Z}$:

A1. Addition is commutative in $\mathbb{Z}$.

A2. Addition is associative in $\mathbb{Z}$.

A3. There is an identity element for addition in $\mathbb{Z}$.

A4. Every member of $\mathbb{Z}$ has an inverse element under addition.

Binary operations with these properties appear in many different areas of mathematics, so it is useful to introduce some notation and terminology that can be used across all such areas. Given a set A with a binary operation $*$, we use the notation $\langle A, * \rangle$ when the set with its binary operation are to be considered together as an entity. Likewise, a set A with two binary operations $*$ and $\vee$ may be denoted as $\langle A, *, \vee \rangle$. Such **algebraic structures** are classified according to the properties of the binary operation(s). We start with a classification involving just one binary operation.

Definition 4.3.1. If a set A with a binary operation $*$ has the four properties:
(i) $*$ is commutative;
(ii) $*$ is associative;
(iii) there exists an identity element for $*$ in A;
(iv) every member of A has an inverse element under $*$;
then $\langle A, * \rangle$ is called an **Abelian group** or **commutative group**.
If the binary operation does not have the commutative property but has all the others, then $\langle A, * \rangle$ is simply a **group**.

In the context of number systems, we are nearly always concerned with commutative operations, so (non-commutative) groups have little relevance to the discussion in this book; but they are very important in many other contexts. In any case, we can now say that $\langle \mathbb{Z}, + \rangle$ is an Abelian group, whereas $\langle \mathbb{Z}, \times \rangle$ is not a group because we do not have inverse elements under multiplication. So we can now prove theorems on addition in $\mathbb{Z}$ by arguments based only on the properties of Abelian groups: no knowledge of Group Theory is required, beyond the meanings of "commutative", "associative", "identity element", and "inverse element". The theorems will also involve subtraction, the inverse operation of addition: recall that doing the inverse operation $\bar{*}$ with an element a is the same as doing the operation $*$ with the inverse element $\bar{a}$ (see Theorem 4.2.13). Whereas the concept of inverse operation is used extensively in number systems, it is not generally used in more abstract work with groups, but we shall use it in abstract proofs below. The overbar notation used in this book for inverse elements and inverse operations is also *not* in general use in books on abstract algebra.

We now state a theorem in two equivalent forms: firstly in terms of addition and subtraction of integers, and secondly in terms of an operation $*$ and its inverse operation in an Abelian group.

Theorem 4.3.2. *For all $x, y \in \mathbb{Z}$:*
 (i) $-(x + y) = (-x) - y$;
 (ii) $-(x - y) = (-x) + y$.

Theorem 4.3.3. *For all $x, y \in A$ where $\langle A, * \rangle$ is an Abelian group:*
 (i) $\overline{(x * y)} = \bar{x} \bar{*} y$;

 (ii) $\overline{(x \bar{*} y)} = \bar{x} * y$.

First satisfy yourself that the two theorems are equivalent: to do this, recall that the abstract operation $*$ is equivalent to the arithmetic operation $+$, their respective inverse operations are $\bar{*}$ and $-$, and an overbar over a member of A, indicating inverse element, is equivalent to a minus sign in front of a number, indicating additive inverse. Next, recall that if an operation is both commutative and associative and we need to do k instances of the operation to combine $k + 1$ elements (for any $k \geq 2$), then we can rearrange the elements and do the operations in any order we like: see the discussion following Definition 3.2.7. This will be useful as we now prove part (i) of the abstract theorem.

Proof of Theorem 4.3.3 (i). The theorem states that the inverse element of $(x * y)$ is $\bar{x} \bar{*} y$. By definition of "inverse element", this is equivalent to saying that

$$(x * y) * (\bar{x} \bar{*} y) = i_* \quad \text{and} \quad (\bar{x} \bar{*} y) * (x * y) = i_*, \tag{4.4}$$

where i_* is the identity element for the operation $*$. So we shall prove the theorem by verifying the first equation in (4.4); the second equation follows immediately since $*$ is commutative.

Now,

$$
\begin{aligned}
(x * y) * (\bar{x} \bar{*} y) &= (x * y) * (\bar{x} * \bar{y}) && [\textit{Theorem 4.2.13}] \\
&= (x * \bar{x}) * (y * \bar{y}) && [\textit{* is commutative and associative}] \\
&= i_* * i_* && [\textit{Def. 4.2.9}] \\
&= i_*. && [\textit{Def. 3.2.9}] \quad \square
\end{aligned}
$$

In case you find it difficult to follow those four lines of algebra because of the abstract symbols, rewrite them using the usual symbols of arithmetic: $+$ instead of $*$, $-$ instead of $\bar{*}$, $(-x)$ instead of $\bar{x}$ and 0 instead of i_* (since zero is the additive identity in $\mathbb{Z}$). It is also interesting to contrast the above proof with one based on the arithmetic of integers defined as ordered pairs of natural numbers:

Proof of Theorem 4.3.2 (i). Represent x by (t, s) and y by (p, n). Then

$$
\begin{aligned}
-(x + y) &= -((t, s) + (p, n)) \\
&= -(t + p, s + n) &&[Def.\ 4.2.1] \\
&= (s + n, t + p) &&[Theorem\ 4.2.11] \\
&= (s, t) + (n, p) &&[Def.\ 4.2.1] \\
&= -(t, s) - (p, n) &&[Theorem\ 4.2.11\ and\ Corollary\ 4.2.15] \\
&= (-x) - y. &&\square
\end{aligned}
$$

Decide for yourself which proof is neater.

Exercise 4.3.4. *(a) Prove part (ii) of Theorem 4.3.2 using the arithmetic of integers defined as ordered pairs of natural numbers.*
(b) Prove part (ii) of Theorem 4.3.3 using the Abelian group properties.

4.3.2 A Commutative Ring

The multiplication operation in $\mathbb{Z}$ has properties similar to three of those listed for addition at the start of the previous subsection:

M1. Multiplication is commutative in $\mathbb{Z}$.

M2. Multiplication is associative in $\mathbb{Z}$.

M3. There is an identity element for multiplication in $\mathbb{Z}$.

We have one further algebraic property which connects the two binary operations:

D. Multiplication distributes over addition in $\mathbb{Z}$.

Combined with the properties of addition, this motivates the following classification of an algebraic structure with two binary operations.

Definition 4.3.5. A set A with binary operations $*$ and $\vee$ is called a **commutative ring** if:
   (i) $\langle A, * \rangle$ is an Abelian group;
   (ii) $\vee$ is commutative
   (iii) $\vee$ is associative
   (iv) there exists an identity element for $\vee$ in A;
   (v) $\vee$ distributes over $*$.
   If $\vee$ is not commutative, but the other properties all hold, then $\langle A, *, \vee \rangle$ is simply a **ring**.[iii]

Clearly the algebraic structure $\langle \mathbb{Z}, +, \times \rangle$ is a commutative ring.

We shall now prove several theorems on arithmetic in $\mathbb{Z}$ using only the commutative ring properties (so the theorems will be valid for any subsequent number system which is a commutative ring); although of course they can also be proved using the arithmetic of integers defined as ordered pairs of natural numbers. The first theorem is so familiar that it seems obvious, but in fact it is quite remarkable. We state it first in a familiar arithmetic form, and then in an abstract form.

[iii]Some authors exclude the requirement for an identity element for $\vee$ from the definition of a ring; for the case where the identity element is present, they would refer to a "ring with identity" or some similar phrase for what we call a "ring".

Theorem 4.3.6. $x \times 0 = 0$ *and* $0 \times x = 0$ *for all* $x \in \mathbb{Z}$.

Theorem 4.3.7. $x \vee i_* = i_*$ *and* $i_* \vee x = i_*$ *for all* x *in any ring* $\langle A, *, \vee \rangle$.

Why is it so remarkable that multiplying any number by zero should result in zero? Because zero is the *additive* identity element, and yet it has this universal property with respect to *multiplication*. This fact must be connected with the one property in Definition 4.3.5 that connects the two operations, namely the distributive law. We will see this in the abstract proof of Theorem 4.3.7, but we first provide a proof of Theorem 4.3.6 using ordered-pair arithmetic, since this is one instance where such a proof is simpler than the algebraic proof.

Proof of Theorem 4.3.6. The number 0 is represented by the ordered pair (s, s), and the arbitrary integer x by (p, n). Then

$$(p, n) \times (s, s) = (ps + ns, ps + ns) \qquad [Def.\ 4.2.2]$$
$$\equiv (s, s), \qquad [Def.\ 4.1.1]$$

i.e. $x \times 0 = 0$. The verification that $0 \times x = 0$ is similar. □

Proof of Theorem 4.3.7. Let x and y be any elements of the ring $\langle A, *, \vee \rangle$, with i_* as the identity element for $*$. Overbars indicate inverse elements under $*$, which exist because $\langle A, * \rangle$ is an Abelian group; in particular, we will use $\overline{(x \vee y)}$, the inverse of $(x \vee y)$ under $*$. Now,

$$(x \vee i_*) * (x \vee y) = x \vee (i_* * y) \qquad [\vee\ distributes\ over\ *]$$
$$= x \vee y \qquad [i_*\ is\ identity\ element]$$
$$\therefore ((x \vee i_*) * (x \vee y)) * \overline{(x \vee y)} = (x \vee y) * \overline{(x \vee y)}$$
$$\therefore (x \vee i_*) * ((x \vee y) * \overline{(x \vee y)}) = (x \vee y) * \overline{(x \vee y)} \qquad [*\ is\ associative]$$
$$\therefore (x \vee i_*) * i_* = i_* \qquad [operating\ with\ inverse\ elements]$$
$$\therefore x \vee i_* = i_*.$$

If the ring is commutative, it follows immediately that $i_* \vee x = i_*$ as well; but this remains true even if the ring is not commutative, as can be verified by repeating the above proof with appropriate quantities written on opposite sides of the $\vee$ operation. □

The amount of effort required to prove the abstract equivalent of "$x \times 0 = 0$" confirms that this theorem should not be considered obvious or trivial. It will now be used in proofs of further theorems which relate multiplication to additive inverses and subtraction; the theorems will be stated in familiar arithmetic forms, but proved in abstract algebraic forms so that it is clear that they apply to any number system which is a commutative ring.

Theorem 4.3.8. *For all* $x, y, z \in \mathbb{Z}$:
  (i) $(-x) \times y = -(x \times y)$;
  (ii) $x \times (-y) = -(x \times y)$;
  (iii) $(-x) \times (-y) = x \times y$.

Proof of (i). In the notation that we have been using, (i) can be written in abstract form as

$$\overline{x} \vee y = \overline{(x \vee y)},$$

i.e. the claim is that the inverse under $*$ of $(x \vee y)$ is $\overline{x} \vee y$. So we need to verify that

$$(x \vee y) * (\overline{x} \vee y) = i_* \quad \text{and} \quad (\overline{x} \vee y) * (x \vee y) = i_*.$$

We verify the first of these equations, using properties of commutative rings; the second then follows since $*$ is commutative.

Now,

$$(x \vee y) * (\overline{x} \vee y) = (x * \overline{x}) \vee y \qquad\qquad [\vee \text{ distributes over } *]$$
$$= i_* \vee y \qquad\qquad [\text{operating with inverse element}]$$
$$= i_*. \qquad\qquad [\textit{Theorem 4.3.7}] \qquad \square$$

Exercise 4.3.9. *(a) Verify Theorem 4.3.8 (i) using the arithmetic of ordered pairs. [Start by representing x as (p, n) and y as (t, s).]*
(b) Write down abstract versions of parts (ii) and (iii) of the theorem, and prove them using properties of commutative rings.

Theorem 4.3.10. *Multiplication distributes over subtraction in ℤ; i.e. for all $x, y, z \in ℤ$:*

$$x(y - z) = xy - xz \quad and \quad (x - y)z = xz - yz.$$

This should not come as any surprise, since subtraction is the same as addition of an additive inverse, and we know that multiplication distributes over addition.

Exercise 4.3.11. *Rewrite the equations in Theorem 4.3.10 in our notation involving abstract operations $*$ and $\vee$, and then verify them using the properties of commutative rings. [Hint: you will need Definition 3.8.1 and Theorem 4.2.13 on inverse operations, as well as the law that multiplication distributes over addition.]*

There are some noteworthy consequences of Theorem 4.3.8. Firstly, setting $x = 1$ in part (i) of the theorem, we have the following corollary.

Corollary 4.3.12.
$$(-1) \times y = -(1 \times y) = -y :$$
multiplying an integer by -1 yields the additive inverse of the integer.

This is of interest because we obtain the *additive* inverse of any number by *multiplying* by a particular number.

Exercise 4.3.13. *Write the equation in Corollary 4.3.12 in abstract form, for an element y of a ring $\langle A, *, \vee \rangle$. Note that since the number 1 is the identity element for multiplication, which is the operation $\vee$ in our abstract notation, we can represent 1 as $i_\vee$ in abstract form. Use overbars for inverses under $*$.*

Next by setting $y = x$ in part (iii) of Theorem 4.3.8, we obtain:

Corollary 4.3.14. *For all $x \in ℤ$, $(-x)^2 = x^2$.*

Here we have used the notation of exponentiation, following the idea that exponentiation by positive integers can be defined inductively as for natural number exponents. We can take this further:

Exercise 4.3.15. *Prove by induction on m that*

$$(-x)^{2m} = x^{2m} \quad and \quad (-x)^{2m+1} = -x^{2m+1}$$

for all $x \in ℤ, m \in ℤ_+$ (treating $ℤ_+$ as being the same as ℕ).

Finally, it is important to observe that, just as 0 and 1 are *distinct* elements of ℤ, the identity elements for the two operations must be distinct in any non-trivial ring:

Exercise 4.3.16. *Prove that if $i_\vee = i_*$ in a ring $\langle A, *, \vee \rangle$, then the set A consists only of the single member i_* (so that if the identity elements for the two operations were the same, the ring would be trivial).*

4.4   Order in $\mathbb{Z}$

We wish to define order relations on $\mathbb{Z}$, satisfying the requirements of trichotomy and transitivity. To do this, we need to refer to the original definition of integers as equivalence classes of ordered pairs of natural numbers, under the equivalence relation (Definition 4.1.1) which we can write as:

$$(t, s) \equiv (p, n) \quad \text{if and only if} \quad t + n = p + s.$$

As when defining arithmetic operations on $\mathbb{Z}$, we define order relations between integers in terms of ordered pairs $(t, s), (p, n) \in \mathbb{N} \times \mathbb{N}$ chosen arbitrarily from the equivalence classes representing two integers:-

Definition 4.4.1.

$$(t, s) \oslash (p, n) \quad \text{if and only if} \quad t + n < p + s.$$

$$(t, s) \ominus (p, n) \quad \text{if and only if} \quad t + n > p + s.$$

These new relations $\oslash$ and $\ominus$ are *well-defined*: if $(t_1, s_1) \equiv (t_2, s_2)$ and $(p_1, n_1) \equiv (p_2, n_2)$, then the same relations $\oslash$ or $\ominus$ will apply between the subscript-2 pairs as between the subscript-1 pairs. The trichotomy of $\oslash$ and $\ominus$ follows immediately from their definition in terms of the trichotomous relations $<$ and $>$ in $\mathbb{N}$, and transitivity is easy to verify from Definition 4.4.1 by considering relations between three ordered pairs.

Exercise 4.4.2. *Verify that the operation $\oslash$ (see Definition 4.4.1) is transitive. [You need to show that if $(t, s) \oslash (p, n)$ and $(p, n) \oslash (q, m)$, then $(t, s) \oslash (q, m)$, where you know all the properties of the relation $<$ in $\mathbb{N}$.]*

Now recall that we have a bijection between $\mathbb{Z}_+$ and $\mathbb{N}$, with each natural number r corresponding to a positive integer which can be represented by the ordered pair $(r + s, s)$.

Theorem 4.4.3. *There is **order isomorphism** between the relations $\oslash, \ominus$ in $\mathbb{Z}_+$ and the relations $<, >$ in $\mathbb{N}$; i.e.*

$$(r + s, s) \oslash (m + s, s) \quad \text{if and only if} \quad r < m$$

and

$$(r + s, s) \ominus (m + s, s) \quad \text{if and only if} \quad r > m.$$

Proof.

$$r < m \Leftrightarrow (r + s) + s < (m + s) + s \qquad [\textit{Theorem 3.5.7, adding or cancelling s twice}]$$
$$\Leftrightarrow (r + s, s) \oslash (m + s, s) \qquad\qquad\qquad\qquad\qquad [\textit{Def. 4.4.1}]$$

and similarly with $>$ and $\ominus$. $\square$

We already have the isomorphisms between the arithmetic operations in $\mathbb{Z}_+$ and $\mathbb{N}$, which allow us to use the same symbols $+$ and $\times$ in $\mathbb{Z}$ as in $\mathbb{N}$; the order isomorphism now means that we no longer need separate symbols $\oslash$ and $\ominus$ for the order relations in $\mathbb{Z}$, but can use $<$ and $>$ (and the terminology, "less than" and "greater than") in $\mathbb{Z}$ as in $\mathbb{N}$. The isomorphisms have established that positive integers behave identically to natural numbers in every way; we can treat them as being the same objects, so effectively $\mathbb{N} = \mathbb{Z}_+$ and $\mathbb{N} \subset \mathbb{Z}$.

But what happens when order relations are applied to zero and the negative integers? We first consider some fundamental properties involving zero.

Theorem 4.4.4. *For $x, y \in \mathbb{Z}$:*
   $x > 0$ *if and only if* $x \in \mathbb{Z}_+$.
   $y < 0$ *if and only if* $y \in \mathbb{Z}_-$.

This statement that positive integers are greater than zero and negative integers are less than zero is not trivial, since the definitions of positive and negative integers are not in terms of order; but it does mean that henceforth we will be able to use the words "positive" and "negative" to mean respectively "> 0" and "< 0".

Proof of Theorem 4.4.4. A positive integer x is represented by an ordered pair (p, n) where $p > n$; a negative integer y is represented by an ordered pair (q, m), where $q < m$ [see Definitions 4.1.5 and 4.1.6 and note Definition 3.5.1 for order in $\mathbb{N}$]. The integer 0 may be represented by the ordered pair (s, s).
    Then

$$x \in \mathbb{Z}_+ \Leftrightarrow p > n$$
$$\Leftrightarrow p + s > s + n \qquad\qquad [\textit{Theorem 3.5.7}]$$
$$\Leftrightarrow (p, n) > (s, s) \qquad\qquad [\textit{Def. 4.4.1}]$$
$$\Leftrightarrow x > 0$$

and

$$y \in \mathbb{Z}_- \Leftrightarrow q < m$$
$$\Leftrightarrow q + s < s + m \qquad\qquad [\textit{Theorem 3.5.7}]$$
$$\Leftrightarrow (q, m) < (s, s) \qquad\qquad [\textit{Def. 4.4.1}]$$
$$\Leftrightarrow y < 0$$

where in the last line of each proof the order isomorphism allows us to use the standard symbols $>, <$ rather than $\oslash, \oslash$. $\square$

Next, a theorem relating multiplication to order with respect to zero:

Theorem 4.4.5. *If $x > 0$ and $y > 0$, then $xy > 0$.*

Proof. This follows immediately from the identification of being a positive integer with being greater than zero [*Theorem 4.4.4*], together with the isomorphism of positive integers with natural numbers, in which multiplication is closed. $\square$

In the Natural Numbers, order relations are preserved by addition (Theorem 3.5.7). The isomorphisms imply that this applies to positive integers; we now show that it extends to all integers.

Theorem 4.4.6. *For $x, y, z \in \mathbb{Z}$:*

$$x > y \Leftrightarrow x + z > y + z.$$

Proof. The theorem states that order is preserved by both addition and cancellation of addition; we prove the former, and leave the latter as an exercise.
    Writing the integers in ordered pair form, $x = (p, n), y = (q, m), z = (r, l)$, Definition 4.4.1 means that the order relation $x > y$ becomes

$$p + m > q + n. \qquad\qquad (4.5)$$

We also have the ordered-pair forms for the additions,

$$x + z = (p + r, n + l), \qquad y + z = (q + r, m + l). \tag{4.6}$$

Noting that (4.5) is a relation between natural numbers so that we can use Theorem 3.5.7 to add $r + l$ to both sides, we have

$$(p + m) + (r + l) > (q + n) + (r + l)$$
$$\therefore (p + r) + (m + l) > (q + r) + (n + l),$$

which according to (4.6) and Definition 4.4.1 means that $x + z > y + z$. □

Exercise 4.4.7. *Show that if $x + z > y + z$, then $x > y$.*

Since subtraction in $\mathbb{Z}$ is the same as addition of an additive inverse, we immediately have that order is also preserved by subtraction and cancellation of subtraction:

Corollary 4.4.8. *For $x, y, z \in \mathbb{Z}$:*

$$x > y \Leftrightarrow x - z > y - z.$$

In Section 4.3 we defined a *commutative ring* to be a set with two binary operations having the basic properties displayed by addition and multiplication in $\mathbb{Z}$, and we showed that further properties of these operations in $\mathbb{Z}$ (Theorem 4.3.6 to the end of Section 4.3) were consequences of the commutative ring structure. We can now define an **ordered commutative ring** as being a commutative ring which additionally has order relations with the basic properties discussed so far in this Section. We shall use the standard symbols and terminology for addition, multiplication, and zero rather than the corresponding symbols $*, \vee, i_*$ used in the abstract definitions of groups and rings in Section 4.3.

Definition 4.4.9. An **ordered commutative ring** is a set A with two binary operations, addition and multiplication, satisfying the requirements of $*$ and $\vee$ in Definition 4.3.5, and a pair of order relations $<$ and $>$ satisfying the following requirements for all $x, y, z \in A$:
(i) $x > y \Leftrightarrow y < x$
(ii) Trichotomy: either $x = y$ or $x < y$ or $x > y$
(iii) Transitivity: if $x < y$ and $y < z$, then $x < z$
(iv) If $x > y$, then $x + z > y + z$
(v) If $x > 0$ and $y > 0$, then $x \times y > 0$.

We will now derive the remaining properties of order in $\mathbb{Z}$ using just the properties in Definition 4.4.9, rather than the definitions of order and arithmetic operations originally given for $\mathbb{Z}$. This will emphasise that the subsequent properties are consequences of the ordered ring structure, and so will certainly be present in any further ordered rings that we may define (in particular the Rational and Real number systems), as well as helping us to identify that certain rings cannot be ordered (see Section 5.6 and Chapter 8). So in the theorems and proofs below, the symbols x, y, z represent integers or elements of any other ordered ring.

First, the observation (following Theorem 4.2.11) that positive and negative integers form additive inverse pairs can be generalised for ordered rings as follows:

Theorem 4.4.10. *Additive inverse elements have opposite orders relative to zero; i.e. $x > 0$ if and only if $-x < 0$, and $x < 0$ if and only if $-x > 0$.*

Proof. Adding $-x$ to each side of the relation $x > 0$, property (iv) in Definition 4.4.9 yields $0 > -x$, so that $-x < 0$ by property (i). Similarly for the case $x < 0$. □

Combining this with Theorem 4.3.8 (on multiplication of additive inverses) and property (v) of ordered rings, we obtain:

Corollary 4.4.11. *(i) If $x > 0$ and $y < 0$, or if $x < 0$ and $y > 0$, then $xy < 0$.*
(ii) If $x < 0$ and $y < 0$, then $xy > 0$.

Now recall Definition 4.2.12 for the modulus of an integer; this was in terms of whether the integer is positive or negative, which according to Theorem 4.4.4 can be interpreted in terms of order relative to zero. Theorem 4.4.5 and Corollary 4.4.11 can then be used together with Theorem 4.3.8 to obtain:

Corollary 4.4.12. *The modulus preserves multiplication:* $|x \times y| = |x| \times |y|$.

This is easily verified by considering the cases where each of x and y is positive, negative, or zero.

Setting $y = x$ in part (ii) of Corollary 4.4.11 and in property (v) of ordered rings (Definition 4.4.9), we obtain the important result:

Theorem 4.4.13. $x^2 \geq 0$ *for all x in any ordered ring, with $x^2 = 0$ only if $x = 0$.*

We can continue by induction as in Exercise 4.3.15 to find that even powers of any non-zero integer are positive, while odd powers are positive or negative according to whether the original integer is positive or negative:

Corollary 4.4.14. *For any $m \in \mathbb{Z}_+$:*
$x^{2m} > 0$ *unless $x = 0$.*
$x^{2m+1} > 0$ *if $x > 0$, and $x^{2m+1} < 0$ if $x < 0$.*

A further important property of additive inverses is:

Theorem 4.4.15. *If $x < y$, then $-x > -y$. If $x > y$, then $-x < -y$.*

Proof. Add $-x - y$ to each side of the relations $x < y$ and $x > y$ to obtain the required results, again using properties (iv) and (i) in Definition 4.4.9. $\square$

A consequence of this theorem is that order is reversed when both sides of an order relation are subtracted *from* the same number:

Corollary 4.4.16. *If $x > y$, then $z - x < z - y$. If $x < y$, then $z - x > z - y$.*

Proof. Simply add z to each side of the relations between $-x$ and $-y$ in Theorem 4.4.15. $\square$

The effect of multiplication on order relations is more interesting in ordered rings than in $\mathbb{N}$ where Theorem 3.5.12 applies:

Theorem 4.4.17.
$$\text{If } x > y \text{ and } z > 0, \text{ then } xz > yz;$$
$$\text{If } x > y \text{ and } z < 0, \text{ then } xz < yz.$$

This theorem says that multiplying both sides of an order relation by a positive number preserves the order relation, while multiplying by a negative number reverses it, i.e. changing a $>$ relation to $<$; here we have used the terms "positive/negative" for "greater/less than zero", following Theorem 4.4.4.

Proof of Theorem 4.4.17. Adding $-y$ to both sides of the relation $x > y$, we have $x - y > 0$ from property (iv) of Definition 4.4.9.

With $z > 0$, property (v) then yields $(x - y)z > 0$; hence $xz - yz > 0$ [*Theorem 4.3.10*], and adding yz to both sides yields $xz > yz$.

For the case where $z < 0$, Theorem 4.4.10 gives $-z > 0$, so property (v) yields $(x - y)(-z) > 0$, so that $-xz + yz > 0$ [*Theorems 4.3.8 and 4.3.10*]. Adding xz to both sides then gives $yz > xz$, so $xz < yz$. $\qquad\square$

An obvious corollary is that if $x < y$ (rather than $x > y$), multiplication by a positive number preserves the relation $<$ while multiplication by a negative number reverses it to $>$.

To see the advantage of using the basic properties of ordered rings for the proof, you might like to try proving part of the above theorem using the ordered-pair definitions of multiplication and order in $\mathbb{Z}$:

Exercise 4.4.18. *Use ordered-pair representations of the integers $x, y, z \in \mathbb{Z}$ to show that if $x > y$ and $z < 0$, then $xz < yz$. You should write $x = (p, n)$ and $y = (r, l)$, and note that from Theorem 4.4.4 and Definition 4.1.6, if $z < 0$ its ordered-pair representation has the form $(t, q + t)$.*

Example 4.4.19. (i) Given $4 > 2$, $3 > 0$ and $-3 < 0$, Theorem 4.4.17 yields

$$4 \times 3 > 2 \times 3 \quad \text{and} \quad 4 \times (-3) < 2 \times (-3) \quad [\text{i.e. } 12 > 6 \text{ and } -12 < -6].$$

(ii) Given $4 > -2$, $3 > 0$ and $-3 < 0$, Theorem 4.4.17 yields

$$4 \times 3 > (-2) \times 3 \quad \text{and} \quad 4 \times (-3) < (-2) \times (-3) \quad [\text{i.e. } 12 > -6 \text{ and } -12 < 6].$$

Because the result of multiplying an order relation depends on whether the multiplier is positive or negative, there is no simple cancellation law for multiplication in order relations in ordered rings such as $\mathbb{Z}$. However, we do have the following useful theorem.

Theorem 4.4.20.

(i)   If $xy > 0$,   then either: $x > 0$ and $y > 0$,   or: $x < 0$ and $y < 0$.

(ii)   If $xy < 0$,   then either: $x > 0$ and $y < 0$,   or: $x < 0$ and $y > 0$.

(iii)   If $xy = 0$,   then $x = 0$ or $y = 0$.

Exercise 4.4.21. *Verify Theorem 4.4.20 using proof by exhaustion: work through the nine cases in which each of x and y may be positive, negative, or zero, evaluating whether the product xy is positive, negative, or zero in each case.*

4.4.1   How to Solve Inequalities

Theorem 4.4.6, Corollary 4.4.8, and Theorem 4.4.20 are the basic tools for solving **algebraic inequalities**, where we want to determine the set of numbers x satisfying some order relation between two expressions involving the unknown x. These theorems hold in any ordered commutative ring (in particular the Rational and Real Numbers as well as the Integers), so methods for solving inequalities in Integers will also be valid in those number systems. The range of inequalities that can be solved in integer arithmetic is rather limited (in particular because we cannot do the inverse operation of multiplication), but nevertheless, since we have established the necessary tools it is opportune to give some examples to illustrate the use of those tools. But first we need to establish one further property of order that is specific to $\mathbb{Z}$: an extension to the Integers of Theorem 3.5.9 on the "gaps" between natural numbers.

Theorem 4.4.22.

$$\nexists k \in \mathbb{Z} \text{ such that } n < k < n+1, \text{ for any } n \in \mathbb{Z}.$$

Proof. The order isomorphism between $\mathbb{Z}_+$ and $\mathbb{N}$, together with Theorem 3.5.9, means that there is no positive integer k between positive integers n and $n+1$. There is a bijection between $\mathbb{Z}_-$ and $\mathbb{N}$ which, together with Theorem 4.4.15, ensures a "reverse ordering" of the negative integers, with no negative integer between $-r$ and $-(r+1)$ for any $r > 0$ (i.e. between n and $n+1$ where $n+1 \in \mathbb{Z}_-$). Finally, $-1 < 0 < 1$, with all positive integers being greater than 1 and all negative integers less than -1. $\square$

Just as Corollary 3.5.11 follows from Theorem 3.5.9, we can state essentially the same Corollary in a form appropriate to integers as a consequence of Theorem 4.4.22:

Corollary 4.4.23. *For* $n, m \in \mathbb{Z}$:

     (i) $n < m \Leftrightarrow n+1 \leq m$.   *(ii)* $n > m \Leftrightarrow n \geq m+1$.

     (iii) $n < m+1 \Leftrightarrow n \leq m$.   *(iv)* $n+1 > m \Leftrightarrow n \geq m$.

Before embarking on some examples, it is important to understand the logic of the process of solving inequalities (or indeed equations). We are answering the question: what are the values of the unknown (typically denoted x) which will guarantee that the inequality (or equation) is satisfied? However, the solution process starts with the inequality (or equation) and ends with the specification of x. So we are trying to ascertain that the last line in this process implies that the first line is true, whereas our calculation proceeds in logical steps from the first to the last line. Thus the solution is only valid if each step is a two-way implication, as indicated by the phrase "if and only if" or the symbol $\Leftrightarrow$. You may like to check that every step in the arguments in the examples below is a two-way implication, even where this is not made explicit with a $\Leftrightarrow$ symbol.

Our first example of solving an inequality in $\mathbb{Z}$ is an elementary example requiring only Corollary 4.4.8.

Example 4.4.24. Find all $x \in \mathbb{Z}$ satisfying: *(i)* $x + 3 > -2$,   *(ii)* $x + 3 \geq -2$.

Solution. Simply subtracting 3 from both sides of (i) yields $x > (-2) - 3 = -5$. The solution set can be written $\{x \in \mathbb{Z} : x > -5\}$ or $\mathbb{Z}_+ \cup \{-4, -3, -2, -1, 0\}$.

Since equalities obey theorems of the same forms as order relations regarding addition, subtraction, and cancellation of these operations, the solution of (ii) proceeds as for (i) but with $>$ replaced with $\geq$. So the solution set for (ii) can be written $\{x \in \mathbb{Z} : x \geq -5\}$ or $\mathbb{Z}_+ \cup \{-5, -4, -3, -2, -1, 0\}$. $\square$

More interesting cases involve **quadratic** expressions, of the form $x^2 + bx + c$. We have not formally discussed factorisation; however, the distributive and commutative laws are sufficient to show that

$$(x + m)(x + n) = x^2 + (m + n)x + mn,$$

so that an expression of the form $x^2 + bx + c$ may be written as $(x+m)(x+n)$ if by inspection we can find integers m, n such that $m + n = b$ and $mn = c$. If we have an order relation between a product of the form $(x + m)(x + n)$ and zero, we can use Theorem 4.4.20, which will require considering two cases.

Example 4.4.25. Find all $x \in \mathbb{Z}$ satisfying $x^2 > 3x + 4$.

Solution. We first need to write this as an order relation with zero on one side; by subtracting $3x + 4$ we obtain

$$x^2 - 3x - 4 > 0.$$

Observing that $1 + (-4) = -3$ and $1 \times (-4) = -4$, this becomes

$$(x + 1)(x - 4) > 0.$$

Theorem 4.4.20 then implies that

$$\text{either: } x + 1 > 0 \text{ and } x - 4 > 0, \quad \text{or: } x + 1 < 0 \text{ and } x - 4 < 0,$$

in which each order relation is of the type in Example 4.4.24, easily solved by addition or subtraction of a number.

Case 1: $x + 1 > 0$ and $x - 4 > 0 \Leftrightarrow x > -1$ and $x > 4$. Now, $4 > -1$, so transitivity implies that if $x > 4$, then $x > -1$. So the condition $x > -1$ is redundant, and our solution in this case is simply $x > 4$.

Case 2: $x + 1 < 0$ and $x - 4 < 0 \Leftrightarrow x < -1$ and $x < 4$. A transitivity argument similar to that in Case 1 yields that $x < 4$ is redundant, and we are left with $x < -1$.

Combining the cases, our solution set is $\{x \in \mathbb{Z} : x > 4 \text{ or } x < -1\}$, which can also be written as $\mathbb{Z} \backslash \{-1, 0, 1, 2, 3, 4\}$. $\square$

Quadratic expressions are polynomials of degree 2. Formally:

Definition 4.4.26. A **polynomial** of **degree** K in a number system $\mathbb{S}$ is an expression of the form

$$a_0 + a_1 x + a_2 x^2 + \cdots + a_{K-1} x^{K-1} + a_K x^K, \tag{4.7}$$

where $a_0, a_1, \ldots, a_K$ are given numbers in $\mathbb{S}$, with $a_K \neq 0$, and x is a **variable**, which may take any value from $\mathbb{S}$. Such a polynomial may be written more concisely using the sum notation of Definition 3.7.7, as

$$\sum_{k=0}^{K} a_k x^k.$$

[Note that the starting index is $k = 0$ here, rather than $k = 1$ as in the inductive Definition 3.7.7; but the meaning should be obvious. However, this extension of the notation implies that $x^0 = 1$ (to make $a_0 x^0 = a_0$); since exponentiation by zero will not be defined formally until Chapter 6, we may take the representation of 1 by x^0 here to be a convenience that makes the notation $\sum_{k=0}^{K} a_k x^k$ work.]

Given a polynomial of degree $K > 2$, expressing it as a product of linear factors of the form $(x - n)$ is not usually possible by inspection; where it is possible, solving an inequality involving such a polynomial generally requires more than one application of Theorem 4.4.20.

Example 4.4.27. Find all $x \in \mathbb{Z}$ satisfying $x^4 - 3x^2 - 4 \leq 0$.

Solution. Noting that the polynomial in this inequality is the same as that in Example 4.4.25 but with x replaced by x^2, we can write the inequality as

$$(x^2 + 1)(x^2 - 4) \leq 0.$$

Theorem 4.4.20 then implies that

$$\text{either: } x^2 + 1 \geq 0 \text{ and } x^2 - 4 \leq 0, \quad \text{or: } x^2 + 1 \leq 0 \text{ and } x^2 - 4 \geq 0,$$

Case 1: $x^2 + 1 \geq 0$ and $x^2 - 4 \leq 0 \Leftrightarrow x^2 \geq -1$ and $x^2 \leq 4$. Now, we have $x^2 \geq 0 \; \forall x \in \mathbb{Z}$

(Theorem 4.4.13), and since $0 \geq -1$ transitivity gives $x^2 \geq -1 \; \forall x \in \mathbb{Z}$. So we only need to consider the condition $x^2 \leq 4$, which can be written as

$$x^2 - 4 \leq 0.$$

Observing that $2 + (-2) = 0$ and $2 \times (-2) = -4$, this becomes

$$(x+2)(x-2) \leq 0$$

and we again use Theorem 4.4.20 to give

$$\text{either: } x + 2 \geq 0 \text{ and } x - 2 \leq 0, \quad \text{or: } x + 2 \leq 0 \text{ and } x - 2 \geq 0.$$

We now have two sub-cases:

Case 1(i): $x + 2 \geq 0$ and $x - 2 \leq 0 \Leftrightarrow x \geq -2$ and $x \leq 2$.

Case 1(ii): $x + 2 \leq 0$ and $x - 2 \geq 0 \Leftrightarrow x \leq -2$ and $x \geq 2$. Since $2 > -2$, the relation $x \geq 2$ together with transitivity implies $x > -2$, which contradicts the condition $x \leq -2$ according to trichotomy. So this case does not yield any solutions.

Case 2: $x^2 + 1 \leq 0$ and $x^2 - 4 \geq 0 \Leftrightarrow x^2 \leq -1$ and $x^2 \geq 4$. We have already noted that $x^2 \geq 0 \; \forall x \in \mathbb{Z}$, and since $0 > -1$ transitivity yields $x^2 > -1$, contradicting the condition $x^2 \leq -1$. So there are no solutions in this case.

The only solutions have been found in Case 1(i): the solution set is $\{x \in \mathbb{Z} : -2 \leq x \leq 2\}$, which can be clarified by writing as $\{-2, -1, 0, 1, 2\}$. □

Exercise 4.4.28. *Find all $x \in \mathbb{Z}$ satisfying each of the following inequalities, writing your final answers as a solution set (as in Examples 4.4.24–4.4.27).*

(i) $x^2 + 5x \leq 6$ *(ii)* $x^2 + 5x < 6$ *(iii)* $8 - 2x < x^2$
(iv) $x^4 + 5x^2 > 36$ *(v)* $x^4 + 36 > 13x^2$ *(vi)* $x^3 + 3x^2 < 10x.$

The next example using Theorem 4.4.20 is so important and useful that we shall designate it as a theorem itself:

Theorem 4.4.29. *If $x^2 < y^2$ and $y > 0$, then $x < y$.*

Informally, you might think of this theorem as giving a condition for when an order relation is preserved by taking the square root of both sides (although of course we have not yet defined "square root"!) – the condition being that the greater quantity is known to be positive.

Proof of Theorem 4.4.29.

$$x^2 < y^2 \Leftrightarrow x^2 - y^2 < 0$$
$$\Leftrightarrow (x-y)(x+y) < 0$$
$$\Leftrightarrow \text{either: } x - y > 0 \text{ and } x + y < 0,$$
$$\text{or: } x - y < 0 \text{ and } x + y > 0.$$

Case 1: $x - y > 0$ and $x + y < 0 \Leftrightarrow x > y$ and $x < -y$ so, by transitivity, $y < -y$. However, given $y > 0$, we have $-y < 0$, so transitivity yields $y > -y$. This contradiction means that no solution is possible in this case.

Case 2: $x - y < 0$ and $x + y > 0 \Leftrightarrow x < y$ and $x > -y$ so, by transitivity, $y > -y$. This is permissible since $y > 0$ and $-y < 0$, and we have shown that $x < y$ as required. □

Exercise 4.4.30. *Prove that if $x^2 < y^2$ and $y < 0$, then $x > y$.*

The order isomorphism between positive integers and natural numbers means that Theorem 3.5.15 applies in $\mathbb{Z}$, in the form that

$$x > y \Rightarrow x^n > y^n \quad \text{for } x, y \in \mathbb{Z}_+, n \in \mathbb{N}.$$

Taking the case $n = 2$ and combining with Theorem 4.4.29, we can say (informally) that an ordering is preserved when we take the square or square root of both sides, provided that all quantities involved are positive. More formally:

Corollary 4.4.31. *$x^2 > y^2$ if and only if $x > y$, provided that $x > 0$ and $y > 0$.*

This statement is valid (and very useful!) in all number systems which are ordered rings. The following interpretation of order relations involving the *modulus* also applies in $\mathbb{Z}$ and in other ordered rings:

Theorem 4.4.32. *For any $b > 0$, $|x - a| < b$ if and only if $a - b < x < a + b$.*

Proof. Case 1: $x \geq a$. From Definition 4.2.12 and Theorem 4.4.4, $|x - a| = x - a$ in this case, so

$$|x - a| < b \Leftrightarrow x - a < b \Leftrightarrow x < a + b.$$

Thus we have

$$a \leq x < a + b.$$

Case 2: $x < a$. From Definition 4.2.12 and Theorem 4.4.4, $|x - a| = -(x - a) = a - x$ in this case, so

$$|x - a| < b \Leftrightarrow a - x < b \Leftrightarrow x > a - b.$$

Thus we have

$$a - b < x < a.$$

Combining the two cases yields $a - b < x < a + b$. $\qquad\square$

Example 4.4.33. If $|x + 3| < 2$, we have $a = -3, b = 2$ in the notation of Theorem 4.4.32, so $-3 - 2 < x < -3 + 2$, i.e. $-5 < x < -1$. [If $x \in \mathbb{Z}$, this means that $x \in \{-4, -3, -2\}$.]

From trichotomy, Theorem 4.4.32 implies:

Corollary 4.4.34. *For any $b > 0$, $|x - a| > b$ if and only if $x < a - b$ or $x > a + b$.*

4.5   Finite, Infinite, and Bounded Sets in $\mathbb{Z}$

Many of the theorems relating to subsets of $\mathbb{Z}$ are similar to those on subsets of $\mathbb{N}$, but it is important to note the differences.

Theorem 4.5.1. *If a non-empty subset of $\mathbb{Z}$ is bounded below, then it has a least member; if bounded above, it has a greatest member.*

Whereas non-empty subsets of $\mathbb{N}$ *unconditionally* have a least member, in $\mathbb{Z}$ the existence of a least member is conditional on being bounded below, in the same way as the existence of a greatest member is conditional on being bounded above (in both $\mathbb{Z}$ and $\mathbb{N}$). So $\mathbb{Z}$ is *not well-ordered*; recall that when introducing the Well-Ordering of $\mathbb{N}$, we mentioned that this, like the Principle of Induction, was a distinctive property of $\mathbb{N}$.

In proving this theorem, we will treat positive integers as being identical to natural numbers (as justified by the isomorphisms with respect to arithmetic operations and order), so any theorem on $\mathbb{N}$ can be applied to $\mathbb{Z}_+$.

Proof of Theorem 4.5.1. Given $A \subseteq \mathbb{Z}$, the simplest case is where $A \subseteq \mathbb{Z}_+$; since $\mathbb{Z}_+ = \mathbb{N}$, Theorems 3.6.9 and 3.6.10 apply: A has a least member, and if bounded above it has a greatest member.

Suppose now that $A \nsubseteq \mathbb{Z}_+$ but A is bounded below. Let h be any lower bound, so that $a \geq h$ for each $a \in A$. Then $a - h + 1 \geq 1$ for each $a \in A$, so the set

$$A_l := \{n \in \mathbb{Z} : n = a - h + 1, a \in A\}$$

consists only of positive integers (≥ 1), i.e. $A_l \subseteq \mathbb{N}$. So A_l has a least member [*Theorem 3.6.9*]; let $l = \min A_l$. Now, by reversing the procedure that produced A_l from A, we have

$$A = \{a \in \mathbb{Z} : a = n + h - 1, n \in A_l\}.$$

Since $n \geq l \; \forall n \in A_l$ [*Definition of* $l = \min A_l$], we have $a \geq l + h - 1 \; \forall a \in A$ [*Order preserved when adding* $h-1$]. Also, since $l \in A_l$, we have $l + h - 1 \in A$. So $l + h - 1$ is a lower bound for A and is also a member of A, and is therefore the *least member* of A [*Definition 3.6.3 (ii)*].

Similarly if A is bounded above, let k be an upper bound. Then the set

$$A_u := \{m \in \mathbb{Z} : m = k - a + 1, a \in A\}$$

consists only of positive integers, so has a least member u [*Theorem 3.6.9*]. We then find $\max A = k - u + 1$ by an argument similar to that used above for $\min A = l + h - 1$. $\square$

The next theorem also differs from the corresponding theorem for $\mathbb{N}$ (Theorem 3.7.20) because $\mathbb{Z}$ is not well-ordered.

Theorem 4.5.2. *A non-empty subset of* $\mathbb{Z}$ *is finite if and only if it is bounded above and below.*

Exercise 4.5.3. *Prove Theorem 4.5.2, making use of Theorem 3.7.20 by applying a procedure similar to that used in proving Theorem 4.5.1.*

Finally, a theorem that may surprise you:

Theorem 4.5.4. $\mathbb{Z}$ *is countably infinite.*

Intuitively, we would think of $\mathbb{Z}$ as being more than twice the size of $\mathbb{N}$: for each natural number, there is one positive integer and one negative integer, and $\mathbb{Z}$ also includes 0. But the theorem says that $\mathbb{Z}$ is the same size as $\mathbb{N}$, or more precisely, has the same cardinality as $\mathbb{N}$. To prove the theorem, we simply need to find a bijection from $\mathbb{Z}$ to $\mathbb{N}$. Such a bijection is

$$f(n) = \begin{cases} 2n & \text{if } n \in \mathbb{Z}_+ \\ 1 - 2n & \text{if } n \in \mathbb{Z}_- \cup \{0\}. \end{cases}$$

To see how this works, note how f maps integers into the first few natural numbers: $f(0) = 1, f(1) = 2, f(-1) = 3, f(2) = 4, f(-2) = 5$, etc.. Considering the natural numbers to be identical to the positive integers, so that $\mathbb{N} \subset \mathbb{Z}$, this is a bijection from $\mathbb{Z}$ to a proper subset of itself, which we know to be possible with infinite sets (Theorem 3.7.23). At Hilbert's Hotel (see following Definition 3.7.26), we imagine a new guest and a coach containing countably infinitely many guests all turning up when the hotel is full. Each of the guests already in the hotel is moved to a room with a number twice their original room number (i.e. from room n to room $2n$); the single new guest (representing the number 0) is accommodated in room 1, while the guests from the coach are placed in the remaining empty rooms: coach passenger number m going to room $1 + 2m$ (where m takes the role of $-n$ for $n \in \mathbb{Z}_-$ in our bijection above).

Investigations

1. This question concerns a number system $\widetilde{\mathbb{Z}}$ with binary operations $\downarrow$ and $+$, defined as follows.

   $\widetilde{\mathbb{Z}} := \mathbb{Z} \cup \{\omega\}$, i.e. $\widetilde{\mathbb{Z}}$ consists of the integers $\mathbb{Z}$ together with one further element, ω.

   If $a \in \mathbb{Z}$ and $b \in \mathbb{Z}$ (i.e. if a and b are any members of $\widetilde{\mathbb{Z}}$ other than ω) then:

   (a) $a \downarrow b := \min\{a, b\}$

   (b) $a \downarrow \omega = \omega \downarrow a := a$

   (c) $\omega \downarrow \omega := \omega$

   (d) $a + b$ is ordinary addition of integers

   (e) $a + \omega = \omega + a := \omega$

   (f) $\omega + \omega := \omega$.

   In the following, you may assume all the theorems about addition and order in the integers; recall that $\min\{a, b\}$ in (i) above means the least member of the set $\{a, b\}$.

   (i) Prove that $\downarrow$ is commutative on $\widetilde{\mathbb{Z}}$.

   (ii) Prove that $\downarrow$ is associative on $\widetilde{\mathbb{Z}}$.

   (iii) Prove that $+$ distributes over $\downarrow$ on $\widetilde{\mathbb{Z}}$.

   (iv) What is the identity element for $\downarrow$ on $\widetilde{\mathbb{Z}}$?

   (v) Show that not every element of $\widetilde{\mathbb{Z}}$ has an inverse element under $\downarrow$.

   (vi) Noting that 0 is the identity element for $+$ in $\widetilde{\mathbb{Z}}$, does every element of $\widetilde{\mathbb{Z}}$ have an inverse element under $+$?

   (vii) Is $\langle \widetilde{\mathbb{Z}}, \downarrow, + \rangle$ a ring?

   [Hint: In (a), (b), and (c) you are considering binary operations combining two or three elements of $\widetilde{\mathbb{Z}}$: any of these elements can be either an integer or ω, and you need to consider all possible combinations.]

2. (i) Let x, y, z be integers with $x \geq y$ and $y \geq z$. Show that

   $$x^2 - y^2 + z^2 \geq (x - y + z)^2.$$

   [Hint: start by considering the quantity $x^2 - y^2 + z^2 - (x - y + z)^2$.]

   (ii) For any $n \in \mathbb{N}$, let $x_1, x_2, \ldots, x_n, x_{n+1}$ and $y_1, y_2, \ldots, y_n$ be integers with $x_k \geq y_k$ and $y_k \geq x_{k+1}$ for each $k = 1, 2, \ldots, n$. Prove that

   $$\sum_{k=1}^{n+1} x_k^2 - \sum_{k=1}^{n} y_k^2 \geq \left(\sum_{k=1}^{n+1} x_k - \sum_{k=1}^{n} y_k \right)^2.$$

   [Hint: in a proof by induction the result of part (i) provides your anchor. The summation symbol was defined in Definition 3.7.7.]

Chapter 5

Foundations of Number Theory

This chapter contains a modest further exploration of the properties of Natural Numbers and Integers, in particular those properties that will be useful as we develop the theory of other number systems. The idea of Integer Division is of fundamental importance, and we shall see that the familiar way that we write down numbers in base ten (decimal form) is based on the Division Theorem, which actually enables us to write down numbers in any base. We then introduce prime numbers and prime factorisation, which are major themes of the area of mathematics called Number Theory. Congruence is an important tool of Number Theory, and leads naturally to modular arithmetic, in which we define finite number systems and find a connection between prime numbers and algebraic structure.

5.1 Integer Division

If asked, "What is 9 divided by 4?", you might answer, "$2\frac{1}{4}$", or you might say, "2 with remainder 1". The first answer introduces rational numbers, which we shall consider in Chapter 6; but the second answer is also correct and is an example of what we call **Integer Division**. The possibility of doing integer division is guaranteed by the Division Theorem (Theorem 5.1.5), but in order to prove this theorem we first require Lemma 5.1.1 and its Corollary 5.1.3. We shall also state another similar lemma and corollary, which will come in useful later on.

Lemma 5.1.1. *For any $n \in \mathbb{Z}$ and $m \in \mathbb{Z}_+$, $\exists k \in \mathbb{N}$ such that $km > n$.*

Lemma 5.1.2. *For any $n \in \mathbb{Z}$ and $m \in \mathbb{Z}_+\backslash\{1\}$, $\exists k \in \mathbb{N}$ such that $m^k > n$.*

These lemmas state that if we have a positive integer m and another integer n, we can always find a natural number multiple of m and a natural number power of m to yield integers greater than n, however large n may be (except in the case $m = 1$ for powers).

Recall that we have established that natural numbers and positive integers are essentially the same objects (due to isomorphisms with respect to arithmetic operations and order), so natural numbers and integers can be combined in arithmetic formulae and order relations as in the statements of these lemmas. The reason for writing $k \in \mathbb{N}$ here (rather than $k \in \mathbb{Z}_+$) is so that we can invoke the Well-Ordering Principle to obtain:

Corollary 5.1.3. *Given $n \in \mathbb{Z}$ and $m \in \mathbb{Z}_+$, there is a least value of $k \in \mathbb{N}$ such that $km > n$.*

Corollary 5.1.4. *Given $n \in \mathbb{Z}$ and $m \in \mathbb{Z}_+\backslash\{1\}$, there is a least value of $k \in \mathbb{N}$ such that $m^k > n$.*

The lemmas seem fairly obvious, but do need to be proved:

Proof of Lemmas 5.1.1 and 5.1.2. If $n \leq 0$, let $k = 1$. Then $km = m$ and $m^k = m$. Since $m > 0$ and $0 \geq n$, we have the required results by transitivity.

If $n > 0$, let $k = n + 1$, so $k > n$. Since $m > 0$, we have $m \geq 1$ [*Corollary 4.4.23*], so by Theorem 4.4.17, $km \geq k \times 1 = k > n$, yielding $km > n$ by transitivity.

To obtain $m^k > n$ with $n > 0$ and $m > 1$, we can let $k = n$ and use induction on n (since $n > 0$ means that $n \in \mathbb{N}$) to show that $m^n > n$:

(a) For $n = 1$: $m^n = m^1 = m > 1 = n$, so $m^n > n$.

(b) Given the inductive hypothesis, $m^n > n$,

$$m^{S(n)} = m \times m^n$$
$$> m \times n \qquad\qquad [\textit{Inductive hypothesis and Theorem 3.5.12}]$$
$$\geq 2n \qquad\qquad [\textit{Theorem 3.5.12, noting that } m > 1 \Rightarrow m \geq 2]$$
$$= n + n$$
$$\geq n + 1$$
$$= S(n),$$

i.e. $m^{S(n)} > S(n)$, as required.   □

We can now prove our central result:

Theorem 5.1.5 (The Division Theorem). *Given $n \in \mathbb{Z}$ and $d \in \mathbb{N}$, there is a unique **quotient** $q \in \mathbb{Z}$ and a unique **remainder** $r \in \mathbb{Z}$ such that*

$$n = qd + r \quad \textit{with} \quad 0 \leq r < d.$$

It is important to note that the uniqueness of the quotient and remainder are consequences of the restriction $0 \leq r < d$. For example, with $n = 9$ and $d = 4$, we satisfied the equation $n = qd + r$ with $q = 2$ and $r = 1$ in the discussion at the beginning of this section; but we could also satisfy $n = qd + r$ with $q = 1$ and $r = 5$, or $q = 3$ and $r = -3$, or even $q = -2$ and $r = 17$, none of which would obey the restriction on the remainder in the Theorem.

The proof is in two parts: first we prove the *existence* of q and r satisfying the requirements of the theorem, and then their *uniqueness*.

Proof of the Division Theorem. We consider three cases:

(i) If $0 \leq n < d$:

Let $q = 0$ and $r = n$; then $n = qd + r$ and $0 \leq r < d$.

(ii) If $n \geq d$:

From Corollary 5.1.3, there is a least value of $k \in \mathbb{N}$ such that $kd > n$. Let k_l be this least value of k; so $k_l d > n$ but $(k_l - 1)d \not> n$, which becomes $(k_l - 1)d \leq n$ by trichotomy. Now let $q = k_l - 1$, so we have

$$(q + 1)d > n \quad \text{and} \quad qd \leq n.$$

Also let $r = n - qd$, to satisfy $n = qd + r$.

From $qd \leq n$, we have $n - qd \geq 0$, i.e. $r \geq 0$.

From $(q + 1)d > n$, we have

$$qd + d > n$$
$$\therefore d > n - qd$$
$$\therefore r < d.$$

(iii) If $n < 0$:

Let k_l be the least value of $k \in \mathbb{N}$ such that $kd \geq -n$ (it should be obvious that the above lemmas and corollaries are true with the relation $>$ replaced by $\geq$); so $k_l d \geq -n$ but $(k_l - 1)d \ngeq -n$, which becomes $(k_l - 1)d < -n$ by trichotomy. Now let $q = -k_l$, so we have

$$(-q)d \geq -n \quad \text{and} \quad (-q-1)d < -n.$$

Let $r = n - qd$, to satisfy $n = qd + r$.

From $(-q)d \geq -n$, we have $n - qd \geq 0$, i.e. $r \geq 0$.

From $(-q-1)d < -n$, we have

$$n - qd < d$$
$$\therefore r < d.$$

In all three cases, we have satisfied $n = qd + r$ with $r \geq 0$ and $r < d$. We now show that the q and r we have obtained are unique; as usual, we do so by supposing non-uniqueness, which in this case means that $\exists q', r' \in \mathbb{Z}$ with $n = q'd + r'$ and $0 \leq r' < d$, and either or both $q' \neq q$ and $r' \neq r$.

From $qd + r = n = q'd + r'$ we obtain

$$(q - q')d = r' - r,$$

from which it is clear (since $d \neq 0$) that $q' \neq q$ if and only if $r' \neq r$, i.e. one of the inequalities cannot be true without the other. If $q' \neq q$ we can take $q' < q$ without loss of generality; this is simply labelling the lesser of the two quotients as q', since by trichotomy one of them has to be less than the other. Then

$$q - q' > 0$$
$$\therefore q - q' \geq 1 \qquad\qquad [\text{Corollary 4.4.23}]$$
$$\therefore (q - q')d \geq d \qquad\qquad [d > 0]$$
$$\therefore r' - r \geq d$$
$$\therefore r' \geq r + d.$$

But we require $r \geq 0$, so $r' \geq r + d$ implies that $r' \geq d$, which contradicts the requirement that $r' < d$. Hence we cannot find $r' \neq r$ with both r and r' satisfying the requirements of the Theorem: there is a *unique* remainder and a *unique* quotient satisfying the Theorem. $\square$

Definition 5.1.6. If $n = qd$ with $n, q, d \in \mathbb{Z}$, then d is a **divisor** or **factor** of n, and we also say that n is **divisible** by d.

Notation: $d|n$ if d is a divisor of n. $d \nmid n$ if d is *not* a divisor of n.

We could say that d is a divisor of n if and only if there is zero remainder when n is divided by d (in integer division). The words "divisor" and "factor" mean the same, and the choice of which word to use is a matter of convention depending on the context.

Exercise 5.1.7. *Show that if d is a divisor of both m and n (for some $m, n \in \mathbb{Z}$), then d is also a divisor of $am + bn$, $\forall a, b \in \mathbb{Z}$.*

An important case is where $d = 2$, so that the only possible remainders are 0 and 1. We then have the familiar definitions:

Definition 5.1.8. An integer n is **even** if it is divisible by 2; it is **odd** if it has remainder 1 on division by 2.

In particular, the results on powers of integers in Exercise 4.3.15 can be reframed as: $(-x)^n = x^n$ if n is even, while $(-x)^n = -(x^n)$ if n is odd.

5.2   Expressing Integers in Any Base

If we write, for example, "4287", we recognise this sequence of symbols as representing an integer. The symbols 4, 2, etc. are known as **digits**, and the value of each digit depends on its position in the sequence as well as on the digit itself: this is the **place-value** system. Indeed, we are so familiar with this system that it seems to be the obvious, indeed the only, way to write numbers; and it certainly facilitates numerical calculations. But the system is far from obvious: the ancient Greeks, for all their prowess in geometry and logical reasoning, did not write numbers in this way; neither did the Romans, whose great feats of engineering were achieved *despite* not having a system for writing numbers that was convenient for calculations. The decimal place-value system that we are familiar with originated in India in or before the 6th century, was then taken up by the Arabs, and only arrived in Europe with the publication of *Liber Abaci* by Leonardo Pisano (better known by his nickname Fibonacci) in 1202.

The word **decimal** means that there are ten digit-symbols (0, 1, 2, 3, 4, 5, 6, 7, 8, 9). The number of symbols is called the **base**, and place-value systems can in principle work with any base $b \in \mathbb{N}\backslash\{1\}$; there is a trade-off between the need to have a large font of symbols if the base is large, and the greater length of the sequence of symbols required to represent a given integer if the base is smaller. Bases other than ten are or have been used in some applications. The ancient Babylonians had a place value system in base sixty, although there was no zero symbol and each of the other fifty-nine digit symbols was a somewhat cumbersome assemblage of up to five symbols for tens and up to nine symbols for units. A modern computer is essentially a network of switches which can each be in one of two positions, off or on, usually represented by the digits 0 and 1; so the state of a component may be expressed by a number written in base two (binary), although it is often found more convenient to use a base that is a power of two, in particular base eight (octal) or base sixteen (hexadecimal). In living organisms, genetic information is stored digitally in base four: DNA consists of sequences of chemical components of four types, denoted A, C, G, T, so a DNA molecule could be thought of as a number written in base four.[i]

The representation of any positive integer n in any base $b \in \mathbb{N}\backslash\{1\}$ in a place-value system is founded on the Division Theorem. [For a negative integer n, we simply put a minus sign in front of the base-b representation of the positive integer $-n$.] Integer division by b yields remainders in the set $\{r \in \mathbb{Z} : 0 \leq r < b\}$, and we require a symbol for each of the b integers in this set. The procedure starts by dividing n by b:

$$n = q_0 b + r_0 \,.$$

If $0 \leq n < b$ we have $q_0 = 0$ (case (i) in the proof of the Division Theorem) and n is simply represented by a single digit, the symbol for the value of r_0. If $n \geq b$ then $q_0 > 0$ (case (ii) of the Division Theorem) and we divide q_0 by b:

$$q_0 = q_1 b + r_1,$$

so that

$$n = (q_1 b + r_1)b + r_0 = q_1 b^2 + r_1 b + r_0.$$

The new quotient is then divided by b, and after $k + 1$ iterations of this process (for $k \geq 2$)

---

[i]The initials A, C, G, T stand for adenine, cytosine, guanine, thymine, respectively; these components are actually known as *bases* in chemical terminology, a completely different use of the word "base" from the mathematical one.

we have

$$q_{k-1} = q_k b + r_k$$

and

$$n = q_k b^{k+1} + r_k b^k + r_{k-1} b^{k-1} + \cdots + r_1 b + r_0$$

(which can be formally verified by induction on k). Here $q_k = 0$ if and only if $b^{k+1} > n$ (i.e. in case (i) of the Division Theorem, considering the last equation as a division of n by b^{k+1}). But Lemma 5.1.2 and Corollary 5.1.4 ensure that there does exist $k \in \mathbb{N}$ such that $b^{k+1} > n$; in particular there exists a least such k, which we denote K; so $q_K = 0$ but $q_{K-1} > 0$ and hence $r_K > 0$. We now have

$$n = r_K b^K + r_{K-1} b^{K-1} + \cdots + r_1 b + r_0.$$

This is the base-b representation of n, and is normally shortened by simply writing the symbols for the values of the remainders as a string of digits: "$r_K r_{K-1} \ldots r_1 r_0$", with the base being understood. For example, in base ten, if we write "2503", that is a shorthand for $2 \times 10^3 + 5 \times 10^2 + 0 \times 10 + 3$; whereas in base 6, "2503" is a shorthand for $2 \times 6^3 + 5 \times 6^2 + 0 \times 6 + 3$. Clearly there is a need for clarity if different bases are under consideration, and we may use subscripts to indicate bases: 2503_{ten} and 2503_6 for the examples in the previous sentence. Note the use of the word "ten" rather than the digit form "10": this is because in *every* base b, "10" is the base-b representation of the number b.[ii] The established symbols $0, 1, 2, 3, 4, 5, 6, 7, 8, 9$ are unambiguous, but for bases $b \geq$ ten we shall use the English words to avoid any ambiguity when specifying a base. Note also that if $b >$ ten we require more digit symbols than those in common use in base ten: we shall adopt the convention of using capital letters, $A = \text{ten}, B = \text{eleven}, C = \text{twelve}, \ldots$.

A familiar rule in base ten is that appending a zero at the end of a number multiplies it by ten, e.g. $25030 = 10 \times 2503$. The general rule is that appending a zero at the end of a number in base b multiplies it by b, since

$$b \times (r_K b^K + r_{K-1} b^{K-1} + \cdots + r_1 b + r_0) = r_K b^{K+1} + r_{K-1} b^K + \cdots + r_1 b^2 + r_0 b + 0,$$

written as "$r_K r_{K-1} \ldots r_1 r_0 0$". The example, $25030 = 10 \times 2503$, is true in every base $b \geq 6$.

The procedure for obtaining a base-b representation has been described as if the integer n is known in some undefined format, whereas in practice we will nearly always start with a number already written in some base; so the practical problem is usually of conversion from one base to another. We are used to doing all our arithmetical calculations in base ten, so converting a number from another base to base ten is simply a matter of recalling what the "shorthand" base-b representation means, for example,

$$2503_6 = 2 \times 6^3 + 5 \times 6^2 + 0 \times 6 + 3 = 615_{\text{ten}}.$$

Converting from base ten to another base b simply involves the procedure described above: repeated integer division by b to find the string of remainders that constitute the base-b representation. The computations can all be written in base ten, but if $b >$ ten then any remainders greater than ten must then be expressed using base-b digits A, B, etc..

Example 5.2.1. Find the representations of the number 2503_{ten} in: *(i)* base 6, *(ii)* base fifteen.

[ii] Hence the joke: "There are 10 kinds of people in the world: those who understand binary, and those who don't".

Solution. (i) In base 6:

$$
\begin{aligned}
2503 &= 417 \times 6 + 1: & q_0 &= 417, r_0 = 1 \\
417 &= 69 \times 6 + 3: & q_1 &= 69, r_1 = 3 \\
69 &= 11 \times 6 + 3: & q_2 &= 11, r_2 = 3 \\
11 &= 1 \times 6 + 5: & q_3 &= 1, r_3 = 5 \\
1 &= 0 \times 6 + 1: & q_4 &= 0, r_4 = 1.
\end{aligned}
$$

So $2503_{\text{ten}} = 15331_6$.

(ii) In base fifteen, using the digits A, B, C, D, E for ten, eleven, twelve, thirteen, fourteen, respectively:

$$
\begin{aligned}
2503 &= 166 \times 15 + 13: & q_0 &= 166, r_0 = D \\
166 &= 11 \times 15 + 1: & q_1 &= 11, r_1 = 1 \\
11 &= 0 \times 15 + 11: & q_2 &= 0, r_2 = B.
\end{aligned}
$$

So $2503_{\text{ten}} = B1D_{\text{fifteen}}$. $\qquad\qquad\qquad\qquad\qquad\qquad\qquad\qquad\qquad\qquad$ $\square$

Exercise 5.2.2. *Write the number* 9376_{ten} *in (i) base 7, (ii) base fifteen. For (ii), use the symbols* A, B, C, D, E *for ten, eleven, twelve, thirteen, fourteen, respectively.*

Doing computations in any other base is in principle no more difficult than in base ten; it is only familiarity that makes base ten seem easier, but the methods are the same regardless of the base. To start with, we need addition and multiplication tables, displaying sums and products of all pairs of single-digit numbers from 1 to $(b-1)$ (addition and multiplication by zero being trivial).[iii] For example, here are tables for base 6.

+	1	2	3	4	5
1	2	3	4	5	10
2	3	4	5	10	11
3	4	5	10	11	12
4	5	10	11	12	13
5	10	11	12	13	14

×	1	2	3	4	5
1	1	2	3	4	5
2	2	4	10	12	14
3	3	10	13	20	23
4	4	12	20	24	32
5	5	14	23	32	41

To construct such tables, recall the inductive definitions of addition and multiplication in the Natural Numbers: going along a row in the addition table, you are simply adding 1 for each successive entry, while going along row m in the multiplication table you are adding m for each successive entry. But always bear in mind which base you are in: for example in base 6, adding 1 to 5 yields 10 (not "6"); and adding 5 to 23 yields 32 (i.e. $5 + (2b + 3) = 3b + 2$ with $b = 6$). Note that the tables are symmetric about the leading diagonal: this is because addition and multiplication are commutative.

Addition, subtraction and multiplication of multi-digit integers then proceeds on the basis of these tables in the same way that these operations are done in base ten. For long multiplication, the most transparent (if somewhat long-winded) way to set out a calculation is the "grid" method: to multiply a $(K+1)$-digit number $r_K \ldots r_0$ by a $(L+1)$-digit number $s_L \ldots s_0$, recall that we are actually doing the multiplication

$$
\sum_{i=0}^{K} r_i b^i \times \sum_{j=0}^{L} s_j b^j, \text{[iv]} \tag{5.1}
$$

[iii]There is a tradition in schools of teaching multiplication tables up to 12×12 in base ten, but in principle it is only necessary to learn tables up to 9×9.

[iv]Recall the convention noted in Definition 4.4.26 that $b^0 = 1$, where the sum notation is used with a starting index of 0.

in which each term $r_i b^i$ is written in base b as the digit r_i followed by i zeroes, and similarly for $s_j b^j$. Now, (5.1) is the sum of $(K+1)(L+1)$ products of terms, $r_i b^i \times s_j b^j = r_i s_j b^{i+j}$ $(i = 0, 1, \ldots, K; j = 0, 1, \ldots, L)$, where the base-$b$ multiplication table yields each product $r_i s_j$ and the factor b^{i+j} entails $i + j$ zeroes after the digit(s) for $r_i s_j$. These multiplications are set out in a grid, and then all the $(K+1)(L+1)$ products are added.

Example 5.2.3. In base 6: find $415_6 \times 52_6$.

Solution.

	400	10	5
50	32000	500	410
2	1200	20	14

Here we have multiplied 400 by 50, by first finding $4 \times 5 = 32$ in the base-6 multiplication table above, and then noting that $i = 2$ zeroes of 400 with $j = 1$ zero of 50 yields a total of $i + j = 3$ zeroes. So $400 \times 50 = 32000$ in base 6, as shown in the appropriate cell in the grid above. Together with the results in the other cells, we then add:

$$
\begin{array}{r}
3 \; 2 \; 0 \; 0 \; 0 \\
+ \quad\quad 5 \; 0 \; 0 \\
+ \quad\quad 4 \; 1 \; 0 \\
+ \quad 1 \; 2 \; 0 \; 0 \\
+ \quad\quad\quad 2 \; 0 \\
+ \quad\quad\quad 1 \; 4 \\
\hline
3 \; {}^1 4 \; 5 \; 4 \; 4
\end{array}
$$

Here, a superscript number in the addition results line indicates carrying from the column to the right, in the usual way; so in the third column from the right, $0 + 5 + 4 + 2 = 15_6$, with 5 in the results line in the same column and 1 carried to the left.

So $415_6 \times 52_6 = 34544_6$. □

Being the inverse operation of addition, subtraction $s_j - r_j$ uses the addition table by seeking the column in which an addition result s_j can be found in the row r_j.

Example 5.2.4. In base 6: find $415_6 - 52_6$.

Solution.

$$
\begin{array}{r}
4 \; 1 \; 5 \\
- \quad 5 \; 2 \\
\hline
{}^1 3 \; 2 \; 3
\end{array}
$$

In the units column we have $5 - 2 = 3$ (as in any base $b \geq 6$), but in the next column to the left we require $1 - 5$. Since $1 < 5$, we find $11 - 5$ by noting that the result 11 in row 5 of the base-6 addition table is found in column 2, and carry 1 for subtraction in the next column to the left. Finally, $415_6 - 52_6 = 323_6$. □

Exercise 5.2.5. *Write out addition and multiplication tables for base 7 and do the following calculations in base 7 (all numbers being written in base 7).*
  (i) $524 + 56$ (ii) $524 - 56$ (iii) 524×56
  Check your calculations by converting the numbers 524_7 and 56_7 and your answers to base ten.

5.3   Prime Numbers and Prime Factorisation

The concept of prime numbers is usually defined within the context of the Natural Numbers, and we accordingly devote most attention here to primes in $\mathbb{N}$; but we shall also consider generalisations to other number systems, which will be explored in more detail in Chapter 8.

5.3.1   Prime Numbers and Prime Factorisation in $\mathbb{N}$

The number 1 has only one factor: 1 itself. Every other natural number n has the factors, 1 and n, since $1 \times n = n$; and it may or may not have other factor(s).

Definition 5.3.1. A number $n \in \mathbb{N}$ is **prime** if it has only two distinct factors, 1 and n. A number $n \in \mathbb{N}$ is **composite** if it has more than two factors.

This definition partitions $\mathbb{N}$ into three subsets:

$$\mathbb{N} = \{1\} \cup \{\text{Prime numbers}\} \cup \{\text{Composite numbers}\}.$$

If k is a factor of a composite number n with $k \neq 1$ and $k \neq n$, then $kl = n$ for some $l \in \mathbb{N}$ with both $1 < k < n$ and $1 < l < n$. Formal verification of these bounds on k and l is easy using Theorems 3.5.12 and 3.5.14, given that 1 is the least natural number.

Prime numbers have been of interest for thousands of years: both Euclid and Eratosthenes, whose contributions to our understanding of them are included below, worked in Alexandria in the third century BC when that city was the dominant centre of ancient Greek philosophy and science. The importance of prime numbers is first apparent in the following theorem, which underlies all of Number Theory.

Theorem 5.3.2 (The Fundamental Theorem of Arithmetic). *Every $n \in \mathbb{N}\backslash\{1\}$ can be expressed uniquely as the product of one or more prime factors.*

This theorem is often expressed informally by saying that prime numbers are the "building blocks" from which all natural numbers are made. It asserts that for any given natural number (apart from 1), there is a *unique* set of prime factors that can be used to build it. This may seem obvious, but in Chapter 8 we shall define number systems in which more than one prime factorisation can be found for certain numbers, with an entirely reasonable generalisation of the definition of "prime". So we do need to prove the Fundamental Theorem of Arithmetic for $\mathbb{N}$.

Note that we need to prove two things: firstly, the existence of a prime factorisation for every $n \in \mathbb{N}\backslash\{1\}$, and then its uniqueness. Both proofs make use of the Well-Ordering Principle to construct a proof by contradiction: we suppose that the theorem is false for some $m \in \mathbb{N}$, and well-ordering then implies that there is a *least* natural number m_0 for which it is false. The contradiction is obtained by showing that there must then be a number less than m_0 for which it is also false. Such an argument is often referred to as a proof by **infinite descent**.

Proof of existence of prime factorisation. If $m \in \mathbb{N}$ is prime, it satisfies the theorem: from Definition 5.3.1, it can be written uniquely as the product of one prime, itself. So if there exist any natural numbers which do not satisfy the theorem, they must be composite. We can denote the least number not having a prime factorisation as m_0, and since it is composite we can write $m_0 = k_0 l_0$, where $1 < k_0 \leq l_0 < m_0$. Now, k_0 and l_0 cannot both have prime factorisations, since then m_0 could be expressed as the product of the prime factor(s) of k_0 multiplied by the prime factor(s) of l_0, so m_0 would in fact have a prime factorisation. But if either k_0 or l_0 do not have a prime factorisation, we have contradicted the definition of m_0 as the *least* such number. □

Proof of uniqueness of prime factorisation. Non-uniqueness would mean the existence of number(s) with two or more prime factorisations; we again denote by m_0 the least such number. We can write

$$m_0 = p_1 \times p_2 \times \cdots \times p_k = q_1 \times q_2 \times \cdots \times q_l,$$

where $p_1, p_2, \ldots, p_k$ and $q_1, q_2, \ldots, q_l$ are the primes in two distinct factorisations of m_0.

Now, *none* of the p_i $(i = 1, \ldots, k)$ can equal any of the q_j $(j = 1, \ldots, l)$; for if any of them were equal, we could write $p_1 = q_1$ (i.e. labelling the equal primes with subscript 1) and define

$$m_1 := p_2 \times p_3 \times \cdots \times p_k$$

so that $m_0 = m_1 p_1 = m_1 q_1$. Cancellation of the factor q_1 then yields

$$m_1 := q_2 \times q_3 \times \cdots \times q_l.$$

We now have two prime factorisations of m_1; clearly $m_1 < m_0$, so we have contradicted the definition of m_0 as the least number with non-unique prime factorisation.

We can now suppose that $p_1 < q_1$ without loss of generality (i.e. labelling the lesser of the two unequal primes as p_1). Define

$$r := (q_1 - p_1) q_2 q_3 \cdots q_l \tag{5.2}$$
$$= q_1 q_2 q_3 \cdots q_l - p_1 q_2 q_3 \cdots q_l$$
$$= m_0 - p_1 q_2 q_3 \cdots q_l \tag{5.3}$$
$$< m_0.$$

Now, $p_1 \nmid (q_1 - p_1)$, because if p_1 was a divisor of $(q_1 - p_1)$ we would have $q_1 - p_1 = s p_1$ for some $s \in \mathbb{N}$, so $q_1 = (1 + s) p_1$, i.e. q_1 would not be prime. So, when $(q_1 - p_1)$ is written as a product of prime factors, expression (5.2) yields a prime factorisation of r that *does not include* p_1. But from (5.3) we have

$$r = p_1 p_2 p_3 \cdots p_k - p_1 q_2 q_3 \cdots q_l$$
$$= p_1 (p_2 p_3 \cdots p_k - q_2 q_3 \cdots q_l),$$

so that there exists a prime factorisation of r that *does include* p_1. Thus r has (at least) two distinct prime factorisations; since $r < m_0$, this contradicts the definition of m_0 as the *least* natural number with non-unique prime factorisation. Hence there cannot exist any $m \in \mathbb{N}$ with non-unique prime factorisation. □

The next theorem, first written down and proved by Euclid, is also of fundamental importance.

Theorem 5.3.3. *There exist infinitely many primes.*

Proof. The proof is by contradiction: we suppose that P, the set of all primes, is finite with some cardinality $n \in \mathbb{N}$: $P = \{p_1, p_2, \ldots, p_n\}$. Now consider Π, the product of all the primes:

$$\Pi := \prod_{i=1}^{n} p_i.$$

Each prime is a divisor of Π, so if we do integer division of the number $\Pi + 1$ by any prime p_i, we obtain

$$\Pi + 1 = q_i p_i + 1$$

for some $q_i \in \mathbb{N}$. This satisfies the requirements of the Division Theorem, since $1 < p_i$ for each prime p_i (recalling that 1 is not prime). So it means that $p_i \nmid (\Pi + 1)$ for each prime p_i. Furthermore, $\Pi + 1 > p_i$ for each prime p_i. Hence either $\Pi + 1$ is a prime not in the set P, or else it is a composite with prime factors not in the set P; either way, we have contradicted P being the set of *all* primes. $\square$

We now turn to the practical business of determining which numbers are prime and of finding prime factorisations of composite numbers. This is a theme which is still the subject of research, as the difficulty of factorising very large numbers is crucial to modern cryptography; but in this book we shall only consider the most elementary methods. We start with a lemma which will help us avoid doing unnecessary work.

Lemma 5.3.4. *If p is a prime factor of n and $p^2 > n$, all other prime factors of n are less than p.*

Proof. Since $p|n$ we have $n = pq$ for some $q \in \mathbb{N}$. If $p^2 > n$ we then have $p^2 > pq$, so $p > q$ [*Theorem 3.5.14*]. Now, q may be prime or composite; in either case, all its prime factors are certainly less than or equal to q, and so less than p. $\square$

Theorem 5.3.3 implies that we cannot hope to list all primes; but we can construct a list of all primes less than or equal to some (arbitrary) upper bound N by means of a **sieve algorithm**. This involves first writing down all natural numbers from 2 to N (omitting 1 because it is not prime) and then going through a systematic procedure to remove composite numbers from the list. The simplest such algorithm was devised by Eratosthenes, and may be set out as an inductive procedure for finding the i'th prime p_i in an increasing list.

Algorithm 5.3.5. The sieve of Eratosthenes

(a) For $i = 1$: $p_1 = 2$.
(b) Having found p_i (for $i = 1, 2, \ldots$): if $p_i^2 \leq N$, delete all multiples of p_i starting from p_i^2; then $p_{S(i)}$ is the next remaining (not deleted) number after p_i, and the procedure continues. But if $p_i^2 > N$, the procedure stops, and the remaining numbers are all the primes less than or equal to N.

Notes on the algorithm. The first prime is 2 because 1 is not a prime, and 2 is the next number in $\mathbb{N}$. Lemma 5.3.4 ensures that any multiples of p_i which are less than p_i^2 will have been deleted before the i'th iteration, since they are also multiples of a smaller prime. Similarly, if $p_i^2 > N$, any composite numbers less than or equal to N will have been deleted as multiples of smaller primes than p_i, so no further iterations are required. The algorithm is computationally simple: the only multiplication required is to evaluate each p_i^2, with further multiples of p_i then being found by successive addition.

To illustrate the algorithm, we apply it with the rather small upper bound, $N = 15$. First list the natural numbers from 2 to 15:

$$2 \quad 3 \quad 4 \quad 5 \quad 6 \quad 7 \quad 8 \quad 9 \quad 10 \quad 11 \quad 12 \quad 13 \quad 14 \quad 15$$

$p_1 = 2$ and $2^2 = 4$, so delete multiples of 2 starting with 4, leaving

$$2 \quad 3 \quad \quad 5 \quad \quad 7 \quad \quad 9 \quad \quad 11 \quad \quad 13 \quad \quad 15$$

The next remaining number after $p_1 = 2$ is $p_2 = 3$. Now, $3^2 = 9$, so delete multiples of 3 starting from 9; as we add 3 successively, we note that 12 has already been deleted, and then move on to delete 15. We are left with

$$2 \quad 3 \quad 5 \quad 7 \qquad 11 \quad 13$$

The next remaining number after $p_2 = 3$ is $p_3 = 5$. But $5^2 = 25 > 15 = N$, so the procedure stops, and the last list of numbers consists of all primes ≤ 15.

Even from this short list of just the first six primes, it is apparent that the primes are distributed in a rather irregular manner. We might ask whether there is some formula to generate all primes; but no computationally simple formula exists. For example, a polynomial of some degree $K \in \mathbb{N}$ in a single variable n, i.e. an expression of the form

$$a_0 + a_1 n + \cdots + a_{K-1} n^{K-1} + a_K n^K$$

or more concisely

$$\sum_{i=0}^{K} a_i n^i,$$

in which a_i $(i = 0, 1, \ldots, K)$ are fixed integers, would constitute a computationally simple formula; but it is shown in Exercise 5.3.6 that no such formula will generate only primes as n ranges through natural number values.[v] A particularly famous case relates to polynomials of form

$$a_0 - n + n^2 :$$

clearly this will be composite when $n = a_0$, since then the polynomial reduces to a_0^2; but Euler noticed that with $a_0 = 41$, the polynomial yields prime values for every natural number $n < 41$. This stands as a warning to anyone tempted to deduce "theorems" from a large number of examples: finding 40 successive primes from this formula does not imply that the formula will always yield primes!

Exercise 5.3.6. *Prove that no polynomial formula in a single variable will yield only primes, as follows.*
(i) Consider first linear formulae,

$$P(n) = a_0 + a_1 n \qquad (a_0, a_1 \in \mathbb{Z}; a_1 \neq 0).$$

This is supposed to yield only primes, i.e. $P(n)$ is a prime for every $n \in \mathbb{N}$. In particular, $P(1)$ is prime. Let $q = P(1)$ and then show that q is a factor of $P(1+q)$ (so that the formula fails to yield a prime when $n = 1 + q$).
(ii) Now extend the proof to general polynomials,

$$P(n) = a_0 + a_1 n + a_2 n^2 + \cdots + a_K n^K$$
$$(K \in \mathbb{N}; \ a_i \in \mathbb{Z} \ (i = 0, 1, 2, \ldots, K); a_K \neq 0).$$

You should again let $q = P(1)$ and show that q is a factor of $P(1 + q)$. If you cannot immediately do the proof for general polynomials, try quadratic functions first.

---

[v]Nevertheless, more advanced Number Theory has shown the existence of polynomials in multiple variables which generate all primes when the variables are given all possible combinations of non-negative integer values. The first concrete example of such a polynomial was found by Jones, Sato, Wada and Wiens in 1976, and is of degree 25 in 26 variables. Further such formulae have been found, with a trade-off between reducing the number of variables while increasing the degree and *vice versa*. The computational inefficiency of these formulae lies not only in the need to systematically go through all combinations of choices of $0, 1, 2, 3, \ldots$ for each of the variables and to compute a messy formula for each combination, but in the fact that the formulae only yield a prime when the result is positive, while for most combinations the result turns out negative!

A list of primes (generated for example by the sieve of Eratosthenes) will be useful if one wants to determine the prime factorisation of a given natural number M. More precisely, we will want a list of primes p_i such that $p_i^2 \leq M$ in order to use the **trial division** method to factorise M. We present this method as an inductive procedure.

Algorithm 5.3.7. Trial Division

(a) For $i = 1$: let $q_1 = M$.

(b) Having found q_i: if $p_i \nmid q_i$, then $q_{S(i)} = q_i$; but if $p_i | q_i$, divide q_i by p_i as often as possible (i.e. as long as the result has zero remainder) to yield $q_{S(i)}$. So if α_i divisions by p_i are possible, then $q_i = p_i^{\alpha_i} q_{S(i)}$ and $p_i \nmid q_{S(i)}$.

When $q_i < p_i^2$ the procedure stops: either q_i is the final prime factor of M, or $q_i = 1$. The prime factorisation of M then consists of this final q_i (if not equal to 1) together with the list of $p_i^{\alpha_i}$ that you have divided by.

Notes on the algorithm. Lemma 5.3.4 ensures that if $q_i < p_i^2$ when we have proceeded to the i'th iteration, then q_i cannot be composite: if it was, one of its prime factors would be less than p_i, and so have already been divided out. In particular, this applies to q_1, which is why the list of known primes does not need to include any with a square greater than M. The algorithm is computationally simple, but is not efficient for very large numbers M as required in cryptography, for which more sophisticated algorithms have been devised.

Example 5.3.8. Find the prime factorisation of 1288.

Solution.

$$
\begin{array}{llll}
i = 1: & p_1 = 2, q_1 = 1288; & 1288 = 644 \times 2 = 322 \times 2^2 = 161 \times 2^3; \ 2 \nmid 161; & \text{Factor } 2^3 \\
i = 2: & p_2 = 3, q_2 = 161; & 3 \nmid 161 & \\
i = 3: & p_3 = 5, q_3 = 161; & 5 \nmid 161 & \\
i = 4: & p_4 = 7, q_4 = 161; & 161 = 23 \times 7; \ 7 \nmid 23; \quad \text{Factor } 7 & \\
i = 5: & p_5 = 11, q_5 = 23; & 23 < 11^2, \text{ so } 23 \text{ is the final prime factor.} &
\end{array}
$$

So $1288 = 2^3 \times 7 \times 23$. □

The primes used in this example were known from our implementation (above) of the sieve of Eratosthenes. We might have anticipated needing all primes whose square is less than or equal to 1288; since $35^2 = 1225$ and $36^2 = 1296$, this would have meant using $N = 35$ in the sieve before starting to factorise 1288. However, this is a "worst case", and the division by factors of 2 and 7 left us needing a much shorter list of primes. In fact, the process could have been stopped during iteration $i = 4$ by observing that $7^2 > 23$.

The uniqueness of prime factorisations (Theorem 5.3.2) means that a divisibility statement $x|y$ is equivalent to saying that the prime factors of x are a subset of the prime factors of y. This includes the special cases: (a) If x is prime, then $x|y$ means that x is a prime factor of y; (b) If $x = 1$, the prime factors of x are the empty set, which is a subset of every other set; (c) If $x = y$, the prime factors of x and of y are equal sets, which are subsets of each other.

We conclude this subsection with a definition and some theorems on factorisation which will be very useful later.

Definition 5.3.9. Two natural numbers are **coprime** if their prime factorisations have no primes in common.

Equivalently, we could simply say that x and y are coprime if they have no common factors; since if the prime factors of x are all different from those of y, any composite factor of x (a product of some its prime factors) must be different from any composite factor of y.

Note that it does not make sense to talk about *a number* being coprime; coprimeness is a property of a comparison between two numbers. We can say "x and y are coprime" or "x is coprime with y". Neither number need be prime itself. For example, 15 and 28 are coprime: their respective prime factorisations are 3×5 and $2^2 \times 7$, which do not have any factors in common. For the case of the number 1 (which has no prime factorisation), it will be convenient to define 1 as being coprime with every $n \in \mathbb{N}$.

The property of prime numbers stated in the next theorem is so important that in more advanced work it is taken as the *definition* of being prime:

Theorem 5.3.10. *Suppose p is prime and $y, z \in \mathbb{N}$. If $p|yz$, then either $p|y$ or $p|z$, or both.*

Proof. Let the prime factorisations of y and z be

$$y = p_1 p_2 \ldots p_k \quad \text{and} \quad z = q_1 q_2 \ldots q_l$$

(in which, rather than writing powers of primes, we allow repetitions in the list of primes: for example, if $y = 1288$, we have $p_1 = p_2 = p_3 = 2, p_4 = 7, p_5 = 23$). Then the prime factorisation of yz is

$$yz = p_1 p_2 \ldots p_k q_1 q_2 \ldots q_l.$$

Thus any prime which appears in the prime factorisation of yz has come from the prime factorisation of y or that of z, so either $p|y$ or $p|z$. $\qquad\square$

Note the importance of p in the theorem being prime: for example, with $p = 6$ (composite), $y = 3$, and $z = 4$, we would have the product $yz = 12$ and $p|yz$, but 6 is not a factor of either 3 or 4. The factors of the composite number 6 can be split between y and z, but this cannot be done with a prime p. It is also worth noting that the proof relies on the uniqueness of prime factorisations: if another, completely different, prime factorisation of yz was available, a prime in that factorisation need not have come from y or z.

Theorem 5.3.11. *Suppose $x, y, z \in \mathbb{N}$ with x and y being coprime. If $x|yz$, then $x|z$.*

Proof. We again write the prime factorisations of y and z as

$$y = p_1 p_2 \ldots p_k \quad \text{and} \quad z = q_1 q_2 \ldots q_l$$

where repetitions are again allowed, so that

$$yz = p_1 p_2 \ldots p_k q_1 q_2 \ldots q_l.$$

If $x|yz$, the prime factors of x are a subset of these prime factors of yz (because prime factorisations are unique). But if x and y are coprime, the prime factors of x do *not* include any from $\{p_i : i = 1, \ldots, k\}$, the set of prime factors of y. So the prime factors of x can only be a subset of $\{q_j : j = 1, \ldots, l\}$, i.e. of the prime factors of z. Hence $x|z$. $\qquad\square$

Note again the importance of the conditions of the theorem: it is not valid without the requirement for x and y to be coprime. For example, with $x = 6, y = 3, z = 4$ (so x and y have the common factor 3), we have $yz = 12$ and $x|yz$ without $x|z$.

Theorem 5.3.12. *Suppose $x, y, z \in \mathbb{N}$ with x and y being coprime. If $x|z$ and $y|z$, then $xy|z$.*

Proof. First, $x|z$ means that the prime factorisation of z includes all the prime factors of x. Next, $y|z$ means that it also includes those prime factors of y that have not already been accounted for among those of x. But that means *all* the prime factors of y, since x and y are coprime. So the prime factorisation of z includes all prime factors of x together with all prime factors of y, which together constitute the prime factors of xy. Hence $xy|z$. $\qquad\square$

Again, we see the necessity of the coprimeness condition in the theorem. For example, with $x = 6$, $y = 4$ and $z = 12$, we have $x|z$ and $y|z$ but $xy = 24 \nmid 12$. Referring to the proof of the theorem, we have all prime factors of 6 being among those of 12; but when we consider the prime factors of 4, one factor 2 has already been accounted for among the factors of 6, and so does not have to be included again among the factors of 12, even though it does appear among the factors of 6×4.

5.3.2   Primes in $\mathbb{Z}$ and Other Number Systems

What happens if we try to extend our definitions and theorems on primes and prime factorisations to the Integers, $\mathbb{Z}$?

If p is prime in $\mathbb{N}$, then in $\mathbb{Z}$ it has factors -1 and $-p$ as well as the factors 1 and p specified in Definition 5.3.1; and the number $-p$ also has the same four factors in $\mathbb{Z}$. But it seems sensible to still consider p to be prime in $\mathbb{Z}$, and then $-p$ should also be prime in $\mathbb{Z}$. In any factorisation of p or $-p$ as the product of two factors, one of those factors will be 1 or -1, and the other factor will be p or $-p$. So if we classify the numbers 1 as -1 as **units**, we can generalise Definition 5.3.1 by saying that a number is prime if, when it is written as the product of two factors, one of those factors must be a unit. Furthermore, if we describe p and $-p$ as **associates**[vi] of each other, we can say that the other factor of a prime must be either the prime itself or its associate. This then allows us to circumvent the problem that the prime factorisation of a composite number appears not to be unique in $\mathbb{Z}$ because, for example, if n is the product of just two primes in $\mathbb{N}$, $n = p_1 \times p_2$, then we also have $n = (-p_1) \times (-p_2)$ in $\mathbb{Z}$: as long as the factors in one factorisation of n are all associates of those in any other factorisation, we shall consider that uniqueness is not violated.

The informal definitions in the previous paragraph allow us to extend the concepts of prime numbers and prime factorisations from $\mathbb{N}$ to $\mathbb{Z}$, which is all that is required for the theory of Rational Numbers in the next chapter. But it is important to have formal definitions and theorems for units, associates and primes, to enable us to develop the theory for further number systems, in particular the quadratic extensions of $\mathbb{Z}$ which we shall meet in Chapter 8. Since these number systems are all commutative rings, we shall assume the associative and commutative properties of multiplication throughout the development below.

Definition 5.3.13. A **unit** is a number with a multiplicative inverse element.

From Definition 4.2.9, if a number system has a multiplicative identity element $i_\times$ (where in particular $i_\times = 1$ in $\mathbb{N}$ and $\mathbb{Z}$), numbers $a, \overline{a}$ are multiplicative inverse elements of each other (and are therefore units) if they satisfy $a \times \overline{a} = \overline{a} \times a = i_\times$. Since $i_\times \times i_\times = i_\times$, a multiplicative identity element is automatically a unit; and it is easy to verify that $\mathbb{N}$ has no units other than 1. But in $\mathbb{Z}$, -1 is also a unit since $(-1) \times (-1) = 1$.

Exercise 5.3.14.

Prove that 1 and -1 are the only numbers in $\mathbb{Z}$ with multiplicative inverses.[vii] *You may use any of the properties of multiplication and order in $\mathbb{Z}$.*

In Chapter 8 we will encounter number systems with several units, and the following general theorem is then important:

Theorem 5.3.15. *A product of any two units is itself a unit.*

[vi]Do not be confused by the terminology: "associate" here is nothing to do with the property of operations being "associative".

[vii]Thus in $\mathbb{N}$ and $\mathbb{Z}$ all units are their own multiplicative inverses (i.e. $a = \overline{a}$ in the equation defining the inverse element); but in other number systems this need not be the case.

Proof. Suppose u_1, u_2 are units, with multiplicative inverses $\overline{u_1}, \overline{u_2}$, respectively. Then the product $u_1 u_2$ has the multiplicative inverse $\overline{u_2}\,\overline{u_1}$ since

$$
\begin{aligned}
(u_1 u_2) \times (\overline{u_2}\,\overline{u_1}) &= u_1 (u_2 \overline{u_2}) \overline{u_1} && [\times \text{ is associative}] \\
&= u_1 i_\times \overline{u_1} && [\overline{u_2} \text{ is inverse of } u_2] \\
&= u_1 \overline{u_1} && [i_\times \text{ is multiplicative identity}] \\
&= i_\times, && [\overline{u_1} \text{ is inverse of } u_1]
\end{aligned}
$$

with a similar calculation verifying that $(\overline{u_2}\,\overline{u_1}) \times (u_1 u_2) = i_\times$. $\qquad\Box$

It then follows immediately that:

Corollary 5.3.16. *In any number system which is a commutative ring, the set of units with the operation of multiplication forms an Abelian group.*

Proof. Theorem 5.3.15 ensures that multiplication is closed on the set of units. Then, considering the four requirements in Definition 4.3.1: the commutative and associative properties are ensured by the number system being a commutative ring; the identity element for multiplication is its own inverse and so is a unit; and Definition 5.3.13 ensures that every member of the set has an inverse. $\qquad\Box$

Before generalising the definition of primes to number systems which have unit(s) other than the multiplicative identity element, it is useful to have the formal definition of an associate.

Definition 5.3.17. A number m is an **associate** of another number n if there is a unit u such that $m = nu$.

Thus in $\mathbb{Z}$, we have that $-n$ is an associate of n (for any $n \in \mathbb{Z}$), since $-n = n \times (-1)$, where -1 is a unit.

Theorem 5.3.18. *Being an associate is an equivalence relation.*

Proof. We need to show that the relation of being an associate is reflexive, symmetric, and transitive.

Reflexive: n is an associate of itself because $n = n \times i_\times$, where the multiplicative identity element is a unit.

Symmetric: If $m = nu$ and the multiplicative inverse of u is $\overline{u}$, then $m\overline{u} = (nu)\overline{u} = n(u\overline{u}) = ni_\times = n$; i.e. $n = m\overline{u}$, where $\overline{u}$ is a unit.

Transitive: If $m = nu_1$ and $n = ku_2$, then $m = (ku_2)u_1 = k(u_2 u_1)$, where $u_2 u_1$ is a unit according to Theorem 5.3.15.

[Note that we have used the property that multiplication is associative in both the "symmetric" and "transitive" parts of the proof.] $\qquad\Box$

This equivalence relation generates a partition of a number system into **associate classes**: subsets of numbers in which all members of a subset are associates of each other. In particular, the units in a number system form a single associate class: for if u_1, u_2 are units, then $u_2 = i_\times u_2 = (u_1 \overline{u_1}) u_2 = u_1 (\overline{u_1} u_2)$, where $\overline{u_1} u_2$ is a unit by Theorem 5.3.15. In the case of $\mathbb{Z}$, the partition consists of an associate class $\{n, -n\}$ for each $n \in \mathbb{Z}_+$, including the units class $\{1, -1\}$, and there is also the class $\{0\}$; the additive identity i_+ $(= 0$ in $\mathbb{Z})$ will always form its own associate class, due to its property that $i_+ \times n = i_+$ for every n in a ring (Theorem 4.3.7).

In general we can notate the associate class containing a number n as $[n]$. We can then define a multiplication operation on associate classes:

Definition 5.3.19.

$$[m] \otimes [n] := [k] \quad \text{if} \quad m \times n = k.$$

For the operation $\otimes$ to be well-defined, we require that $m_* \times n_* \in [k]$ if $m_* \in [m]$ and $n_* \in [n]$ with $m \times n = k$. But if $m_* \in [m]$ and $n_* \in [n]$, we have $m_* = mu_1$ and $n_* = nu_2$ for some units u_1, u_2, so that $m_* \times n_* = (mu_1)(nu_2) = (mn)(u_1u_2) = k(u_1u_2) \in [k]$ since u_1u_2 is a unit by Theorem 5.3.15. [We have used both the associative and commutative properties of multiplication here.]

Turning now to primes, our formal generalisation of Definition 5.3.1 is:

Definition 5.3.20. A number is **prime** if it can be written as the product of two factors *only* with one of those factors being a unit.

So the special status of the number 1 in Definition 5.3.1 for primes in $\mathbb{N}$ derives not from its being the multiplicative identity element, but from its being the only unit in $\mathbb{N}$.

The importance of identifying associates is now seen in the following theorem:

Theorem 5.3.21. *If p is prime, then all members of its associate class $[p]$ are also prime.*

Note that this theorem allows us to refer to a **prime class**: the associate class of a prime.

Proof. If q and p are associates of each other, then by Definition 5.3.17 we can write $p = qu$, where u is a unit. Suppose p is prime. If q was *not* prime, we could write $q = rs$, where neither r nor s are units. But then $p = rs \times u = r \times su$, where neither r nor su are units (since su is in the associate class of a non-unit), which would contradict p being prime. Hence q must be prime. $\qquad\square$

If p is prime in $\mathbb{N}$, then p is also prime in $\mathbb{Z}$: the isomorphism between $\mathbb{N}$ and $\mathbb{Z}_+$ under multiplication ensures that the only factorisation of p in positive integers is $p \times 1$, and the properties of additive inverses under multiplication (see Theorem 4.3.8) may be used to verify that the only other factorisation of p is $(-p) \times (-1)$, i.e. involving an associate of p and a unit. [The latter verification is important because in some of the extensions of $\mathbb{Z}$ that we shall study in Chapter 8, numbers corresponding to p under an isomorphism may not be prime even when p is prime in $\mathbb{Z}$.] Theorem 5.3.21 then shows that $-p$ is prime in $\mathbb{Z}$, with factorisations $(-p) \times 1$ and $p \times (-1)$.

Now consider prime factorisations of composite numbers in a number system containing a unit u other than the multiplicative identity element $i_\times$. We have already observed that if $n = p_1 \times p_2$ in $\mathbb{N}$, we also have the factorisation $n = (-p_1) \times (-p_2)$ in $\mathbb{Z}$. More generally, if $n = p_1 \times p_2$ in a number system with a unit $u \neq i_\times$, we also have the factorisation $n = (p_1 u) \times (p_2 \bar{u})$; and this apparent non-uniqueness of prime factorisations obviously extends to numbers with more than two prime factors. But in writing $n = p_1 \times p_2 = (p_1 u) \times (p_2 \bar{u})$, we have simply replaced the primes in one factorisation with associates of those same primes in the other factorisation, so we do not regard the two factorisations as being substantively different: the Fundamental Theorem of Arithmetic (uniqueness of prime factorisations) is not violated in such a case. We can make a precise definition of uniqueness in the context of prime factorisation by considering how associate classes of numbers may be factorised using the multiplication operation $\otimes$ in Definition 5.3.19:

Definition 5.3.22. The **Fundamental Theorem of Arithmetic** applies (i.e. prime factorisations are considered **unique**) in a number ring if every associate class of numbers, except for the units class and the additive identity, can be expressed uniquely as the product of one or more prime classes using the operation $\otimes$ in Definition 5.3.19.

Thus the Fundamental Theorem of Arithmetic is true in $\mathbb{Z}$, since if $n = p_1 \times p_2 \times \cdots \times p_k$ the only alternative factorisations of n involve replacing an even number of primes p_i by their additive inverses $-p_i$. For example, whereas in $\mathbb{N}$ we would say that 15 has the unique prime factorisation 3×5, in $\mathbb{Z}$ we say that $[15]$ has the unique prime factorisation $[3] \otimes [5]$, which encompasses the factorisations $15 = 3 \times 5 = (-3) \times (-5)$ and also $-15 = 3 \times (-5) = (-3) \times 5$.

The exclusion of the additive identity (zero in the case of $\mathbb{Z}$) from Definition 5.3.22 did not appear in Theorem 5.3.2 because there is no additive identity in $\mathbb{N}$; but it certainly does need to be excluded, as can be seen from the impossibility of expressing 0 as the product of primes in $\mathbb{Z}$. Thus the partition of $\mathbb{N}$ that followed Definition 5.3.1 may be generalised: a ring of numbers is partitioned into the subsets,

$$\{\text{Additive identity}\} \cup \{\text{Units}\} \cup \{\text{Prime numbers}\} \cup \{\text{Composite numbers}\};$$

or we can consider partitioning the set of associate classes of numbers in a ring as

$$\{\text{Additive identity}\} \cup \{\text{Units class}\} \cup \{\text{Prime classes}\} \cup \{\text{Composite classes}\}.$$

In later chapters, we shall encounter the Rational, Real, and Complex number systems, in each of which every number apart from the additive identity has a multiplicative inverse element; so in such number systems the identification of "units" is rather pointless and there are no such things as primes or composites. Before that, we shall meet finite number systems in the later sections of this chapter, in which there may be some numbers with multiplicative inverses and some without; but it is not usually useful to identify primes and composites in such number systems. So the language of "units", "primes" and "composites" is generally reserved for those number systems in which there are infinitely many numbers without multiplicative inverses. $\mathbb{N}$ and $\mathbb{Z}$ both satisfy this requirement, as do the "extensions" of $\mathbb{Z}$ which we shall meet in Chapter 8.

Finally in this section, we mention that the concept of "coprime" has an obvious generalisation to $\mathbb{Z}$ and other number systems with more than one unit, as long as prime factorisations are unique in the sense of Definition 5.3.22: two numbers are coprime if their prime factorisations do not include any numbers from the same prime classes. For example, having established that 15 is coprime with 28 in $\mathbb{N}$, those numbers are certainly coprime in $\mathbb{Z}$, as is -15 with 28, 15 with -28, and -15 with -28.

5.4 Congruence

Questions of divisibility are a major theme of Number Theory. An important tool which allows us to settle such questions, even involving very large integers, is **congruence**. This also forms the basis of our construction of finite number systems in the next section.

Definition 5.4.1. Given $d \in \mathbb{N}\backslash\{1\}$ and $a, b \in \mathbb{Z}$, we say that "a is congruent to b, modulo d" and we write

$$a \equiv b \pmod{d}$$

if a and b have the *same remainder* when divided by d.

The notation $\equiv$ has previously been used for an equivalence relation; its use again here is no coincidence:

Theorem 5.4.2. *Congruence is an equivalence relation.*

Proof. The "same remainder" criterion makes it obvious that each of the three conditions for an equivalence relation is fulfilled:

Reflexive: Any integer a has the same remainder as itself when divided by d: $a \equiv a \pmod d$.

Symmetric: If a has the same remainder as b when divided by d, then b has the same remainder as a: if $a \equiv b \pmod d$, then $b \equiv a \pmod d$.

Transitive: If a has the same remainder as b and b has the same remainder as c when divided by d, then a has the same remainder as c: if $a \equiv b \pmod d$ and $b \equiv c \pmod d$, then $a \equiv c \pmod d$. □

The next theorem provides an alternative definition of congruence.

Theorem 5.4.3. $a \equiv b \pmod d$ *if and only if* $d|(a - b)$.

Definition 5.4.1 says that we should test the congruence of two integers modulo d by dividing the individual numbers by d and comparing remainders. Theorem 5.4.3 says that an alternative way to test their congruence is by dividing d into the difference between the two integers.

Proof of Theorem 5.4.3. According to Definition 5.4.1, $a \equiv b \pmod d \Rightarrow a = q_1 d + r$ and $b = q_2 d + r$ for some $q_1, q_2, r \in \mathbb{Z}$ with r satisfying the requirements of the Division Theorem. Then

$$a - b = (q_1 d + r) - (q_2 d + r)$$
$$= (q_1 - q_2)d,$$

so that d is a divisor of $(a - b)$.

Conversely, according to Definition 5.1.6, $d|(a - b) \Rightarrow a - b = qd$ for some $q \in \mathbb{Z}$. Now divide a by d:

$$a = q_1 d + r,$$

where $q_1, r \in \mathbb{Z}$ and r satisfies the requirements of the Division Theorem. Then

$$b = a - (a - b)$$
$$= (q_1 d + r) - qd$$
$$= (q_1 - q)d + r,$$

i.e. dividing b by d gives quotient $(q_1 - q) \in \mathbb{Z}$ and the *same* remainder r as when a was divided by d. So if $d|(a - b)$, then $a \equiv b \pmod d$ according to Definition 5.4.1. □

Example 5.4.4. Show that $11 \equiv -94 \pmod 5$.

Solution. We can do this in two ways: using the original definition of congruence or the alternative definition provided by Theorem 5.4.3.

First method: we have $11 = 2 \times 5 + 1$ and $-94 = (-19) \times 5 + 1$, so both 11 and -94 have the same remainder, 1, on division by 5.

Second method: $11 - (-94) = 105 = 21 \times 5$, so 5 is a divisor of the difference between 11 and -94. □

The usefulness of congruence in Number Theory is founded on the remaining theorems in this section.

Theorem 5.4.5. *Suppose* $a \equiv b \pmod d$ *and* $x \equiv y \pmod d$. *Then:*
(i) $a + x \equiv b + y \pmod d$;
(ii) $a \times x \equiv b \times y \pmod d$.

We shall prove the addition theorem (i), and leave the multiplication theorem (ii) as an exercise for the reader.

Proof of (i). By Theorem 5.4.3, we have

$$a - b = q_1 d \quad \text{and} \quad x - y = q_2 d$$

for some $q_1, q_2 \in \mathbb{Z}$. Then

$$
\begin{aligned}
(a + x) - (b + y) &= (a - b) + (x - y) \\
&= q_1 d + q_2 d \\
&= (q_1 + q_2)d,
\end{aligned}
$$

with $(q_1 + q_2) \in \mathbb{Z}$. Thus, according to Theorem 5.4.3, we have

$$a + x \equiv b + y \ (\text{mod } d). \qquad \square$$

Exercise 5.4.6. *Prove that if $a \equiv b$ (mod d) and $x \equiv y$ (mod d), then*

$$a \times x \equiv b \times y \ (\text{mod } d).$$

Since subtraction of an integer is the same as addition of its additive inverse (Corollary 4.2.15), the addition theorem also applies to subtraction.

Corollary 5.4.7. *If $a \equiv b$ (mod d) and $x \equiv y$ (mod d), then*

$$a - x \equiv b - y \ (\text{mod } d).$$

However, there is no inverse operation of multiplication in the Integers, so there is no similar corollary of the multiplication theorem. In particular, it is *not* in general true that if $ax \equiv bx$ (mod d), then $a \equiv b$ (mod d). As a numerical counterexample, take $a = 3$, $b = 5$, $x = 9$, $d = 6$: we have $ax = 27, bx = 45$, so $ax - bx = -18 = (-3) \times 6$ so that $ax \equiv bx$ (mod 6); but $3 \not\equiv 5$ (mod 6). You might like to think about conditions on the relation between x and d under which such an inverse multiplication theorem would be true; although all will become clear later in this chapter.

Example 5.4.8. Show that $137 \times 89 - 73$ is divisible by 12.

Obviously the numbers involved here are small enough that it would not be difficult to answer this by direct calculation. However, the example is offered as a simple illustration of the above theorem and corollary. So imagine that you know the 12 times table (up to 12×12) but do not know how to do long multiplication. The 12 times table is necessary in this example because of the general principle that questions of divisibility by any given $d \in \mathbb{N}$ are solved by considering congruence modulo d.

Solution. The method of working is to look for congruences to small (positive or negative) numbers that are easier to handle than the original numbers in the problem. So in this example we note that

$$137 \equiv 5 \ (\text{mod } 12), \quad 89 \equiv 5 \ (\text{mod } 12) \quad \text{and} \quad 73 \equiv 1 \ (\text{mod } 12).$$

Hence,

$$
\begin{aligned}
137 \times 89 - 73 &\equiv 5 \times 5 - 1 \ (\text{mod } 12) \\
&= 24 \\
&\equiv 0 \ (\text{mod } 12),
\end{aligned}
$$

where the last line is included to emphasise that divisibility by d is established by finding congruence to 0 (mod d) (i.e. zero remainder on division by d); we could have stopped at the previous line, noting from our 12 times table that 24 is divisible by 12. In any case, our calculation has shown that $137 \times 89 - 73 \equiv 0$ (mod 12) (using the transitivity of congruence, and noting that equality is a special case of congruence), and so $12|(137 \times 89 - 73)$. $\qquad\square$

Exercise 5.4.9.

Show that $(79 \times 87 + 53 \times 65)$ is divisible by 7 (without doing any multiplication of numbers larger than 7).

The next theorem is even more useful in considering divisibility of large numbers and also for theoretical developments.

Theorem 5.4.10. *If $a \equiv b$ (mod d), then $a^k \equiv b^k$ (mod d) for any $k \in \mathbb{N}$.*

Proof. Recall that the inductive Definition 3.4.1 of exponentiation applies to natural number powers of integers; so we can prove this theorem by induction.
(a) For $k = 1$: $a^k = a^1 = a$ and $b^k = b^1 = b$; so if $a \equiv b$ (mod d), then $a^k \equiv b^k$ (mod d).
(b) Given the inductive hypothesis $a^k \equiv b^k$ (mod d),

$$
\begin{aligned}
a^{S(k)} &= a \times a^k && [\textit{Def. 3.4.1}] \\
&\equiv b \times b^k \ (\text{mod } d) && [\textit{Inductive hypothesis and Theorem 5.4.5(ii)}] \\
&= b^{S(k)}. && [\textit{Def. 3.4.1}] \qquad\square
\end{aligned}
$$

Note that the congruence does *not* apply to the powers: if $k \equiv l$ (mod d), it is *not* in general true that $a^k \equiv a^l$ (mod d). For example, $4 \equiv 7$ (mod 3) but $2^4 \not\equiv 2^7$ (mod 3), since $2^7 - 2^4 = 128 - 16 = 112$ and $3 \nmid 112$.

Our next example is a rather artificial application of the theorem, but is good for illustrating its use, and also uses a theorem from earlier in this chapter.

Example 5.4.11. Show that 111111 is divisible by 7.

Solution. First note that $9 \times 111111 = 999999 = 10^6 - 1$: since 7 and 9 are coprime, $7|111111$ if and only if $7|(10^6 - 1)$ [*Theorem 5.3.11*].

We now use Theorem 5.4.10 to reduce the fairly large number $10^6 - 1$ progressively to smaller numbers by congruence modulo 7 (since we are considering divisibility by 7).

$$
\begin{aligned}
10^6 - 1 &\equiv 3^6 - 1 \ (\text{mod } 7) && [\textit{Theorem 5.4.10, with } 10 \equiv 3 \ (\text{mod } 7)] \\
&= (3^3)^2 - 1 \\
&= 27^2 - 1 \\
&\equiv (-1)^2 - 1 \ (\text{mod } 7) && [\textit{Theorem 5.4.10, with } 27 \equiv -1 \ (\text{mod } 7)] \\
&\equiv 0 \ (\text{mod } 7). && \qquad\square
\end{aligned}
$$

Notice that in the second-last line of the calculation we used a congruence to a negative integer: -1 is "smaller" (closer to zero) than 6, and there is no restriction in such a calculation to use numbers that satisfy the requirements of a remainder in the Division Theorem.

When using Theorem 5.4.10 to handle large powers, as in the next example, it is useful to recall that $(-1)^n = 1$ if n is even, while $(-1)^n = -1$ if n is odd. This is a correct use of congruence of a power, since evenness and oddness refer to congruence modulo 2; the next example also illustrates more subtle ways to use congruence of a power to deal with divisibility problems involving large powers.

Example 5.4.12. Show that $51^{9753} + 68^{369}$ is divisible by 13.

Solution. We require congruences modulo 13, so we first note that $51 \equiv -1 \pmod{13}$ and $68 \equiv 3 \pmod{13}$. Theorem 5.4.10 then yields

$$51^{9753} + 68^{369} \equiv (-1)^{9753} + 3^{369} \pmod{13}.$$

Now, 9753 is odd so $(-1)^{9753} = -1$; but to deal with 3^{369} we consider smaller powers of 3. We notice that $3^3 = 27 \equiv 1 \pmod{13}$ and that 369 is divisible by 3, so we can write $369 = 3m$ (where we do not need the value of the integer m) and then

$$3^{369} = 3^{3m} = (3^3)^m \equiv 1^m \pmod{13} = 1.$$

So finally
$$51^{9753} + 68^{369} \equiv -1 + 1 \pmod{13} \equiv 0 \pmod{13},$$

i.e. $51^{9753} + 68^{369}$ is divisible by 13. □

Exercise 5.4.13. *(i) Show that $(36^{62} - 62^{36})$ is divisible by 7 (without taking powers of any numbers larger than 7).*
(ii) Find the remainder when $317^{654} - 119^{456}$ is divided by 11 (without taking powers of any numbers larger than 11). [Hint: what is the nearest multiple of 11 to 2^5?]
(iii) Show that $9,999,999,999,999$ (that's 13 nines) is divisible by 53. [Hints: $9,999,999,999,999 = 10^{13} - 1$; note that $10^2 = 2 \times 53 - 6$.]

5.5 Modular Arithmetic

Since congruence modulo d is an equivalence relation on $\mathbb{Z}$, it partitions $\mathbb{Z}$ into equivalence classes, known as **congruence classes**:

Definition 5.5.1. For any $d \in \mathbb{N}$ let r be an allowed remainder on division by d (i.e. $r \in \mathbb{Z}$ with $0 \leq r < d$). Then the **congruence class** $[r]_d$ is the set of all integers with remainder r on division by d:
$$[r]_d := \{n \in \mathbb{Z} : n = qd + r; \, q \in \mathbb{Z}\}.$$

Thus the members of a congruence class are all the integers that are congruent to r modulo d.

Example 5.5.2. (i) $[3]_5 = \{\dots, -12, -7, -2, 3, 8, 13, \dots\}$: there are infinitely many positive and negative integers with remainder 3 on division by 5.
(ii) We can write the set of all even integers as $[0]_2$ and the set of all odd integers as $[1]_2$.

For each $d \in \mathbb{N}\backslash\{1\}$ we can now define a new number system, $\mathbb{Z}_d$, in which each number is a congruence class modulo d. [We have done something similar before: Definition 4.1.4 also defines each integer as an equivalence class containing infinitely many objects.]

Definition 5.5.3. For each $d \in \mathbb{N}\backslash\{1\}$,

$$\mathbb{Z}_d := \{[r]_d : r \in \mathbb{Z}, 0 \leq r < d\}.$$

This is a finite number system, since there are d congruence classes modulo d: we could write Definition 5.5.3 as

$$\mathbb{Z}_d = \{[0]_d, [1]_d, \ldots, [d-1]_d\}.$$

This method of using a finite set of integers to represent quantities that can in principle take infinitely many integer values is common in our daily lives, particularly in our representation of time. The 12-hour clock is an application of $\mathbb{Z}_{12}$, representing each hour by its "remainder" after noon or midnight when the half-day is divided by 12 – although in the case of zero remainder, we refer to "12 o'clock" rather than "0 o'clock". For minutes and seconds we use $\mathbb{Z}_{60}$, but $\mathbb{Z}_{12}$ is used again to represent longer periods of time, albeit with names rather than numbers: the 12 months of the year and, in the Chinese calendar, the 12 animal names for the years. The author of this book was born in the Year of the Dog, and is also writing this sentence in the Year of the Dog: his age in years at this time is a multiple of 12. Another application of $\mathbb{Z}_{12}$ is in music, in the Western 12-tone scale, with notes being named A, B-flat, ..., G-sharp.[viii] All these applications show the cyclic character of $\mathbb{Z}_d$: the next integer after $[d-1]_d$ is $[0]_d$, and d integers after $[r]_d$ takes us back to $[r]_d$; the next hour after 12 o'clock is 1 o'clock, the next minute after 12:59 is 1:00, and 12 hours after 12:59 is again 12:59; the next note after G-sharp is A, and 12 notes higher (or lower) than G-sharp is again G-sharp. The process of moving on n hours or sounding a note n tones higher can be thought of as adding n in the number system $\mathbb{Z}_{12}$, and we can similarly think of moving back n hours or sounding n notes lower as subtracting n in $\mathbb{Z}_{12}$. So we now provide formal definitions of addition and subtraction in $\mathbb{Z}_d$, and also of multiplication in $\mathbb{Z}_d$ which does not have any obvious counterparts in daily life. These processes are called **modular arithmetic** (but are often informally referred to as "clock arithmetic", due to the analogy with the 12-hour clock).

Definition 5.5.4. In $\mathbb{Z}_d$ we define $[m]_d + [n]_d$, $[m]_d - [n]_d$ and $[m]_d \times [n]_d$ respectively to be the congruence classes modulo d of the remainders when $m+n$, $m-n$ and $m \times n$ are divided by d.[ix]

As was done following Definitions 4.2.1 and 4.2.2 for addition and multiplication in $\mathbb{Z}$, we should check that the arithmetic operations in $\mathbb{Z}_d$ are *closed* and *well-defined*. The closed property is immediately apparent from Definition 5.5.4, since "congruence classes modulo d of the remainders . . ." are precisely what the members of $\mathbb{Z}_d$ are defined to be. The well-defined property (that the same result would be obtained whichever member of an equivalence class we take to represent the class) is guaranteed by Theorem 5.4.5 and Corollary 5.4.7, since a remainder on division by d is congruent modulo d.

To show how arithmetical calculations in $\mathbb{Z}_d$ work, we do some examples in $\mathbb{Z}_{12}$.

[viii]Why 12 tones in the scale? A scale in which all combinations of notes sounded harmonically perfect would have t tones where $3^t = 2^s$ for some $s, t \in \mathbb{N}$, which is obviously impossible due to the Fundamental Theorem of Arithmetic. So for a scale to sound good, we try to approximate this ideal: we seek $s, t \in \mathbb{N}$ such that $3^t \approx 2^s$ (with s, t not too large), and it turns out that the difference between $3^{12} = 531441$ and $2^{19} = 524288$ is small enough for a 12-tone scale to sound acceptable. Nevertheless, there are many and diverse **temperaments** (ways of defining the 12 tones to account for the mismatch between 3^{12} and 2^{19}), and the search for temperaments that minimise any dissonance in music has been an important application of mathematics over many centuries.

[ix]When we defined addition and multiplication operations in $\mathbb{Z}$ in Section 4.2, we initially used new symbols $\oplus$ and $\otimes$ since the operations were defined differently from $+$ and $\times$ in $\mathbb{N}$, and we only consented to use the same symbols as in $\mathbb{N}$ once we had established isomorphisms between $\mathbb{N}$ and a subset of $\mathbb{Z}$ with their respective operations. There are no isomorphisms between the infinite set $\mathbb{Z}$ and any finite sets $\mathbb{Z}_d$, but the operations in Definition 5.5.4 are so closely related to the respective operations in $\mathbb{Z}$ that it would be very pedantic to introduce new symbols for them, even temporarily.

Example 5.5.5.

(i) $[9]_{12} + [7]_{12} = [4]_{12}$ (since $9 + 7 = 16 = 1 \times 12 + 4$)
(ii) $[9]_{12} - [7]_{12} = [2]_{12}$ (since $9 - 7 = 2 = 0 \times 12 + 2$)
(iii) $[7]_{12} - [9]_{12} = [10]_{12}$ (since $7 - 9 = -2 = (-1) \times 12 + 10$)
(iv) $[9]_{12} \times [7]_{12} = [3]_{12}$ (since $9 \times 7 = 63 = 5 \times 12 + 3$)
(v) $[9]_{12} \times [4]_{12} = [0]_{12}$ (since $9 \times 4 = 36 = 3 \times 12 + 0$)

In each case we have done the ordinary addition, subtraction, or multiplication of integers and then found the remainder when the result is divided by the modulus 12.[x] The result of (v) may seem rather curious: multiplying two non-zero numbers and obtaining zero, which cannot occur in $\mathbb{Z}$ or in any of the infinite number systems that we shall define later; its significance will become clearer when we discuss $\mathbb{Z}_d$ as an algebraic structure.

Exercise 5.5.6. *Evaluate the following in* $\mathbb{Z}_9$:
   (i) $[3]_9 + [7]_9$ *(ii)* $[3]_9 - [7]_9$ *(iii)* $[3]_9 \times [7]_9$ *(iv)* $[6]_9 \times [6]_9$

If we need to do many modular arithmetic calculations in a particular number system $\mathbb{Z}_d$, it is helpful to have addition and multiplication tables for $\mathbb{Z}_d$. These are produced by the procedure of taking remainders as in Example 5.5.5; but a remainder when a number is divided by d is simply the last digit of the number expressed in base d, so entries in the tables for $\mathbb{Z}_d$ will simply be the last digits of the corresponding entries in the addition and multiplication tables in base d (see Section 5.2). We present here the addition and multiplication tables for $\mathbb{Z}_3$ and $\mathbb{Z}_4$, which will be used in examples below. For completeness we include the columns and rows for zero (which were omitted from the base-6 tables in Section 5.2). To simplify the rather cumbersome notation $[r]_d$, we omit the subscript for the modulus d, which is the same throughout any table.
    For $\mathbb{Z}_3$:

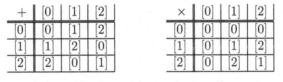

+	[0]	[1]	[2]
[0]	[0]	[1]	[2]
[1]	[1]	[2]	[0]
[2]	[2]	[0]	[1]

×	[0]	[1]	[2]
[0]	[0]	[0]	[0]
[1]	[0]	[1]	[2]
[2]	[0]	[2]	[1]

For $\mathbb{Z}_4$:

+	[0]	[1]	[2]	[3]
[0]	[0]	[1]	[2]	[3]
[1]	[1]	[2]	[3]	[0]
[2]	[2]	[3]	[0]	[1]
[3]	[3]	[0]	[1]	[2]

×	[0]	[1]	[2]	[3]
[0]	[0]	[0]	[0]	[0]
[1]	[0]	[1]	[2]	[3]
[2]	[0]	[2]	[0]	[2]
[3]	[0]	[3]	[2]	[1]

An important application of modular arithmetic is to **Diophantine equations**, which are polynomial equations in two or more unknowns for which integer solutions are sought. In particular, modular arithmetic is useful when it is required to prove that such an equation does not have a solution: if a solution exists in $\mathbb{Z}$, then by Theorem 5.4.5 a solution exists when the equality is replaced by a congruence modulo d (for any $d \in \mathbb{N}$), and hence a solution exists in $\mathbb{Z}_d$ (with any constants in the equation being replaced by their remainders modulo d). The contrapositive of this is that if no solution exists in some $\mathbb{Z}_d$, then no

---

[x]A different use of the word "modulus" from that introduced in Definition 4.2.12. There should not be any confusion: the word is used for the number d in Definitions 5.5.1 and 5.5.3 only in the context of modular arithmetic.

solution exists in $\mathbb{Z}$. So a proof by exhaustion becomes feasible, searching through finitely many possible values in $\mathbb{Z}_d$ for each unknown, rather than needing to consider infinitely many values in $\mathbb{Z}$. Note that the converse argument is not valid: the existence of a solution of an equation in $\mathbb{Z}$ cannot be inferred from a solution in some $\mathbb{Z}_d$.

Example 5.5.7. Using $\mathbb{Z}_3$, show that there do not exist any integers m, n such that

$$2m^2 + 11 = 9n^2.$$

Solution. First note the congruence classes of the constants 11 and 9 modulo 3: $11 \in [2]_3$ and $9 \in [0]_3$. So in $\mathbb{Z}_3$ the equation reduces to

$$[2]m^2 + [2] = [0]n^2$$
$$= [0]. \tag{5.4}$$

We need to consider all possible values of m^2 in $\mathbb{Z}_3$:

$$[0]^2 = [0], \quad [1]^2 = [1], \quad [2]^2 = [1]$$

(where we have used the multiplication table for $\mathbb{Z}_3$), so m^2 can take the values $[0]$ or $[1]$ in $\mathbb{Z}_3$. Taking these values in turn in $[2]m^2 + [2]$ on the left-hand side of (5.4), we have $[2] \times [0] + [2] = [2]$ and $[2] \times [1] + [2] = [2] + [2] = [1]$ (using the addition table for $\mathbb{Z}_3$). In neither case do we obtain $[0]$, the required value on the right-hand side of (5.4); so the equation in $\mathbb{Z}_3$ has no solution, and hence the original Diophantine equation has no solution. □

It is of interest to note what happens if we do this calculation in $\mathbb{Z}_4$. For the constants, we have $11 \in [3]_4$ and $9 \in [1]_4$, so our Diophantine equation reduces to

$$[2]m^2 + [3] = [1]n^2 = n^2.$$

For the possible values of m^2 and n^2 in $\mathbb{Z}_4$, the multiplication table for $\mathbb{Z}_4$ gives

$$[0]^2 = [0], \quad [1]^2 = [1], \quad [2]^2 = [0], \quad [3]^2 = [1].$$

So m^2 and n^2 can take the values $[0]$ and $[1]$ in $\mathbb{Z}_4$. With $m^2 = [1]$, the left-hand side of our equation in $\mathbb{Z}_4$ is $[2]m^2 + [3] = [2] \times [1] + [3] = [2] + [3] = [1]$ (using the addition table for $\mathbb{Z}_4$); since n^2 on the right-hand side of the equation can take the value $[1]$, the equation *can* be satisfied in $\mathbb{Z}_4$. This does not imply that the original Diophantine equation can be satisfied, and indeed we have already shown that it does not have any solutions. Unfortunately, there is no simple rule for deciding which modulus d is the best choice for attempting a proof of the type in the above example.

Exercise 5.5.8. *Find numbers $m, n \in \mathbb{Z}_4$ such that $[2]_4 m^2 - [3]_4 n^2 = [1]_4$.*
   Show that there do not exist numbers $m, n \in \mathbb{Z}_8$ such that $[2]_8 m^2 - [3]_8 n^2 = [1]_8$.
   What can you say about the existence of integers $m, n \in \mathbb{Z}$ that satisfy the Diophantine equation $2m^2 - 3n^2 = 1$?

Another example of the use of modular arithmetic concerns sums of squares, an important topic in Number Theory:

Example 5.5.9. Show that $m^2 + n^2 \not\equiv 3 \pmod 4$ for any $m, n \in \mathbb{Z}$.

Solution. This question is framed in terms of (non-)congruence modulo 4, and is equivalent to the question: "Show that there do not exist $m, n \in \mathbb{Z}_4$ such that $m^2 + n^2 = [3]$". We have already shown that the square of an integer in $\mathbb{Z}_4$ can only be $[0]$ or $[1]$. So the possible values of $m^2 + n^2$ are:

$$[0] + [0] = [0], \quad [0] + [1] = [1], \quad [1] + [0] = [1], \quad [1] + [1] = [2],$$

so it is *not* possible to have $m^2 + n^2 = [3]$. □

The set of all primes in $\mathbb{N}$ is partitioned into three subsets according to their remainder modulo 4 (not four subsets: there are obviously no primes in $[0]_4$). According to Example 5.5.9, the primes in $[3]_4$ cannot be expressed as a sum of squares. There is a single prime, 2, in $[2]_4$, with $2 = 1^2 + 1^2$; and for primes in $[1]_4$ there is a remarkable theorem that each such prime can be expressed as a sum of squares of positive integers in a unique manner; but to prove that requires knowledge of Number Theory beyond the scope of this book.

5.6 $\mathbb{Z}_d$ as an Algebraic Structure

The finite number systems $\mathbb{Z}_d$ (for all $d \in \mathbb{N}\backslash\{1\}$) inherit the commutative and associative properties of both addition and multiplication, and also the distributive law, from $\mathbb{Z}$. Furthermore, the numbers $[0]$ and $[1]$ are respectively the identity elements for addition and multiplication in $\mathbb{Z}_d$. Additive inverse elements are also easy to find in $\mathbb{Z}_d$: since $r + (d - r) = (d - r) + r = d \in [0]_d$, we have that $[d - r]$ is the additive inverse element of $[r]$, except when $r = 0$ in which case $[0]$ is the additive inverse of $[0]$ (since $0 + 0 = 0$). So $\mathbb{Z}_d$ has the same algebraic properties, labelled A1–A4, M1–M3, and D in Section 4.3, as $\mathbb{Z}$; thus in the terminology of algebraic structures, we can say that $\langle \mathbb{Z}_d, + \rangle$ is an Abelian group and that $\langle \mathbb{Z}_d, +, \times \rangle$ is a commutative ring: see Definitions 4.3.1 and 4.3.5.

But can we go beyond this? Multiplicative inverse elements have not yet been discussed: since $[1]$ is the multiplicative identity element in $\mathbb{Z}_d$, a multiplicative inverse of $[r]$ in $\mathbb{Z}_d$ would be a number $[s] \in \mathbb{Z}_d$ such that $[r] \times [s] = [s] \times [r] = [1]$. An identity element for any binary operation is its own inverse (since $i_\vee \vee i_\vee = i_\vee$, in the notation of Section 4.3), so $[1]$ is the multiplicative inverse of $[1]$ in $\mathbb{Z}_d$, and indeed 1 is the multiplicative inverse of 1 in $\mathbb{Z}$. Furthermore, in a ring $\langle A, *, \vee \rangle$ we have $i_* \vee a = a \vee i_* = i_* \, \forall a \in A$ [Theorem 4.3.7], and since $i_\vee \neq i_*$ [see at the end of Section 4.3] this means that the identity element for $*$ cannot have an inverse element under $\vee$ (there does not exist $a \in A$ such that $a \vee i_* = i_\vee$). So in the ring $\langle \mathbb{Z}_d, +, \times \rangle$ the number $[0]$ has no multiplicative inverse, and of course in the ring of ordinary integers $\langle \mathbb{Z}, +, \times \rangle$ the number 0 has no multiplicative inverse.

But the abstract properties of rings are not helpful for seeking further multiplicative inverses, for which we need to consider the specific properties of $\mathbb{Z}_d$ for each d. Such inverses are easy to find in a multiplication table for $\mathbb{Z}_d$: if the entry $[1]$ appears in column $[s]$ in row $[r]$, then $[s]$ and $[r]$ are multiplicative inverses of each other. Referring to the multiplication tables for $\mathbb{Z}_3$ and $\mathbb{Z}_4$ in the previous section we find:

- In $\mathbb{Z}_3$, $[2]$ is the multiplicative inverse of $[2]$; so apart from $[0]$ which cannot have a multiplicative inverse, every member of $\mathbb{Z}_3$ does have one;

- In $\mathbb{Z}_4$, $[3]$ is the multiplicative inverse of $[3]$ but $[2]$ does not have a multiplicative inverse: the entry $[1]$ does not appear in the row for $[2]$ in the multiplication table for $\mathbb{Z}_4$. So, unlike $\mathbb{Z}_3$, not every member of $\mathbb{Z}_4\backslash\{[0]\}$ has a multiplicative inverse.

Why this difference between the properties of $\mathbb{Z}_3$ and $\mathbb{Z}_4$? The next theorem tells us when we can find a multiplicative inverse in any $\mathbb{Z}_d$.

Theorem 5.6.1. *A number $[r]$ has a multiplicative inverse in $\mathbb{Z}_d$ if and only if r is coprime with d.*

Proof. Suppose first that r is *not* coprime with d: they share a common factor $q > 1$, so $r = kq$ and $d = lq$ for some $k, l \in \mathbb{N}$. Consider xr, a multiple of r, and divide it by d: $xr = ad + b$ for some $a, b \in \mathbb{Z}$ with $0 \le b < d$. In terms of the factors of r and d, this becomes

$$xkq = alq + b,$$
$$\therefore b = (xk - al)q.$$

We have shown that the remainder when xr is divided by d must be a multiple of q, and so cannot be 1; so $[x] \times [r] \neq [1]$ for any $[x] \in \mathbb{Z}_d$.

Now consider the case where r *is* coprime with d. If $[r]$ has no multiplicative inverse in $\mathbb{Z}_d$, i.e. if there does not exist $[x] \in \mathbb{Z}_d$ with $[r] \times [x] = [1]$, there remain $d-1$ possible values of the product $[r] \times [x]$, with d values of $[x]$ in $\mathbb{Z}_d$. So according to the Pigeonhole Principle there must exist $[x_1], [x_2] \in \mathbb{Z}_d$ (with $x_1 \neq x_2$) such that $[r] \times [x_1] = [r] \times [x_2]$. This means that

$$rx_1 \equiv rx_2 \ (\text{mod } d),$$

i.e. $rx_1 - rx_2 = kd$ for some $k \in \mathbb{Z}$, so $d|r(x_1 - x_2)$. Now, since r is coprime with d this implies $d|(x_1 - x_2)$ [*Theorem 5.3.11*]. But if we label the lesser of $\{x_1, x_2\}$ as x_2, we have $1 \le x_2 < x_1 \le d - 1$ (since $[x_1], [x_2] \in \mathbb{Z}_d$) and hence $0 < x_1 - x_2 < d$; so $d|(x_1 - x_2)$ is impossible. Hence our supposition that $[r]$ has no multiplicative inverse has been contradicted: $[r]$ must have a multiplicative inverse in $\mathbb{Z}_d$. $\qquad\square$

Note that the theorem does apply to the case $r = 1$, recalling that 1 is considered coprime with every $d \in \mathbb{N}\backslash\{1\}$ (see after Definition 5.3.9). In the case of $r = 0$, we already know that $[0]$ has no multiplicative inverse in $\mathbb{Z}_d$, and the concept of coprimeness does not apply to 0 at all.

According to Theorem 4.2.13, the existence of inverse elements under a binary operation implies the possibility of doing the inverse operation.

Definition 5.6.2. The inverse operation of multiplication is **exact division**, for which we use the symbol $\div$:

$$[r] \div [s] = [t] \quad \text{if and only if} \quad [t] \times [s] = [r] \quad \text{for } [r], [s], [t] \in \mathbb{Z}_d.$$

This accords with Definition 3.8.1 of inverse operation, and uses the familiar symbol $\div$ in an unfamiliar context; it will be used in a more familiar situation as the inverse of multiplication in the Rational Numbers in the next chapter. But for now, Theorem 5.6.1 shows when exact division is possible in modular arithmetic:

Corollary 5.6.3. *If and only if r is coprime with d:*
   (i) For any $[y] \in \mathbb{Z}_d$, there exists $[x] \in \mathbb{Z}_d$ such that $[x] \times [r] = [y]$, so that $[y] \div [r] = [x]$.
   (ii) We can cancel a common factor of $[r]$ in $\mathbb{Z}_d$: if $[x] \times [r] = [z] \times [r]$, then $[x] = [z]$.

To do the exact division, i.e. to find the $[x]$ in (i), first find $\overline{[r]}$, the multiplicative inverse of $[r]$ in $\mathbb{Z}_d$; this can be done from the multiplication table for $\mathbb{Z}_d$. Then Theorem 4.2.13 shows that $[y] \div [r] = [y] \times \overline{[r]}$. To verify that cancellation works as in (ii), simply multiply the equation $[x] \times [r] = [z] \times [r]$ on the right by $\overline{[r]}$ and use the associative property. But do not try exact division or cancellation when r and d are not coprime, because then the required multiplicative inverse of $[r]$ won't exist!

Example 5.6.4. In $\mathbb{Z}_3$, we can show that $[1] \div [2] = [2]$ either by inspection of the multiplication table for $\mathbb{Z}_3$ or by noting that the multiplicative inverse of $[2]$ is $[2]$, so that $[1] \div [2] = [1] \times \overline{[2]} = [1] \times [2] = [2]$.

In $\mathbb{Z}_4$, we cannot do the exact division $[1] \div [2]$ because 2 is not coprime with 4.

Exercise 5.6.5. *Evaluate the following in $\mathbb{Z}_9$, or explain why it is not possible to evaluate either of them:*

(i) $[3]_9 \div [7]_9$ (ii) $[7]_9 \div [3]_9$

Following from Theorem 5.6.1 we also have a result which will be important when we classify the algebraic structure of $\mathbb{Z}_d$:

Corollary 5.6.6. *Every number in $\mathbb{Z}_d \backslash \{[0]\}$ has a multiplicative inverse if and only if d is prime.*

Proof. If d is composite, it has some factor q (with $q \neq 1$ and $q \neq d$) with which it is not coprime, so by Theorem 5.6.1 $[q]$ has no multiplicative inverse in $\mathbb{Z}_d$. If d is prime, then r is coprime with d whenever $1 \leq r \leq d-1$, so by Theorem 5.6.1 $[r]$ has a multiplicative inverse for all $[r] \in \mathbb{Z}_d \backslash \{[0]\}$. $\square$

Exercise 5.6.7. *(i) Show that for every $d \in \mathbb{N} \backslash \{1\}$, the multiplicative inverse of $[d-1]$ in $\mathbb{Z}_d$ is $[d-1]$.*

(ii) Show that if $[s]$ is the multiplicative inverse of $[r]$ in $\mathbb{Z}_d$, then $[d-s]$ is the multiplicative inverse of $[d-r]$ in $\mathbb{Z}_d$.

(iii) Deduce that if d is prime and $d \equiv 1 \pmod 4$, then there exists $[r] \in \mathbb{Z}_d$ such that $[d-r]$ is the multiplicative inverse of $[r]$.

Exercise 5.6.8. *(i) Find the multiplicative inverse of every number in $\mathbb{Z}_7 \backslash \{[0]\}$. Hence evaluate $[3]_7 \div [2]_7$ in $\mathbb{Z}_7$.*

(ii) Find the multiplicative inverse of every number in $\mathbb{Z}_{17} \backslash \{[0]\}$ (using the result in Exercise 5.6.7(ii) to reduce the amount of effort required). Hence evaluate $[7]_{17} \div [14]_{17}$ in $\mathbb{Z}_{17}$.

We can now return to the finding in Example 5.5.5 that, whereas in $\mathbb{Z}$ it is only possible for $r \times s = 0$ if either (or both) $r = 0$ or $s = 0$, in $\mathbb{Z}_{12}$ it was possible to have $[r] \times [s] = [0]$ with $[r] \neq [0]$ and $[s] \neq [0]$.

Definition 5.6.9. In $\mathbb{Z}_d$, numbers $[r]$ and $[s]$ are **zero divisors** if $[r] \times [s] = [0]$ with $[r] \neq [0]$ and $[s] \neq [0]$. More generally, members a, b of a ring $\langle A, *, \vee \rangle$ are zero divisors if $a \vee b = i_*$, where i_* is the identity element for the operation $*$, and $a \neq i_*$ and $b \neq i_*$.

So in Example 5.5.5 (v) we found that $[4]$ and $[9]$ are zero divisors in $\mathbb{Z}_{12}$. But in which other number systems $\mathbb{Z}_d$ can we find zero divisors?

Theorem 5.6.10. *If d is prime, then $\mathbb{Z}_d$ has no zero divisors. If d is composite, then the zero divisors in $\mathbb{Z}_d$ are all the numbers $[r]$ such that r is not coprime with d.*

To prove this theorem, refer back to the proof of Theorem 5.6.1 which contains most of the ideas needed.

Exercise 5.6.11. *(i) Prove that if d is composite, then $[r]$ is a zero divisor in $\mathbb{Z}_d$ if r is not coprime with d.*

(ii) Write down all the zero divisors in $\mathbb{Z}_{12}$. For each zero divisor $[r] \in \mathbb{Z}_{12}$, write down all the other numbers $[s] \in \mathbb{Z}_{12} \backslash \{[0]\}$ such that $[r] \times [s] = [0]$.

(iii) Prove that if d is prime, then $\mathbb{Z}_d$ has no zero divisors.

A consequence of Theorem 5.6.10 is that multiplication is closed on $\mathbb{Z}_d \backslash \{[0]\}$ if d is prime: we already know that multiplication is closed on $\mathbb{Z}_d$, and the theorem ensures that multiplying two non-zero members of $\mathbb{Z}_d$ will not give a zero result if d is prime. So for prime d, multiplication is a binary operation on $\mathbb{Z}_d \backslash \{[0]\}$ and has the following properties (in the notation of Section 4.3), the last of which is guaranteed by Corollary 5.6.6:

M1. Multiplication is commutative in $\mathbb{Z}_d \backslash \{[0]\}$.

M2. Multiplication is associative in $\mathbb{Z}_d \backslash \{[0]\}$.

M3. There is an identity element for multiplication in $\mathbb{Z}_d \backslash \{[0]\}$.

M4. Every member of $\mathbb{Z}_d \backslash \{[0]\}$ has an inverse element under multiplication.

Thus according to Definition 4.3.1, $\langle \mathbb{Z}_d \backslash \{[0]\}, \times \rangle$ is an Abelian group. We have already noted that $\langle \mathbb{Z}_d, +, \times \rangle$ is a commutative ring (regardless of whether d is prime), and the inclusion of the new property M4 now defines a new algebraic structure:

Definition 5.6.12. A set A with two binary operations $*$ and $\vee$ constitutes a **field** if $\langle A, *, \vee \rangle$ is a commutative ring and every member of $A \backslash \{i_*\}$ has an inverse element under the operation $\vee$, where i_* is the identity element for the operation $*$.

So the number systems $\mathbb{Z}_d$ with prime d are the first examples of fields that we have encountered (because in our previous number systems, $\mathbb{N}$ and $\mathbb{Z}$, no numbers other than 1, and -1 in $\mathbb{Z}$, have multiplicative inverses). With zero excluded, they also yield our first Abelian groups in which the binary operation is multiplication. Indeed, in general if $\langle A, *, \vee \rangle$ is a field, then both $\langle A, * \rangle$ and $\langle A \backslash \{i_*\}, \vee \rangle$ are Abelian groups.

In $\mathbb{N}$ and $\mathbb{Z}$ we defined the order relations, $<$ and $>$, and it would seem sensible to define order on $\mathbb{Z}_d$ by saying that $[r] < [s]$ if $r < s$ in $\mathbb{Z}$. Such a definition does inherit the essential properties of trichotomy and transitivity from the order relations on $\mathbb{Z}$. However, the next requirement of an ordered ring (see Definition 4.4.9), that order relations are preserved by addition of a number to each side, cannot be satisfied in $\mathbb{Z}_d$. For example, in $\mathbb{Z}_4$ we have $[1] < [2]$; adding $[1]$ to both sides, we get $[1] + [1] < [2] + [1]$ because $[2] < [3]$; but adding $[2]$ to both sides, we have $[1] + [2] > [2] + [2]$ because $[3] > [0]$. Whereas numbers in $\mathbb{Z}$ can be considered as being ordered along an infinite line, numbers in $\mathbb{Z}_d$ are best thought of as lying on a circle, like a clockface.

Investigations

1. (i) Find a simple method for converting an integer written in base 2 (binary) to base 8 (octal), without converting to base ten at any stage. Use this method to convert the binary number 10011101101_2 to octal.

   (ii) By a similar method, convert the binary number 10110111000011_2 to base sixteen (hexadecimal). You should use the symbols A, B, C, D, E, F as digits for ten, eleven, twelve, thirteen, fourteen, fifteen, respectively.

   (iii) Convert the octal number 73144_8 to hexadecimal without converting to base ten at any stage.

2. Investigate the distribution of primes as follows.

(a) Write down all natural numbers from 2 to 200. Use the sieve of Eratosthenes to obtain a list of all the prime numbers less than or equal to 200.

(b) We define the k'th decade D_k by

$$D_k := \{n \in \mathbb{N}; 10(k-1) < n \leq 10k\},$$

e.g. $D_3 = \{21, 22, \ldots, 29, 30\}$.

(i) How many primes are there in each of the decades, D_3 to D_9? Do you see a pattern emerging? Explain this pattern. Now write down the number of primes in each of the decades, D_{10}, D_{11}, D_{12}, and D_{13}, and comment on the results.

(ii) How many primes are there in each of the decades, D_2, D_{11}, and D_{20}? Explain your observations. What is the obvious next member of this sequence of decades? Find how many primes there are in this next decade in the sequence, using the Trial Division method on any numbers in this decade which are not obviously composite.

(iii) Find a list of all primes less than or equal to 10000 (there are many such lists on the internet). From this list, find all the decades containing four primes.

(iv) If you were using Eratosthenes' sieve to find all the primes less than or equal to 10000, what is the largest prime whose multiples you would have to delete? How many multiples of that prime would be remaining for you to delete at that last stage?

(c) **Twin primes** are pairs of primes $\{p_1, p_2\}$ such that $p_2 = p_1 + 2$; so for example, $\{3, 5\}$ are twin primes. **Prime quadruplets** are sets of four primes $\{p_1, p_2, p_3, p_4\}$ such that $p_2 = p_1 + 2, p_3 = p_1 + 6, p_4 = p_1 + 8$; so for example, $\{5, 7, 11, 13\}$ are prime quadruplets.[xi] You have found some prime quadruplets in part (b) above.

(i) Prove that in every pair of twin primes $\{p_1, p_2\}$ *except* $\{3, 5\}$, the smaller number p_1 has remainder 5 when divided by 6.

(ii) Prove that in every set of prime quadruplets $\{p_1, p_2, p_3, p_4\}$ *except* $\{5, 7, 11, 13\}$, the smallest number p_1 has remainder 11 when divided by 30.

3. A prime gap is the difference between two successive prime numbers. For instance, the largest prime gap among the primes less than or equal to 200 (see Investigation 2(a) above) was 14, between the successive primes 113 and 127.

Prove that there exist arbitrarily large prime gaps, i.e. that for any given $g \in \mathbb{N}$ it is always possible to find a prime gap larger than g.

[Hint: consider all the numbers between $n!$ and $(n! + n)$, where $n!$ is the factorial of n, see Definition 3.7.9.]

4. (i) Show that if $2 \mid n^2$, then $2 \mid n$.

(ii) Now consider the more general hypothesis, "if $k \mid n^2$, then $k \mid n$". Is this true for **all** $k \in \mathbb{N}$? If so, prove it. If not, find a counterexample and give conditions on the prime factorisation of k that must be satisfied for the hypothesis to be true.

[xi] At the time of writing it remains unknown whether there are infinitely many twin primes and whether there are infinitely many prime quadruplets. Clearly the existence of infinitely many prime quadruplets would imply the existence of infinitely many twin primes, but the converse would not apply.

5. This investigation concerns **arithmetic progressions** of primes, which are sets of the form $\{p_0, p_0 + d, p_0 + 2d, \ldots, p_0 + (k-1)d\}$ in which there are k numbers, all prime, and differing by a **common difference** d. For example, $\{3, 5, 7\}$ is such a set with $p_0 = 3$, $d = 2$, and $k = 3$; from Investigation 2(c)(i) it follows immediately that this is the only such set of primes with $d = 2$ and $k > 2$.

(i) Show that all other arithmetic progressions of primes with $k \geq 3$ have either d divisible by 6 or $p_0 = 3$.

(ii) Write down an arithmetic progression of primes with $d = 6$ and $k = 5$. Then show that there are no other arithmetic progressions of primes with $d = 6$ and $k > 4$.

(iii) Formulate and explain necessary conditions on the value of d for the existence of arithmetic progressions of primes, for any given $k > 5$: what must d be divisible by? You should give conditions on d that allow the existence of (a) one arithmetic progression, and (b) more than one arithmetic progression, for any given k. Note that your conditions must be *necessary*, but may not be *sufficient*: your values of d will be required in order to obtain a set with a given k, but need not guarantee that such a set actually exists.

Note: It may be useful to refer to the **primorial** function: if p_n is the n'th prime (in the increasing sequence of primes as found by the sieve of Eratosthenes), then its primorial, denoted $p_n\sharp$, is the product of all the primes less than or equal to itself:

$$p_n\sharp := \prod_{k=1}^{n} p_k.$$

6. (a) Prove by induction that $(a^n - 1) = (a-1)(a^{n-1} + a^{n-2} + \cdots + a + 1)$ for all $n \in \mathbb{N}$ and $a \in \mathbb{Z}$.

(b) Hence show that $a^n - 1$ is composite if $a \geq 3$, or if $a = 2$ and n is composite.

(c) A **Mersenne number** is a number of the form $2^p - 1$, where p is prime. Find four Mersenne numbers which are themselves prime, and one Mersenne number which is composite. [This shows that the converse of what you have shown in (b) is not true: if $2^n - 1$ is composite, it does not follow that n is composite.]

(d) A **perfect number** is a number which is equal to the sum of all its divisors apart from itself. For instance, the divisors of 6 (excluding 6 itself) are $1, 2, 3$; and $6 = 1 + 2 + 3$. Show that if the Mersenne number $2^p - 1$ is prime, then the number $2^{p-1}(2^p - 1)$ is perfect.[xii]

(e) Hence find three perfect numbers other than 6.

7. (a) Find the prime factorisation of 18! (18 factorial), writing your answer as a product of powers of primes, $2^{\alpha_1} \times 3^{\alpha_2} \times \cdots$, where $\alpha_1, \alpha_2, \ldots$ are powers to be determined. Do not evaluate 18! or use trial division: recall that $n!$ is the product of all natural numbers less than or equal to n.

(b) Devise a rule or algorithm for determining the power α_i of any given prime p_i in the prime factorisation of $n!$. [It may be useful to introduce the notation $\lfloor n/d \rfloor$ for the quotient obtained when n is divided by d for general natural numbers n and d.]

[xii] It is not too difficult (although it is beyond the scope of this book) to prove that every even perfect number has the form $2^{p-1}(2^p - 1)$, where $2^p - 1$ is a Mersenne prime. At the time of writing no odd perfect numbers have been found, but it has not been proved that no odd perfect numbers exist.

(c) How does the value of α_i relate to the digits of n in base p_i?

(d) When $n!$ is written in a prime base p_i, how many zeroes are there after the last non-zero digit?

(e) How many zeroes are there after the last non-zero digit when $n!$ is written in base ten? Or in base twelve? And what rule or algorithm would you use to answer this for a general composite base?

8. In this investigation we use the symbol $D(n)$ to denote the number of divisors of a positive integer n, including 1 and n itself. For example, $D(10) = 4$, since 10 has 4 divisors $(1, 2, 5,$ and $10)$.

(a) (i) Find $D(n)$ in terms of α for the case where n is a power of a prime number, $n = p^\alpha$.

(ii) Find $D(n)$ in terms of α_1 and α_2 when the prime factorisation of n is of the form $n = p_1^{\alpha_1} p_2^{\alpha_2}$, where p_1 and p_2 are distinct primes.

(iii) Find $D(n)$ in terms of $\alpha_1, \alpha_2, \ldots, \alpha_m$ when the prime factorisation of n is of the form $n = p_1^{\alpha_1} p_2^{\alpha_2} \ldots p_m^{\alpha_m}$, where $p_1, p_2, \ldots, p_m$ are distinct primes.

(b) A prison consists of N cells in a row, numbered consecutively from 1 to N. There is a prisoner locked in each cell. The prison governor decides to release some of the prisoners, as follows. He passes along the row of N cells N times; on the first pass, he turns the locks on every cell door, so that all the doors are unlocked; on the second pass, he turns the locks on every second door, so that the doors of cells $2, 4, 6, \ldots$ are locked again; on the third pass he turns the locks on every third door, (locking those that were unlocked, and unlocking those that were locked); he carries on like this, so that on the k'th pass the locks of cells numbered $k, 2k, 3k, \ldots$ are turned. At the end of the process, the prisoners whose cell doors are unlocked are free to go.

What are the cell numbers of the prisoners who are set free? Write your answer in general terms (rather than giving a list of cell numbers), and explain the connection with $D(n)$ for cell number n.

(c) A **highly composite number** (first defined by Ramanujan in 1915) is a natural number with more divisors than any smaller natural number; so h is highly composite if $D(n) < D(h) \, \forall n < h$.

(i) Prove that there are infinitely many highly composite numbers.

(ii) Find $D(n)$ for each natural number n from 2 to 12, and hence write down a list of all the highly composite numbers less than or equal to 12.

(iii) To find larger highly composite numbers, it would clearly not be feasible to continue computing $D(n)$ for every natural number. Devise a search strategy, based on the formula for $D(n)$ found in part (a)(iii) above, to look for highly composite numbers h with successively higher values of $D(h)$. Use your strategy to find all highly composite numbers less than or equal to 2000.

9. Mersenne numbers were defined in Question 6(c) above: they are numbers of the form $2^p - 1$, where p is prime. Show that no Mersenne number (apart from 3 and 7 themselves) is divisible by 3, 5, or 7.

[Hints: $3 = 2 + 1$; $5 = 2^2 + 1$; $7 = 2^3 - 1$.]

10. For a natural number written in base ten, it is well known that the number is divisible by 9 or by 3 if the sum of its digits is divisible by 9 or by 3 respectively. For example, to show that 783 is divisible by 9 (and by 3), note that $7 + 8 + 3 = 18$ which is divisible

by 9 and by 3. There is also a well known test for divisibility by 11: alternately add and subtract the digits, and if the result is divisible by 11, then so is the original number. For example, to show that 27181 is divisible by 11, note that $2 - 7 + 1 - 8 + 1 = -11$ which is divisible by 11. Finally, the simplest divisibility tests in base ten are that a number is divisible by 2 or by 5 if its final digit is divisible by 2 or by 5, respectively. In the above, recall that 0 is divisible by all natural numbers.

This investigation uncovers the theory behind these tests, and extends it to numbers written in any base. The following notation is used throughout: a number n is written in base b with digits $r_K r_{K-1} \ldots r_1 r_0$ (so the values of the $r_i (i = 0, 1, \ldots, K)$ would change if the base changes, although this is not explicitly shown in the notation for the digits). For example, for the number written 783 in base ten, $r_2 = 7, r_1 = 8, r_0 = 3$.

(a) Recalling that the number written $r_K r_{K-1} \ldots r_1 r_0$ is $n = \sum_{i=0}^{K} r_i b^i$, show that

$$n \equiv \sum_{i=0}^{K} r_i \ (\mathrm{mod} \ (b - 1)).$$

Hence explain the test for divisibility by 9 for numbers written in base ten. Write down a similar divisibility test for numbers written in a general base b.

(b) Show that if q is a divisor of $b - 1$, then

$$n \equiv \sum_{i=0}^{K} r_i \ (\mathrm{mod} \ q).$$

Hence explain the test for divisibility by 3 for numbers written in base ten. Write down a similar divisibility test for numbers written in a general base b.

(c) Show that

$$n \equiv \sum_{i=0}^{K} (-1)^i r_i \ (\mathrm{mod} \ (b + 1)).$$

Hence explain the test for divisibility by 11 for numbers written in base ten. Write down a similar divisibility test for numbers written in a general base b. Also write down a test for divisibility by any divisor of $b + 1$.

(d) Show that if q is a factor of the base b, then n is divisible by q if its final digit r_0 is divisible by q.

(e) Consider now the following numbers written in base sixteen, with the symbols A, B, C, D, E, F as digits for ten, eleven, twelve, thirteen, fourteen, fifteen, respectively.

(i) 57F8   (ii) ACD2D5   (iii) 91FFEC   (iv) 34EAE

Use the above tests to determine whether each of these four numbers is divisible by each of the following numbers (also written in base sixteen): 5, 8, C, 11, 33.

For your convenience, a base-sixteen addition table is provided below.

+	1	2	3	4	5	6	7	8	9	A	B	C	D	E	F
1	2	3	4	5	6	7	8	9	A	B	C	D	E	F	10
2	3	4	5	6	7	8	9	A	B	C	D	E	F	10	11
3	4	5	6	7	8	9	A	B	C	D	E	F	10	11	12
4	5	6	7	8	9	A	B	C	D	E	F	10	11	12	13
5	6	7	8	9	A	B	C	D	E	F	10	11	12	13	14
6	7	8	9	A	B	C	D	E	F	10	11	12	13	14	15
7	8	9	A	B	C	D	E	F	10	11	12	13	14	15	16
8	9	A	B	C	D	E	F	10	11	12	13	14	15	16	17
9	A	B	C	D	E	F	10	11	12	13	14	15	16	17	18
A	B	C	D	E	F	10	11	12	13	14	15	16	17	18	19
B	C	D	E	F	10	11	12	13	14	15	16	17	18	19	1A
C	D	E	F	10	11	12	13	14	15	16	17	18	19	1A	1B
D	E	F	10	11	12	13	14	15	16	17	18	19	1A	1B	1C
E	F	10	11	12	13	14	15	16	17	18	19	1A	1B	1C	1D
F	10	11	12	13	14	15	16	17	18	19	1A	1B	1C	1D	1E

11. An **alphametic** is an arithmetic equation in which each digit is represented by a letter. For instance, in base ten, the alphametic

$$ABC + CDE = BFG$$

would be correct if $A = 2$, $B = 7$, $C = 5$, $D = 1$, $E = 8$, $F = 9$, and $G = 3$, since $275 + 518 = 793$. [Alphametics are often posed as puzzles: find digits corresponding to each of the letters in the alphametic, such that the equation is correct.] Note that this is nothing to do with the use of letters for digits in bases greater than ten.

Show that if

$$ABCD = EFG + HIJ$$

is a correct alphametic in base ten, then ABCD is divisible by 9.

[Hint: the alphametic contains ten distinct letters, each of which must represent a distinct digit in base ten.]

12. There is a theorem that states that every positive integer can be expressed as the sum of squares of four integers. For example, $60 = 7^2 + 3^2 + 1^2 + 1^2$, and $59 = 5^2 + 5^2 + 3^2 + 0^2$. Obviously, 59 can also be expressed as the sum of three squares: $59 = 5^2 + 5^2 + 3^2$. This question concerns those integers that cannot be expressed as the sum of three squares.

(i) By working in $\mathbb{Z}_8$, prove that a positive integer m cannot be expressed as the sum of three squares of integers if $m \equiv 7 \pmod{8}$.

(ii) Prove that a positive integer m cannot be expressed as the sum of three squares of integers if $m = 4^k m_0$ where $m_0 \equiv 7 \pmod{8}$ and $k \in \mathbb{N}$.

[Hint: First prove that if $4m$ can be expressed as the sum of three squares, then so can m. What is the contrapositive of that proposition? Use induction to prove the proposition for all $k \in \mathbb{N}$.]

(iii) Now consider positive integers m of the form $m = s^{2k} m_0$, where $m_0 \equiv 7 \pmod{8}$ and s is any natural number greater than 2. Can they be expressed as the sum of three squares of integers, for any or all $s > 2$? If not, do any values of s yield further instances of positive integers that cannot be expressed as the sum of three squares of integers, that have not already been found in (i) or (ii) above?

[Note: Legendre's Three-Squares Theorem states that a positive integer m cannot be expressed as the sum of three squares of integers *if and only if* $m \equiv 7 \pmod 8$ or $m = 4^k m_0$ with $m_0 \equiv 7 \pmod 8$ and $k \in \mathbb{N}$. You have shown the implication *if*; to show the converse, *only if*, requires knowledge of Number Theory beyond the scope of this book, as does the Four-Squares Theorem mentioned at the start of the question.]

13. For composite d, denote by $\mathbb{Z}_d^\phi$ the set of all numbers $[r] \in \mathbb{Z}_d \backslash \{[0]\}$ such that r is coprime with d. For example,

$$\mathbb{Z}_{24}^\phi = \{[1], [5], [7], [11], [13], [17], [19], [23]\}.$$

We know that multiplication is commutative and associative and has the identity element $[1]$ on $\mathbb{Z}_d^\phi$ (because these properties apply in the set $\mathbb{Z}_d$, of which $\mathbb{Z}_d^\phi$ is a subset). Furthermore, Theorem 5.6.1 shows that every member of $\mathbb{Z}_d^\phi$ has a multiplicative inverse element. However, to verify that $\langle \mathbb{Z}_d^\phi, \times \rangle$ is an Abelian group, it is necessary to show that multiplication is closed on $\mathbb{Z}_d^\phi$.

(i) Show that if $[r], [s] \in \mathbb{Z}_d^\phi$, then $[r] \times [s] \in \mathbb{Z}_d^\phi$ (i.e. that multiplication is closed on $\mathbb{Z}_d^\phi$).

(ii) Given that $\langle \mathbb{Z}_d^\phi, \times \rangle$ is an Abelian group, why is $\langle \mathbb{Z}_d^\phi, +, \times \rangle$ *not* a field when d is composite?

(iii) Verify that each member of $\mathbb{Z}_{24}^\phi$ is its own multiplicative inverse, i.e. that $[r] \times [r] = [1]$ for each $[r] \in \mathbb{Z}_{24}^\phi$.

(iv) Find all $d < 24$ for which this property is true, i.e. that each member of $\mathbb{Z}_d^\phi$ is its own multiplicative inverse. Include prime values of d (for which $\mathbb{Z}_d^\phi = \mathbb{Z}_d \backslash \{[0]\}$) in your survey.

(v) Can you prove that there does not exist $d > 24$ for which each member of $\mathbb{Z}_d^\phi$ is its own multiplicative inverse? [Hint: start by explaining why, if the property is true for some $d \geq 25$, then the prime factorisation of d must include factors of 2, 3, and 5. Then generalise this idea.]

14. Show that the existence of prime factorisation fails in $\mathbb{Z}_d$ (with composite d), by considering the example of $\mathbb{Z}_{12}$ as follows.

(i) Write down the units (numbers with multiplicative inverses) in $\mathbb{Z}_{12}$.

(ii) Write down four numbers in $\mathbb{Z}_{12}$ that are obviously composite because they are composite in $\mathbb{Z}$ with none of the prime factors being units in $\mathbb{Z}_{12}$.

(iii) The numbers which you have not already written down should be the additive identity $[0]$, together with $[2]$, $[3]$, and $[10]$. Show that $[2]$ and $[10]$ are prime in $\mathbb{Z}_{12}$ (i.e. all their factorisations in $\mathbb{Z}_{12}$ involve a unit).

(iv) Find a factorisation of $[3]$ in $\mathbb{Z}_{12}$ that does not involve a unit. Hence $[3]$ is composite. Show that $[3]$ cannot be written as the product of any of the primes in $\mathbb{Z}_{12}$.

Chapter 6

Rational Numbers, $\mathbb{Q}$

6.1 Definition of the Rationals

The finite number systems $\mathbb{Z}_d$ encountered in the last chapter have considerable importance in some areas of mathematics, but may be regarded as a digression from the main purpose of this book, which is to develop the rigorous theory of the five major number systems introduced in Section 1.3. Nevertheless, our discussion of $\mathbb{Z}_d$ does anticipate some of the fundamental ideas behind our next number system, the Rational Numbers. Towards the end of the last chapter, we introduced exact division, the inverse operation of multiplication, with the symbol $\div$. In $\mathbb{Z}_d$, exact division is closed (with the exception of division by zero) if and only if d is prime. In the Integers, exact division is certainly not closed: from the definition of inverse operation, we can obtain $a \div b$ in $\mathbb{Z}$ if and only if there exists $c \in \mathbb{Z}$ such that $c \times b = a$, i.e. if the integer division of a by b leaves zero remainder. So, just as the subtraction $a - b$ in the Natural Numbers was conditional on the relation $a > b$, the exact division $a \div b$ in the Integers is conditional on the relation $b|a$. And just as we removed the conditionality of subtraction by defining a new number system, $\mathbb{Z}$, we shall now use a somewhat similar procedure to define a new number system, $\mathbb{Q}$, to remove the conditionality of exact division; although the exception of division by zero will remain. That exception was necessitated by the ring structure in $\mathbb{Z}_d$; since $\langle \mathbb{Z}, +, \times \rangle$ is also a ring, and in our progression through the five major number systems we want each new system to "include" the previous one (in the sense of having a subset isomorphic to the previous system under the arithmetic operations, so that the algebraic properties of those operations are retained in the new system) we will retain the ring structure in $\mathbb{Q}$. Note that the symbol $\mathbb{Q}$ comes from "quotient", and the name "Rational" for our new number system is not used here in the sense of "reasonable", but derives from "ratio".

Proceeding analogously to our approach to defining Integers in Section 4.1, we start by noting that any $z \in \mathbb{Z}$ can be represented as the result of an exact division, $z = p \div q$, by choosing an arbitrary $q \in \mathbb{Z}_+$ and letting $p = z \times q$ (the restriction of q to *positive* integers simplifies some later definitions). So with two such representations, $z = p_1 \div q_1 = p_2 \div q_2$, we have

$$p_1 = zq_1 \quad \text{and} \quad p_2 = zq_2,$$

so that

$$p_1 q_2 = zq_1 q_2 = p_2 q_1.$$

This motivates the definition of a new equivalence relation (but again using the old symbol $\equiv$) on $\mathbb{Z} \times \mathbb{Z}_+$, the set of ordered pairs of integers in which the second integer must be positive:

Definition 6.1.1.

$$(p_1, q_1) \equiv (p_2, q_2) \quad \text{if and only if} \quad p_1 \times q_2 = p_2 \times q_1.$$

Exercise 6.1.2. *Verify that the relation in Definition 6.1.1 satisfies the requirements of being reflexive, symmetric, and transitive.*

You should find this verification almost as straightforward as for the very similar Definition 4.1.1.

Example 6.1.3. According to Definition 6.1.1,

$$(-8, 4) \equiv (-6, 3)$$

because $(-8) \times 3 = (-6) \times 4$.

We now define Rational Numbers in a similar way to Definition 4.1.4 for Integers:

Definition 6.1.4. A Rational Number is an equivalence class of ordered pairs $(p, q) \in \mathbb{Z} \times \mathbb{Z}_+$ under the equivalence relation $\equiv$ in Definition 6.1.1.

We may characterise the members of such an equivalence class by noting first that $(kp, kq) \equiv (p, q)$ for any $k \in \mathbb{Z}_+$, because $kp \times q = p \times kq$. Thus if p and q have a common factor $d \in \mathbb{Z}_+$, so that $p = ds$ and $q = dt$ for some $s \in \mathbb{Z}$ and $t \in \mathbb{Z}_+$, we have

$$(p, q) = (ds, dt) \equiv (s, t).$$

So for any $(p, q) \in \mathbb{Z} \times \mathbb{Z}_+$, we may cancel out *all* (positive) common factors from the prime factorisations of p and q to obtain a coprime pair (s, t) in the same equivalence class as (p, q); this coprime pair will be unique because of the Fundamental Theorem of Arithmetic and our insistence that the second member of our ordered pairs should be positive. The exception to this argument is when $p = 0$, since zero does not have a prime factorisation; in this case we note that $(0, q) \equiv (0, 1)$ for all $q \in \mathbb{Z}_+$. In all cases, we then have that an equivalence class that constitutes a rational number according to Definitions 6.1.1 and 6.1.4 is a set of ordered pairs,

$$\{(ds, dt) \in \mathbb{Z} \times \mathbb{Z}_+ : d \in \mathbb{Z}_+\},$$

where either $s \in \mathbb{Z}$ and $t \in \mathbb{Z}_+$ are coprime or $(s, t) = (0, 1)$. Each rational number has countably infinitely many representations, (ds, dt), with one such representation for each positive integer d; the representation with $d = 1$, i.e. the pair (s, t) where s and t are coprime or where $s = 0$ and $t = 1$, is called the **lowest terms** representation of the rational number. Whereas after defining Integers as equivalence classes we introduced the temporary notations ζ_r and η_q for those equivalence classes, we shall simply represent the whole equivalence class that constitutes a rational number by its lowest-terms form.

Example 6.1.5. The unique coprime pair in the same equivalence class as $(-8, 4)$ and $(-6, 3)$ (see Example 6.1.3) is $(-2, 1)$. [Note that if we had not insisted on the second member being positive, we could also have found another equivalent coprime pair, $(2, -1)$.] The pair $(-2, 1)$ is the lowest-terms form of a rational number which can also be represented by the pairs $(-8, 4)$ and $(-6, 3)$, and any pair of the form $(-2d, d)$ for $d \in \mathbb{Z}_+$.

Since equivalent ordered pairs are representations of the same, i.e. *equal*, rational numbers, we shall use the symbol $=$ instead of $\equiv$ unless we want to emphasise the use of the equivalence relation 6.1.1. See the similar comment pertaining to integers, below Definition 4.1.4.

To define **positive** and **negative** rational numbers, we first note that because we require $q \in \mathbb{Z}_+$ in ordered pairs (p, q), every pair that is equivalent to (p, q) has its first member positive if p is positive, or its first member negative if p is negative. Thus we can say:

Definition 6.1.6. If $p \in \mathbb{Z}_+$, the pair (p, q) represents a **positive rational**.
If $p \in \mathbb{Z}_-$, the pair (p, q) represents a **negative rational**.

Note that we have not used any concept of order in these definitions. We use the obvious notations $\mathbb{Q}_+$ and $\mathbb{Q}_-$ for the sets of positive and negative rationals, respectively, and observe that there is a single rational which is neither positive nor negative: indeed, $\mathbb{Q} = \mathbb{Q}_+ \cup \mathbb{Q}_- \cup \{(0, 1)\}$.

6.2 Addition and Multiplication on $\mathbb{Q}$

We cannot simply carry over the definitions of addition and multiplication from any previous number system: as when we defined arithmetic operations on $\mathbb{Z}$, new definitions of operations are required for a newly defined number system, and new symbols will be used until we have established that these operations are effectively doing the same as the similarly named operations on the previous number system. So the rational-number versions of addition and multiplication will for now be denoted by $\oplus$ and $\otimes$ respectively:

Definition 6.2.1.
$$(p, q) \oplus (m, n) := (pn + mq, qn).$$

Definition 6.2.2.
$$(p, q) \otimes (m, n) := (pm, qn).$$

Before proceeding further, we should check that the operations $\oplus$ and $\otimes$ are both closed and well-defined. Recall that *closed* means that the output of the operation is within the same set as the inputs, that set being $\mathbb{Z} \times \mathbb{Z}_+$ in Definitions 6.2.1 and 6.2.2. This is clear from the definitions: in particular, qn as the second member of the output ordered pair in both definitions is positive because q and n as second members of input pairs are positive.

Exercise 6.2.3. *Verify that the operations $\oplus$ and $\otimes$ are* well-defined: *you need to show that if you perform the operations on equivalent pairs, the results are equivalent, so that using different representations of the same rational number does not affect the result of an operation.*

The well-defined property justifies our use of a single ordered pair (the lowest-terms form) to represent a rational number in the calculations below.

It is instructive to compare the present definitions with Definitions 4.2.1 and 4.2.2 for operations on integers. The construction of integers was motivated by the need to perform the inverse operation of addition, and started with an equivalence relation involving addition of natural numbers; the definition of addition of integers was then straightforward, while the definition of multiplication was more complicated and was motivated by considering integers to be the result of subtractions of natural numbers and then using familiar arithmetical ideas which had not yet been formally proved. Similarly, the construction of rationals is motivated by the need to perform the inverse operation of multiplication, and was started with an equivalence relation involving multiplication of integers; and now the definition of multiplication of rationals is straightforward (in fact exactly analogous to Definition 4.2.1 for addition of integers), whereas the definition of addition of rationals is more complicated, but can be motivated by recalling (from above Definition 6.1.1) that the ordered pair (p, q) initially arose from considering an integer z as the result of an exact division, $p \div q$; if we now think of the ordered pair on the right of Definition 6.2.1 as representing a division,

$(pn + mq) \div (qn)$, our familiar (but not yet formally derived) ideas about division allow us to reduce this to $(p \div q) + (m \div n)$, corresponding to the expression in ordered-pair form on the left of Definition 6.2.1.

Now consider the subset of $\mathbb{Q}$ consisting of those rationals in which the second member of the pair is 1 when the rational is expressed in lowest terms,

$$\mathbb{Q}_Z := \{(p, q) \in \mathbb{Q} : q = 1\}.$$

This has an obvious bijection to the integers, $f : \mathbb{Q}_Z \to \mathbb{Z}$, given by

$$f : (p, 1) \mapsto p.$$

Combining two members of $\mathbb{Q}_Z$ using the operations $\oplus$ and $\otimes$, Definitions 6.2.1 and 6.2.2 immediately yield

$$(p, 1) \oplus (m, 1) = (p + m, 1)$$

and

$$(p, 1) \otimes (m, 1) = (pm, 1).$$

But our bijection from $\mathbb{Q}_Z$ to $\mathbb{Z}$ gives

$$f : (p + m, 1) \mapsto p + m \quad \text{and} \quad f : (pm, 1) \mapsto pm,$$

so the operations $\oplus$ and $\otimes$ on $\mathbb{Q}_z$ yield results corresponding to the operations $+$ and $\times$ on $\mathbb{Z}$: we have isomorphisms between $\mathbb{Q}_Z$ with $\oplus$ and $\mathbb{Z}$ with $+$, and between $\mathbb{Q}_Z$ with $\otimes$ and $\mathbb{Z}$ with $\times$. Thus we shall be entitled to refer to $\oplus$ and $\otimes$ as "addition" and "multiplication", and to use the usual symbols $+$ and $\times$ on $\mathbb{Q}$ as on $\mathbb{Z}$. We may also identify the members of $\mathbb{Q}_Z$ as integers, and simply write p rather than $(p, 1)$.

Next note that

$$(p, q) \times (q, 1) = (pq, q) \equiv (p, 1)$$

(using Definition 6.2.2 but with the notation $\times$) which, from the definition of exact division as the inverse of multiplication, means that

$$(p, q) = (p, 1) \div (q, 1).$$

This allows us to interpret an arbitrary rational number (p, q) as being formed by the exact division of an integer p by a positive integer q. Since division is often represented by fraction notation, we can therefore dispense with ordered-pair notation and write $\frac{p}{q}$ rather than (p, q). Definitions 6.2.1 and 6.2.2 of addition and multiplication in $\mathbb{Q}$ are then seen to be the familiar rules for adding and multiplying fractions,

$$\frac{p}{q} + \frac{m}{n} = \frac{pn + mq}{qn},$$

$$\frac{p}{q} \times \frac{m}{n} = \frac{pm}{qn}.$$

However, we will still write ordered pairs in proofs and calculations below when we need to refer to our definitions of addition and multiplication on $\mathbb{Q}$.

We want each of our number systems to retain the same arithmetic properties as previous systems. Our next theorem states that addition and multiplication on $\mathbb{Q}$ do indeed have all the properties found on $\mathbb{Z}$, and identifies the identity and inverse elements (in ordered-pair notation in lowest terms) where appropriate.

Theorem 6.2.4. *(i) Addition is commutative in $\mathbb{Q}$.*

(ii) Addition is associative in $\mathbb{Q}$.

(iii) The rational number $(0,1)$ *is the unique identity element for addition in* $\mathbb{Q}$.

(iv) Every rational number has a unique inverse element under addition: the additive inverse of (p,q) *is* $(-p,q)$.

(v) Multiplication is commutative in $\mathbb{Q}$.

(vi) Multiplication is associative in $\mathbb{Q}$.

(vii) The rational number $(1,1)$ *is the unique identity element for multiplication in* $\mathbb{Q}$.

(viii) Multiplication distributes over addition in $\mathbb{Q}$.

To prove any part of this theorem, we need to use the definitions of addition and multiplication in $\mathbb{Q}$, noting that any theorem for $\mathbb{Z}$ is valid for *each member of an ordered pair representing a rational number*; the parts of the theorem cannot be assumed true in $\mathbb{Q}$ just because they are true in $\mathbb{Z}$, except when dealing with rationals in $\mathbb{Q}_Z$ for which the isomorphisms with $\mathbb{Z}$ have been established. We prove two parts of the theorem below, and leave the rest as exercises for the reader.

Proof of Theorem 6.2.4 (ii).

$$
\begin{aligned}
((p,q) + (m,n)) + (k,l) &= (pn + mq, qn) + (k,l) && \text{[Def. 6.2.1]} \\
&= ((pn + mq)l + k(qn), (qn)l) && \text{[Def. 6.2.1]} \\
&= (pnl + mql + kqn, qnl). && (*)
\end{aligned}
$$

Also,

$$
\begin{aligned}
(p,q) + ((m,n) + (k,l)) &= (p,q) + (ml + kn, nl) && \text{[Def. 6.2.1]} \\
&= (p(nl) + (ml + kn)q, q(nl)) && \text{[Def. 6.2.1]} \\
&= (pnl + mlq + knq, qnl).
\end{aligned}
$$

The last expression is equal to the expression marked $(*)$, so completing the proof. □

Note the strategy adopted in this proof: as described in Section 2.3.3, we have proved equality between two expressions by two calculations, showing separately that each of these expressions is equal to a third expression.

Proof of Theorem 6.2.4 (iv)).

$$
\begin{aligned}
(p,q) + (-p,q) &= (pq + (-p)q, q^2) && \text{[Def. 6.2.1]} \\
&= (0, q^2) \\
&\equiv (0,1),
\end{aligned}
$$

and similarly $(-p,q) + (p,q) = (0,1)$. Since $(0,1)$ is the identity element for addition [*Theorem 6.2.4 (iii)*], this verifies that (p,q) and $(-p,q)$ are additive inverses of each other. The uniqueness of the additive inverse is guaranteed by Theorem 4.2.10, since we have proved above that addition is associative in $\mathbb{Q}$. □

Exercise 6.2.5. *Prove the remaining parts of Theorem 6.2.4.*

Recalling that we have found that our ordered pairs are actually a formal way of writing what are more commonly displayed as fractions (or integers when the second member of the pair is 1), we could write the calculation results in the proof of Theorem 6.2.4 (iv) as

$$\frac{p}{q} + \frac{-p}{q} = \frac{-p}{q} + \frac{p}{q} = 0.$$

Here, the minus signs in the numerators denote the additive inverse of the integer p. If we also allow the minus sign to be used in front of a fraction to denote the additive inverse of a rational number, we have

$$-\frac{p}{q} = \frac{-p}{q}.$$

This is not as trivial as it might at first appear: it states that the additive inverse of the rational number $\frac{p}{q}$ is obtained by replacing the integer in the numerator by its additive inverse.

The existence of additive inverses of rational numbers immediately means that subtraction can always be done in $\mathbb{Q}$, according to Theorem 4.2.13: as in $\mathbb{Z}$, we subtract by adding an additive inverse, so

$$\frac{p}{q} - \frac{m}{n} = \frac{p}{q} + \frac{-m}{n}$$
$$= \frac{pn - mq}{qn},$$

in which the last line uses Definition 6.2.1 in fraction notation.

Theorem 6.2.4 can be summed up as saying that $\langle \mathbb{Q}, +, \times \rangle$ is a commutative ring (and this includes that $\langle \mathbb{Q}, + \rangle$ is an Abelian group); hence all the further theorems in Section 4.3 apply in $\mathbb{Q}$ as in $\mathbb{Z}$, including that $x \times 0 = 0$ and $(-1) \times x = -x$, as well as various properties of additive inverses and subtraction. So the rationals have all the same arithmetic properties as the integers; but the object of defining the rationals was to be able to do exact division, the inverse operation of multiplication, which is not in general possible in the integers. The next two theorems establish that we can do this: as in the discussion of subtraction in the integers (Section 4.2), the route to doing an inverse *operation* is via finding inverse *elements*.

Theorem 6.2.6. *Every rational number except* $(0, 1)$ *has a unique inverse under multiplication.*

The requirement that the second member of an ordered pair representing a rational number be positive gives rise to a little awkwardness in the proof, forcing us to consider positive and negative rationals separately; and of course the exceptional case, the zero rational represented as $(0, 1)$, requires separate consideration.

Proof of Theorem 6.2.6. Case (i). If $(p, q) \in \mathbb{Q}_+$, i.e. if $p \in \mathbb{Z}_+$, then $(q, p) \in \mathbb{Q}_+$ and

$$(p, q) \times (q, p) = (pq, qp) \qquad\qquad [Def.\ 6.2.1]$$
$$\equiv (1, 1), \qquad\qquad [Def.\ 6.1.1]$$

and $(q, p) \times (p, q) = (1, 1)$ similarly. Since $(1, 1)$ is the identity element for multiplication in $\mathbb{Q}$ [*Theorem 6.2.4(vii)*], we have established that (q, p) is a multiplicative inverse of (p, q) if $p \in \mathbb{Z}_+$.

Case (ii). If $(p, q) \in \mathbb{Q}_-$, i.e. if $p \in \mathbb{Z}_-$ so that $(-p) \in \mathbb{Z}_+$, then $(-q, -p) \in \mathbb{Q}_-$ and

$$(p, q) \times (-q, -p) = (p(-q), q(-p)) \qquad\qquad [Def.\ 6.2.1]$$
$$= (-pq, -qp)$$
$$\equiv (1, 1), \qquad\qquad [Def.\ 6.1.1]$$

and $(-q, -p) \times (p, q) = (1, 1)$ similarly. We have now shown that $(-q, -p)$ is a multiplicative inverse of (p, q) if $p \in \mathbb{Z}_-$.

Case (iii). If $p = 0$, we have

$$(0, q) \times (n, m) = (0, qm) \not\equiv (1, 1) \quad \text{for any} \quad n, m \in \mathbb{Z},$$

so $(0, q)$ has no multiplicative inverse in $\mathbb{Q}$.

Finally, since multiplication is associative in $\mathbb{Q}$, the inverses found in cases (i) and (ii) are unique according to Theorem 4.2.10 (recalling always that we are representing a rational number by writing a single ordered pair, e.g. (q, p), from its equivalence class). $\quad\square$

Written in fraction notation, the calculation in case (i) is seen to be in accord with familiar ideas on manipulation of fractions:

$$\frac{p}{q} \times \frac{q}{p} = \frac{q}{p} \times \frac{p}{q} = 1.$$

Having obtained inverse elements under multiplication, we can now do the inverse operation:

Theorem 6.2.7. *Exact division is closed on* $\mathbb{Q} \backslash \{(0, 1)\}$.

Proof. According to Theorem 4.2.13, this follows immediately from the existence of multiplicative inverse elements as found in Theorem 6.2.6. In detail, to do the exact division, $(p, q) \div (m, n)$ where $m \neq 0$:

If $m \in \mathbb{Z}_+$, $(p, q) \div (m, n) = (p, q) \times (n, m) = (pn, qm)$;

If $m \in \mathbb{Z}_-$, $(p, q) \div (m, n) = (p, q) \times (-n, -m) = (-pn, -qm)$. $\quad\square$

Writing the first of these in fraction notation, we have

$$\frac{p}{q} \div \frac{m}{n} = \frac{p}{q} \times \frac{n}{m} = \frac{pn}{qm},$$

which expresses the familiar idea that when we divide by a fraction, we "flip the fraction over" and multiply.

Note that fraction notation is also used as a shorthand for exact division, i.e. we can write $\frac{x}{y}$ for $x \div y$ where x and y are *rationals*, whereas we have hitherto only used fraction notation with the numerator and denominator being *integers*; in particular, the multiplicative inverse of $x \in \mathbb{Q} \backslash \{0\}$ may be written $\frac{1}{x}$. The notation is of course consistent, since the result of exact division of integers, $p \div q$ where $p \in \mathbb{Z}$ and $q \in \mathbb{Z}_+$, is the rational number $\frac{p}{q}$.

In terms of algebraic structure, we have already noted that $\langle \mathbb{Q}, +, \times \rangle$ is a *commutative ring*, according to Theorem 6.2.4. But $\mathbb{Q}$ has further structure: Theorem 6.2.6, showing that every element of $\mathbb{Q}$ except for the additive identity element has a multiplicative inverse, identifies $\langle \mathbb{Q}, +, \times \rangle$ as a *field* according to Definition 5.6.12, which also means that both $\langle \mathbb{Q}, + \rangle$ and $\langle \mathbb{Q} \backslash \{0\}, \times \rangle$ are Abelian groups. The fields we encountered in Section 5.6 were finite; $\mathbb{Q}$ is the first infinite field that we have met. The isomorphisms of a subset of $\mathbb{Q}$ with the infinite set $\mathbb{Z}$ make it clear that $\mathbb{Q}$ is infinite; but just how infinite will be the subject of the next section.

6.3 Countability of $\mathbb{Q}$

At the end of Chapter 4, we showed that the cardinality of $\mathbb{Z}$ was the same as that of $\mathbb{N}$, even though a naive estimate would suggest that there are twice as many integers as natural numbers. We might naively expect that the cardinality of $\mathbb{Q}$ is approximately the square of that of $\mathbb{N}$: there are countably infinitely many positive integers, and for each of those there are countably infinitely many coprime integers to combine with, to form fractions in lowest terms. But what we have seen regarding $\mathbb{Z}$ shows that cardinality does not work like that for infinite sets. In this section we will show that the cardinality of $\mathbb{Q}$ is the same as that of $\mathbb{N}$, i.e. the rational numbers are countable. We will do this by first proving two much more general and powerful theorems on countability: the countability of $\mathbb{Q}$ will follow almost immediately from these theorems, and the countability of many more infinite sets will also be easily provable from the general theorems. Recall first that according to Theorem 3.7.25 and Definition 3.7.26, to prove countability requires finding a bijection to some subset of $\mathbb{N}$ (which could be $\mathbb{N}$ itself, or either a finite or infinite subset of $\mathbb{N}$); or equivalently, finding an injection to $\mathbb{N}$.

The following two theorems consider a countable set of sets, $\{A_1, A_2, \ldots\}$, i.e. there are countably many sets $A_i (i = 1, 2, \ldots)$, and each set contains countably many members. So we can denote the j'th member of set A_i as $a_{ij} (i, j \in \mathbb{N})$.

Theorem 6.3.1. *A countable union of countably many sets is itself countable; i.e. the set* $A_1 \cup A_2 \cup \cdots$ *contains countably many members.*

Proof. The Fundamental Theorem of Arithmetic provides a simple way of obtaining the required bijection to a subset of $\mathbb{N}$. Let p_i be the i'th prime number, where primes are listed in increasing order, as found by the Sieve of Eratosthenes (Algorithm 5.3.5). Then the function

$$f : A_1 \cup A_2 \cup \cdots \to \mathbb{N} \quad \text{defined by} \quad f : a_{ij} \mapsto p_i^j$$

is an injection, because the Fundamental Theorem ensures that any number expressible as the j'th power of some prime p_i cannot be expressed as any other prime factorisation. □

Theorem 6.3.2. *The Cartesian product of finitely many countable sets is itself countable; i.e. the set* $A_1 \times A_2 \times \cdots \times A_k$ *is countable, for any* $k \in \mathbb{N}$.

Proof. We again use the Fundamental Theorem of Arithmetic. An arbitrary member of $A_1 \times A_2 \times \cdots \times A_k$ is the ordered k-tuple,

$$(a_{1j_1}, a_{2j_2}, \ldots, a_{kj_k}),$$

in which we select the j_1'th member of A_1, the j_2'th member of A_2, and so on up to the j_k'th member of A_k, to construct the n-tuple. The function,

$$g : A_1 \times A_2 \times \cdots \times A_k \to \mathbb{N}$$

defined by

$$g : (a_{1j_1}, a_{2j_2}, \ldots, a_{kj_k}) \mapsto p_1^{j_1} \times p_2^{j_2} \times \cdots \times p_k^{j_k},$$

is an injection because no two prime factorisations will yield the same number in $\mathbb{N}$. □

We shall now see how either of these theorems can be used to prove the countability of $\mathbb{Q}$; later in this book, they will be used to show that even larger sets of numbers can still be bijected to subsets of $\mathbb{N}$.

Theorem 6.3.3. $\mathbb{Q}$ *is countably infinite.*

Proof. We already know that $\mathbb{Q}$ is infinite, because it has a subset isomorphic to $\mathbb{Z}$, which is infinite. We now offer two alternative proofs of countability.

For each $n \in \mathbb{Z}_+$, let D_n be the set of all rationals represented by ordered pairs (m, n)

in lowest terms; so D_n bijects to the set of all integers m which are coprime with n. This set is countable (since the coprime integers m are a subset of $\mathbb{Z}$), and there are countably many such sets D_n (one for each $n \in \mathbb{Z}_+$). The set of all rationals is the union of these sets, $\mathbb{Q} = D_1 \cup D_2 \cup \cdots$, so is countable according to Theorem 6.3.1.

Alternatively, since each rational number can be represented uniquely by a coprime ordered pair $(m, n) \in \mathbb{Z} \times \mathbb{Z}_+$, we have that $\mathbb{Q}$ bijects to a subset of $\mathbb{Z} \times \mathbb{Z}_+$. Since $\mathbb{Z}$ and $\mathbb{Z}_+$ are countable, Theorem 6.3.2 implies that $\mathbb{Z} \times \mathbb{Z}_+$, and hence $\mathbb{Q}$, is countable. $\qquad\square$

Exercise 6.3.4. *The Library of Babel[i] contains every book which it is possible to print using a font of 75 characters (26 upper-case and 26 lower-case letters, 10 numerals, 12 punctuation marks, and a space). A book may have p pages, for any $p \in \mathbb{N}$. A page may have n lines, for any $n \in \mathbb{N}$ (with each page in a particular book having the same number of lines). A line may have c characters, for any $c \in \mathbb{N}$ (with each line in a particular book having the same number of characters). Each character can be any one of the 75 in the font.*
(i) Prove that the set of books in the Library of Babel is countable.
(ii) Would there be more books in the Library of Babel if a font of 7000 Chinese characters was used instead of just 75 characters?

We have previously illustrated issues of countability by reference to Hilbert's Hotel. Suppose now that every room in the hotel is full, and a countably infinite fleet of coaches arrives, with each coach containing countably infinitely many passengers wanting rooms at the hotel. Using the function in the proof of Theorem 6.3.1, the guest originally in room m may be moved to room 2^m, and passenger number m from coach number n can then occupy room number p_{n+1}^m, for each $m, n \in \mathbb{N}$ (where we take powers of the $(n+1)$'th prime because powers of 2, the first prime, have been used for the original guests). Alternatively, using the function in the proof of Theorem 6.3.2, passenger m from coach n occupies room number $2^m 3^n$.

But how does this relate to countability of the rational numbers? Our two methods for assigning rooms have treated the coachloads of passengers as follows. In the first method, each coach is a copy of $\mathbb{N}$, containing one passenger for each element of $\mathbb{N}$; the total set of passengers is the union of the countably infinitely many coachloads, $\mathbb{N} \cup \mathbb{N} \cup \ldots$. In the second method, the passenger in seat m of coach n is labelled by the ordered pair (m, n), and so is a member of $\mathbb{N} \times \mathbb{N}$. These are similar to, but not the same as, our methods for proving the countability of $\mathbb{Q}$. In the first method we identified $\mathbb{Q}$ with the union of sets D_n of integers coprime to n rather than the union of copies of $\mathbb{N}$; in the second method $\mathbb{Q}$ was identified with a subset of $\mathbb{Z} \times \mathbb{Z}_+$ rather than $\mathbb{N} \times \mathbb{N}$. But easy adjustments allow us to identify guests originally in the hotel as integers, and passengers arriving in the coaches as non-integer rationals (those which in lowest terms have denominator greater than 1). First observe that there are empty seats on the coaches: if m and n are not coprime, the passenger who might have sat in seat m on coach n would have been in seat m_c on an earlier coach n_c, where $mn_c = nm_c$ (our equivalence relation) and m_c is coprime with n_c. Secondly, to accommodate negative rationals (and zero), rooms of the original hotel guests and seats in each coach can be reallocated using the function in the proof of Theorem 4.5.4 (the countability of $\mathbb{Z}$) before any reassignment of hotel rooms is done to accommodate the coach passengers.

Either of our methods of assigning rooms may be regarded as wasteful, leaving many rooms empty after all coach passengers have been accommodated. In the first method, rooms with numbers that are not a power of a prime are empty: that includes rooms $1, 6, 10, 12, \ldots$. In the second method, rooms with prime factorisations including factors other than 2 and

---

[i] *The Library of Babel* is the title of a short story by Jorge Luis Borges; but in Borges' story the number of books in the library is huge but finite.

3 are empty, including rooms $5, 7, 10, 11, \ldots$ as well as room 1. A variety of methods have been devised to fill all rooms with passengers representing the rationals, i.e. to construct a bijection from $\mathbb{Q}$ to $\mathbb{N}$, rather than to a proper subset of $\mathbb{N}$ with Theorem 3.7.25 then confirming that a bijection to $\mathbb{N}$ is possible. One such method is described below, and others are left for you to explore in the Investigations. Note that whereas if we were wanting to biject $\mathbb{N} \times \mathbb{N}$ to $\mathbb{N}$ it would be possible to write down an explicit function $f : \mathbb{N} \times \mathbb{N} \to \mathbb{N}$ as we did when bijecting $\mathbb{Z}$ to $\mathbb{N}$ (see Theorem 4.5.4), the requirement for coprimeness of m and n in the ordered pair (m, n) prevents us from doing this with $\mathbb{Q}$, essentially because of the irregular distribution of primes. So what is given below is an *algorithm*, which will ultimately specify a natural number corresponding to each rational, with no natural numbers being left out; so giving us a bijection $f : \mathbb{Q} \to \mathbb{N}$ but without a "neat" formula.

We identify $\mathbb{Q}$ with the subset of $\mathbb{Z} \times \mathbb{Z}_+$ in which $m \in \mathbb{Z}$ and $n \in \mathbb{Z}_+$ are coprime, not forgetting the special case of zero, i.e. $(0, 1)$ in ordered pair form. Letting $f : (0, 1) \mapsto 1$, all remaining ordered pairs (m, n) to be considered have $m \in \mathbb{Z}_+$ or $m \in \mathbb{Z}_-$. Recall now that $-m$ is coprime with n if and only if m is coprime with n. So consider $m \in \mathbb{Z}_+$ (with $n \in \mathbb{Z}_+$ in any case), and let $q = m + n$. Then $q \geq 2$ and, for any given $q \in \mathbb{Z}_+ \backslash \{1\}$, there are $q - 1$ ordered pairs $(m, n) = (q - n, n) \in \mathbb{Z}_+ \times \mathbb{Z}_+$. So the algorithm works as follows.

- Take successive integer values of q, starting from 2.

- For each q, consider successive values of n, from 1 to $q - 1$, and let $m = q - n$ for each n.

- If m is coprime with n, then the pair (m, n) represents a positive rational. The bijection maps this to the next available natural number, which will be even. Also, $(-m, n)$ represents a negative rational, which is mapped to the next odd natural number.

The results of this algorithm are shown in Table 6.1 up to $q = 6$, with rationals written in standard fraction or integer form (rather than ordered pairs).

	$q = 1$	$q = 2$	$q = 3$	$q = 4$
$n = 1$	$0 \mapsto 1$	$1 \mapsto 2, -1 \mapsto 3$	$2 \mapsto 4, -2 \mapsto 5$	$3 \mapsto 8, -3 \mapsto 9$
$n = 2$			$\frac{1}{2} \mapsto 6, -\frac{1}{2} \mapsto 7$	NC
$n = 3$				$\frac{1}{3} \mapsto 10, -\frac{1}{3} \mapsto 11$

	$q = 5$	$q = 6$
$n = 1$	$4 \mapsto 12, -4 \mapsto 13$	$5 \mapsto 20, -5 \mapsto 21$
$n = 2$	$\frac{3}{2} \mapsto 14, -\frac{3}{2} \mapsto 15$	NC
$n = 3$	$\frac{2}{3} \mapsto 16, -\frac{2}{3} \mapsto 17$	NC
$n = 4$	$\frac{1}{4} \mapsto 18, -\frac{1}{4} \mapsto 19$	NC
$n = 5$		$\frac{1}{5} \mapsto 22, -\frac{1}{5} \mapsto 23$

TABLE 6.1: A bijection from $\mathbb{Q}$ to $\mathbb{N}$. For example, $-\frac{1}{2} \mapsto 7$ indicates that the rational $-\frac{1}{2}$ is mapped to the natural number 7. "NC" indicates that m and n are not coprime (where $m = q - n$), so no new rational is to be accounted for.

You may observe that the bijection used to count the integers in Theorem 4.5.4 is embedded into this bijection for counting the rationals: in both bijections a positive number (integer or rational) is mapped to an even natural number, with the corresponding negative number mapped to the following odd natural number.

6.4 Exponentiation and Its Inverse(s) on $\mathbb{Q}$

6.4.1 Integer Powers

When exponentiation was defined on the Natural Numbers, the inductive definition used multiplication in a similar way to how addition was used in the definition of multiplication. However, it soon became clear that exponentiation behaved very differently from multiplication: there were no commutative or associative laws for exponentiation; instead, we had Theorems 3.4.2 to 3.4.7, which are commonly referred to as the "Laws of Indices". Furthermore, there was no identity element for exponentiation: although $n^1 = n$ for all $n \in \mathbb{N}$, it is not in general true that $1^n = n$. Since there was no identity element, we could certainly not define inverse elements under exponentiation.

When it came to integers, we noted that it was not possible to define an exponentiation operation that was closed on $\mathbb{Z}$ and admitted an isomorphism between $\mathbb{N}$ and $\mathbb{Z}_+$. But it was possible to use the inductive definition of exponentiation to define x^n where $x \in \mathbb{Z}$ and $n \in \mathbb{Z}_+$, with the positive integer n treated as a natural number to allow induction. Because the definition involves multiplication, which has the same properties on $\mathbb{Z}$ as on $\mathbb{N}$, the inductive proofs of the Laws of Indices remain valid for these positive integer powers of integers. In $\mathbb{Q}$, multiplication again has the same arithmetic properties (and more) as in $\mathbb{N}$, so we can again use Definition 3.4.1 to define x^n where $x \in \mathbb{Q}$ and $n \in \mathbb{N}$ (or, more pedantically, where n is a member of the subset of $\mathbb{Q}$ which is isomorphic to $\mathbb{N}$ under the respective multiplication operations[ii]); and we can again be sure that the Laws of Indices remain valid. But we now have a new property of multiplication in $\mathbb{Q}$: Theorem 6.2.7, the ability to do exact division, the inverse operation of multiplication, except with zero. And since exact division is the same as multiplication by a multiplicative inverse element, theorems involving division can be written down as corollaries of theorems on multiplication. In particular, from Theorem 3.4.5 we have:

Corollary 6.4.1.

$$(x \div y)^n = x^n \div y^n \quad for \quad x \in \mathbb{Q}, y \in \mathbb{Q}\backslash\{0\}, n \in \mathbb{N}.$$

If readers are unsure about the argument leading to this corollary, they can prove this theorem using induction on n together with the theory of multiplication and division in $\mathbb{Q}$, presented in Section 6.2.

Corollary 6.4.1 is only a minor advance in our knowledge of exponentiation. A significant advance would entail being able to define x^y with $x \in \mathbb{Q}$ and $y \in \mathbb{Q}\backslash\mathbb{N}$, i.e. to be able to widen the scope of exponentiation beyond natural number powers. As when defining addition and multiplication operations in a new number system, we want to retain the arithmetic properties that were found in the previous number system: for exponentiation, this means that Theorems 3.4.2 to 3.4.7 should still be valid. Indeed, it is the requirement to satisfy these Laws of Indices that determines how x^y is defined for various categories of rational number y. Below, we present a sequence of definitions followed by "proofs"; a definition cannot be proved (!), but the proofs justify the definitions by deriving them from the Laws of Indices. Our starting point is that x^n is defined according to Definition 3.4.1 for all $x \in \mathbb{Q}$ and $n \in \mathbb{N}$.

Definition 6.4.2.

$$x^0 = 1 \ \forall x \in \mathbb{Q}\backslash\{0\}.$$

[ii]For the remainder of this chapter, we shall ignore the distinction between $\mathbb{N}$ and the subset of $\mathbb{Q}$ which is isomorphic to $\mathbb{N}$ under addition and multiplication; so, for example, it will be acceptable to write $\mathbb{N} \subset \mathbb{Q}$.

"*Proof*".

$$x^0 \times x^n = x^{0+n} \qquad\qquad [\textit{Theorem 3.4.4}]$$
$$= x^n$$
$$\therefore x^0 = x^n \div x^n$$
$$= 1. \qquad\qquad\qquad\qquad \square$$

Note that 0^0 is undefined; since $0^n = 0$ for all $n \in \mathbb{N}$, setting $x = 0$ in the above proof would entail dividing by zero. So zero creates difficulties with exponentiation as it does with division.

Definition 6.4.3.
$$x^{-n} = 1 \div x^n \ \forall x \in \mathbb{Q}\backslash\{0\} \ \text{ and } \ n \in \mathbb{N}.$$

"*Proof*".

$$x^{-n} \times x^n = x^{(-n)+n} \qquad\qquad [\textit{Theorem 3.4.4}]$$
$$= x^0$$
$$= 1 \qquad\qquad\qquad\qquad [\textit{Def. 6.4.2}]$$
$$\therefore x^{-n} = 1 \div x^n. \qquad\qquad\qquad\qquad \square$$

Note that 0^{-n} is undefined (for $n \in \mathbb{N}$) for similar reasons to 0^0 being undefined.

These two theorems have defined zero powers and negative integer powers of all non-zero rationals, so we can now take all integer powers. Before proceeding to define non-integer powers, we need to consider the inverse operation of exponentiation.

6.4.2   Roots and Fractional Powers

Recalling Definition 3.8.1 for the inverse of a binary operation and using the symbol $\wedge$ for exponentiation, the inverse of exponentiation, for which we shall (temporarily) use the symbol $\overline{\wedge}$, is defined by

$$x = z \,\overline{\wedge}\, y \quad \text{if and only if} \quad x \wedge y = z.$$

Whereas there are simple symbols, $-$ and $\div$, for the respective inverses of addition and multiplication, there is no comparable notation for the inverse of exponentiation (the author's $\overline{\wedge}$ is not used outside this book!). Instead, if $z = x^n$ with $n \in \mathbb{N}\backslash\{1\}$, the accepted notation for the inverse operation is $x = \sqrt[n]{z}$ (which would be written $x = z \,\overline{\wedge}\, n$ in our earlier notation), and we say that x is the n'th **root** of z when z is the n'th power of x.[iii] However, bear in mind that, whereas x^0 and x^{-n} are guaranteed to exist in $\mathbb{Q}$ (for $x \neq 0$ and $n \in \mathbb{N}$) because the proofs of Definitions 6.4.2 and 6.4.3 use division which is closed on $\mathbb{Q}\backslash\{0\}$, we have no knowledge of the existence of roots in $\mathbb{Q}$; indeed, given our experience with inverse operations requiring the definition of new number systems, it will not be surprising to find that n'th roots of rational numbers may not exist in $\mathbb{Q}$.

But how should we notate the inverse operation if $z = x^y$ with $y \in \mathbb{Q}\backslash\mathbb{N}$? The following definitions will show that there is ultimately no need for *any* notation for the inverse of exponentiation, regardless of whether the power is a natural number or any other rational number, so we could even dispense with the root notation (although it is often useful).

[iii]If $n = 1$ and $z = x^n$, then $z = x$ so that $\sqrt[1]{z} = z$; thus there is never any need to refer to "1'st root" or to use the root notation in this case. If $n = 2$, it is common practice to omit the number 2 on the root symbol, writing $\sqrt{z}$ rather than $\sqrt[2]{z}$ for the **square root**; however, we shall retain the 2 in $\sqrt[2]{z}$ to emphasise that this is just one case of the general operation symbolised by $\sqrt[n]{z}$.

Definition 6.4.4.

$$x^{1/n} = \sqrt[n]{x} \quad \text{for } n \in \mathbb{N}, \quad \text{if } \sqrt[n]{x} \text{ exists in } \mathbb{Q}.$$

"Proof".

$$(x^{1/n})^n = x^{(1/n) \times n} \qquad\qquad\qquad\qquad [Theorem\ 3.4.7]$$
$$= x^1$$
$$= x,$$
$$\therefore x^{1/n} = \sqrt[n]{x} \qquad\qquad\qquad\qquad [n\text{'th root is inverse of } n\text{'th power.}] \qquad \square$$

Corollary 6.4.5.

$$x^{m/n} = (\sqrt[n]{x})^m \quad \text{for } m \in \mathbb{Z}, n \in \mathbb{N}, \quad \text{if } \sqrt[n]{x} \text{ exists in } \mathbb{Q}.$$

This follows from Definition 6.4.4, noting that $x^{m/n} = x^{(1/n) \times m} = (x^{1/n})^m$ according to Theorem 3.4.7.

We have now defined all rational powers of rational numbers, since every rational can be represented as m/n with $m \in \mathbb{Z}$ and $n \in \mathbb{N} = \mathbb{Z}_+$; but since the definition of non-integer rational powers is in terms of the inverse of exponentiation (i.e. roots), we still need to explore their existence, and indeed their uniqueness, in $\mathbb{Q}$. Considering uniqueness first: we observed in Exercise 4.3.15 that $(-z)^n = z^n \in \mathbb{Z}_+$ when $z \in \mathbb{Z}$ and n is an *even* natural number. Since multiplication in $\mathbb{Q}$ obeys the same theorems as in $\mathbb{Z}$, it is also true that $(-z)^n = z^n \in \mathbb{Q}_+$ when $z \in \mathbb{Q}$ and n is an even natural number. So if $x = z^n$ with even n, both z and $-z$ satisfy the definition of $\sqrt[n]{x}$. To avoid this non-uniqueness, we adopt the convention that $\sqrt[n]{x}$ means the *positive* number z satisfying $x = z^n$ when n is even and $x \in \mathbb{Q}_+$; and hence from Definition 6.4.4 and Corollary 6.4.5, $x^{1/n}$ and $x^{m/n}$ are taken to be positive when n is even. Furthermore, since $z^n \geq 0$ for all $z \in \mathbb{Q}$ when n is even, $\sqrt[n]{x}$ does not exist in $\mathbb{Q}$ when $x \in \mathbb{Q}_-$ and n is even. On the other hand, when n is odd we have $(-z)^n = -z^n$ for all $z \in \mathbb{Q}$, so if $\sqrt[n]{x}$ exists in $\mathbb{Q}$ then $\sqrt[n]{-x} = -\sqrt[n]{x}$. When $x = 0$, we have $0^n = 0$ and so $\sqrt[n]{0} = 0$ for all $n \in \mathbb{N}$, but the exclusions in Definitions 6.4.2 and 6.4.3 mean that when applying Corollary 6.4.5, we have $0^{m/n} = 0$ when $\frac{m}{n} \in \mathbb{Q}_+$ but $0^{m/n}$ is undefined otherwise.

The above considerations mean that we now only need to consider the existence of $x^{m/n}$ for $x \in \mathbb{Q}_+$. At this stage, many authors would offer a proof that $\sqrt[2]{2}$ is not a rational number; and it is sometimes said that this verifies the *existence of irrational numbers*. It does no such thing; it merely shows that not all roots of rational numbers exist within $\mathbb{Q}$. The existence of irrational numbers requires the construction of a new number system to accommodate numbers that do not exist in $\mathbb{Q}$, which will be done in the next chapter. The use of a single counterexample, usually $\sqrt[2]{2}$, is certainly valid as a way to show that roots of rationals may not exist within $\mathbb{Q}$;[iv] but it would be more useful to have a theorem which actually allows us to distinguish which roots, and more generally non-integer powers, of which rational numbers exist within $\mathbb{Q}$. We shall first prove a theorem for roots of positive integers, and then a corollary generalising to rational powers of rational numbers.

[iv]Legend has it that Hippasus of Metapontum, a member of the Pythagorean school of mathematicians, was the first to discover that the length of the diagonal of a square was not a rational multiple of the length of the side (the ratio of diagonal:side being $\sqrt[2]{2}$ according to Pythagoras' theorem); and that he was drowned at sea as a consequence! But while the ancient Greeks certainly knew that $\sqrt[2]{2}$ and certain other square roots were not rational numbers, the history of the discovery is somewhat uncertain. In any case, there remains a tradition to the present day of using the case of $\sqrt[2]{2}$ to demonstrate that roots of rational numbers may not themselves exist in $\mathbb{Q}$, and a wide variety of proofs may be found.

Theorem 6.4.6. *If $x \in \mathbb{Z}_+$ and $n \in \mathbb{N}$, then $\sqrt[n]{x} \in \mathbb{Q}$ if and only if $\exists r \in \mathbb{Z}_+$ such that $x = r^n$.*

Proof. The fact that $\sqrt[n]{x}$ exists in $\mathbb{Q}$ if $x = r^n$ for some $r \in \mathbb{Z}_+ \subset \mathbb{Q}$ is simply the definition of the n'th root. So the theorem is saying that in all other cases it is impossible to find a rational number z such that $x = z^n$. The Fundamental Theorem of Arithmetic ensures that "other cases" do exist: for if r has the prime factorisation, $r = p_1^{\alpha_1} p_2^{\alpha_2} \ldots p_k^{\alpha_k}$ where $p_1, p_2, \ldots, p_k$ are distinct primes raised to powers $\alpha_1, \alpha_2, \ldots, \alpha_k \in \mathbb{N}$, then r^n has the unique prime factorisation, $r^n = p_1^{n\alpha_1} p_2^{n\alpha_2} \ldots p_k^{n\alpha_k}$, i.e. with each prime raised to a power which is a multiple of n. For any $n \geq 2$, it is certainly possible to find numbers x with prime factorisations not of this form; simple cases are when x itself is prime.

If we suppose that $\sqrt[n]{x} \in \mathbb{Q}_+$ (recalling that we only need to consider positive rationals), we can certainly write $\sqrt[n]{x}$ as a fraction in lowest terms:

$$\sqrt[n]{x} = \frac{r}{s} \quad \text{where } r \text{ and } s \text{ are coprime positive integers.}$$

Then

$$
\begin{aligned}
x &= \left(\frac{r}{s}\right)^n \\
&= \frac{r^n}{s^n} \qquad\qquad [\textit{Theorem 6.4.1}] \\
\therefore r^n &= s^n x.
\end{aligned}
$$

So $x \mid r^n$. But since r and s are coprime, r^n and s^n are also coprime, and applying Theorem 5.3.11 to the equation $r^n = s^n x$ then yields $r^n \mid x$. Since a divisor cannot be greater than the number it divides, $x \mid r^n$ and $r^n \mid x$ together imply $x = r^n$. So we have shown that a necessary condition for $\sqrt[n]{x}$ to be rational is that x is the n'th power of an integer. $\qquad\square$

Corollary 6.4.7. *If u and v are coprime positive integers, and m and n are coprime positive integers, then*

$$\left(\frac{u}{v}\right)^{m/n}$$

is rational if and only if $\exists y, z \in \mathbb{Z}_+$ such that $u = y^n$ and $v = z^n$.

Proof. Proceeding as in the proof of Theorem 6.4.6, we suppose that

$$\left(\frac{u}{v}\right)^{m/n} = \frac{r}{s}$$

where r and s are coprime positive integers so that $r/s \in \mathbb{Q}$, and we obtain that

$$r^n = u^m \quad \text{and} \quad s^n = v^m$$

as necessary conditions for our supposition to be true, where the details of the calculation are left for the reader in Exercise 6.4.8.

Now let r have the prime factorisation $r = p_1^{\alpha_1} p_2^{\alpha_2} \ldots p_k^{\alpha_k}$, so that

$$r^n = p_1^{n\alpha_1} p_2^{n\alpha_2} \ldots p_k^{n\alpha_k}.$$

The Fundamental Theorem of Arithmetic implies that since $u^m = r^n$, the prime factorisation of u^m must be identical to this. Clearly this requires the prime factorisation of u to involve precisely the same primes as that of r^n, i.e. $u = p_1^{\beta_1} p_2^{\beta_2} \ldots p_k^{\beta_k}$ for some powers $\beta_1, \beta_2, \ldots, \beta_k \in \mathbb{N}$, so that

$$u^m = p_1^{m\beta_1} p_2^{m\beta_2} \ldots p_k^{m\beta_k}.$$

The power of each prime must be the same in the factorisations of r^n and u^m:

$$n\alpha_i = m\beta_i \quad (i = 1, 2, \ldots, k).$$

So $n \mid m\beta_i$ and since m and n are coprime, this implies $n \mid \beta_i$ so that we can write $\beta_i = n\gamma_i$ for some $\gamma_i \in \mathbb{N}$, for each i. Thus

$$u = p_1^{n\gamma_1} p_2^{n\gamma_2} \cdots p_k^{n\gamma_k} = (p_1^{\gamma_1} p_2^{\gamma_2} \cdots p_k^{\gamma_k})^n :$$

u is the n'th power of a positive integer.

Similarly the condition $s^n = v^m$ implies that v is also the n'th power of a positive integer. $\qquad\square$

Exercise 6.4.8. *Show that if*

$$\left(\frac{u}{v}\right)^{m/n} = \frac{r}{s}$$

where r is coprime with s, u is coprime with v, and m is coprime with n, then

$$r^n = u^m \quad and \quad s^n = v^m.$$

Corollary 6.4.7 covers all positive rational powers of all positive rational numbers. The proof may appear a little cumbersome, but the important point to notice is that the failure of most non-integer powers of rational numbers to be rational is a consequence of the uniqueness of prime factorisations in $\mathbb{N}$. Note that whereas the inadequacies of $\mathbb{N}$ and $\mathbb{Z}$ related to the *inverse* operations of addition and multiplication not being closed in the respective number systems, here it is exponentiation itself, not merely its inverse, that fails to be closed in $\mathbb{Q}$.

6.4.3 Logarithms

When we introduced the concept of the inverse of a binary operation in Section 3.8, we pointed out the possibility of defining a "second inverse". Given an operation $*$ on a set A and two elements $y, z \in A$, the (first) inverse seeks a first input element $x \in A$ such that $x * y = z$, while the second inverse seeks a second input element $x \in A$ such that $y * x = z$. If the operation $*$ is commutative, so that $x * y = y * x$ for all $x, y \in A$, the first and second inverse operations clearly give the same result. This is why we have not had to consider second inverses in relation to addition or multiplication; but exponentiation is not commutative. So far we have considered the first inverse problem for exponentiation, finding a number x which, when raised to the y'th power, yields a given number z: when $y \in \mathbb{N}$ we refer to this x as the y'th root of z, although more generally for any $y \in \mathbb{Q}$, Definition 6.4.4 and Corollary 6.4.5 show that if $x^y = z$, then $x = z^{1/y}$ (although we have seen that with most choices of y and z, no such x may be found in $\mathbb{Q}$). But the second inverse problem for exponentiation is: given $y, z \in \mathbb{Q}$, find a power x such that $y^x = z$. Such a number x is called the **logarithm to base y of** z, and the notation is even more inelegant than the root notation:

Definition 6.4.9.

$$x = \log_y z \quad \text{if} \quad y^x = z.$$

It is useful to observe an immediate consequence of this definition:

Corollary 6.4.10.

$$y^{\log_y z} = z.$$

Note that $\log_y z$ is only defined when both y and z are *positive*: since y^x can only be positive if y is positive, no logarithm to a positive base of a negative number z can ever exist in $\mathbb{Q}$. The reason for not allowing the base y to be negative will become apparent below, but one further restriction on y is immediately apparent: because $1^x = 1$ for all $x \in \mathbb{Q}$, we cannot take a logarithm to base $y = 1$.

Whereas with the first inverse of exponentiation we found that the root notation could be avoided because the result of the inverse operation was in fact expressible in terms of the original operation (see Definition 6.4.4), there is no comparable way of avoiding the use of logarithms. However, we can derive useful theorems, the **Laws of Logarithms**, from the Laws of Indices, noting that although the latter were proved in Section 3.4 for Natural Numbers, we have defined exponentiation in $\mathbb{Q}$ so that it satisfies the same theorems as in $\mathbb{N}$. As with roots and fractional powers, it is not clear whether $\log_y z$ exists in $\mathbb{Q}$ for any given $y, z \in \mathbb{Q}$; but we shall address the issue of existence after we have derived the laws that logarithms must obey when they do exist in $\mathbb{Q}$.

Theorem 6.4.11.
$$\log_n(uv) = \log_n u + \log_n v.$$

Proof. In Theorem 3.4.4 let $u = n^m$ and $v = n^p$ so that $m = \log_n u$ and $p = \log_n v$. Then

$$\begin{aligned}
\log_n(uv) &= \log_n(n^m \times n^p) \\
&= \log_n(n^{m+p}) & [\textit{Theorem 3.4.4}] \\
&= m + p \\
&= \log_n u + \log_n v.
\end{aligned}$$ $\square$

Corollary 6.4.12.
$$\log_n\left(\frac{u}{v}\right) = \log_n u - \log_n v.$$

Exercise 6.4.13. *Prove Corollary 6.4.12.*

Before proceeding further, note that from Definition 6.4.2 we have

$$\log_n 1 = 0$$

for arbitrary $n \in \mathbb{Q}_+$, and then from Corollary 6.4.12,

$$-\log_n u = \log_n\left(\frac{1}{u}\right).$$

The final law of logarithms derives from Theorem 3.4.7 for exponentiation:

Theorem 6.4.14.
$$\log_n(u^v) = v \log_n u.$$

Exercise 6.4.15. *Prove Theorem 6.4.14.*

The above theorems relate logarithms all to the same base. There are also useful formulae for changing base in exponentiation and logarithms:

Theorem 6.4.16.
$$n^v = m^{v \log_m n}.$$

Proof.

$$\begin{aligned}
m^{v \log_m n} &= (m^{\log_m n})^v & [\textit{Theorem 3.4.7}] \\
&= n^v. & [\textit{Corollary 6.4.10}]
\end{aligned}$$ $\square$

Letting $u = n^v$ so that $v = \log_n u$, and taking logarithms to base m of both sides of the last theorem, we obtain:

Corollary 6.4.17.
$$\log_m u = (\log_n u) \times (\log_m n).$$

Exercise 6.4.18. *Prove that*
$$\log_m n = \frac{1}{\log_n m}.$$

Following our findings with roots and non-integer powers, it will come as no surprise that there are similarly restricted circumstances in which logarithms to rational bases of rational numbers exist in $\mathbb{Q}$.

Theorem 6.4.19. *If $x \in \mathbb{Z}_+$ and $z \in \mathbb{Z}_+$, then $\log_x z$ exists in $\mathbb{Q}$ if and only if there are positive integers y, r, s such that $x = y^s$ and $z = y^r$.*

Exercise 6.4.20. *Prove Theorem 6.4.19. Start by supposing that $\log_x z = \frac{r}{s}$ where r and s are coprime positive integers, and use arguments similar to those in the proofs of Theorem 6.4.6 and Corollary 6.4.7.*

Observe that x and z are required to be positive integer powers of the *same* number y. Once we have found the required y, r, s, it is easy to evaluate the logarithm:

Example 6.4.21. (i) $\log_8 32$ is rational: $8 = 2^3$ and $32 = 2^5$. To evaluate $\log_8 32$ we require x such that $32 = 8^x$; this can now be written as $2^5 = (2^3)^x = 2^{3x}$, from which $5 = 3x$, so that $\log_8 32 = 5/3$. Similarly, we could show that $\log_{32} 8 = 3/5$.
(ii) $\log_{18} 216$ does not exist in $\mathbb{Q}$. Consider the prime factorisations of 18 and 216: $18 = 2 \times 3^2, 216 = 2^3 \times 3^3$. Even though 216 is a multiple of 18, and their prime factorisations both include powers of 2 and 3 only, the required condition cannot be satisfied. On the other hand, $324 = 2^2 \times 3^4 = 18^2$, so $\log_{18} 324 = 2$.

We have already mentioned that $\log_y z$ cannot exist in $\mathbb{Q}$ if z is negative and y is positive; referring to the above example, $\log_8(-32)$ cannot exist in $\mathbb{Q}$ because 8^x can only be positive if it exists in $\mathbb{Q}$ for some given $x \in \mathbb{Q}$. But since $(-8)^{5/3} = (\sqrt[3]{-8})^5 = (-2)^5 = -32$, one might argue that we can take a logarithm to a negative base, $\log_{-8}(-32) = \frac{5}{3}$. However, $\log_{-8} 32$ would not exist in $\mathbb{Q}$: there is no $x \in \mathbb{Q}$ such that $(-8)^x = 32$. The situation is reversed if we try to take logarithms to base -8 of -16 or 16: here we have $(-8)^{4/3} = 16$, so we could say that $\log_{-8} 16 = \frac{4}{3}$, but $\log_{-8}(-16)$ would not exist. So if we allow logarithms to negative bases, the general situation is that (with positive y and z) either but not both of $\log_{-y} z$ or $\log_{-y}(-z)$ may exist in $\mathbb{Q}$; and if one of them does exist it is simply equal to $\log_y z$. So it is sensible to just say that logarithms to negative bases are undefined, at least in $\mathbb{Q}$.

Logarithms to rational, non-integer bases do not present such difficulties as long as the base is positive, although they are not commonly discussed. The restriction on their existence in $\mathbb{Q}$ is similar to that for logarithms to positive integer bases:

Corollary 6.4.22. *If u and v are coprime positive integers, and m and n are coprime positive integers, then*
$$\log_{m/n}\left(\frac{u}{v}\right)$$
is rational if and only if $\exists y \in \mathbb{Q}_+$ and $\exists r, s \in \mathbb{Z}_+$ such that $\frac{u}{v} = y^r$ and $\frac{m}{n} = y^s$.

Exercise 6.4.23. *Prove Corollary 6.4.22.*

Example 6.4.24. $\log_{8/9}(32/27) \notin \mathbb{Q}$, even though $\log_8 32$ and $\log_9 27$ are rational. But $\log_{8/9}(81/64) = -2$, since $81/64 = (8/9)^{-2}$.

Exercise 6.4.25. *Find the following logarithms, if they exist in* $\mathbb{Q}$:

$$(i) \ \log_3 6 \qquad (ii) \ \log_3 \frac{1}{27} \qquad (iii) \ \log_{27} \frac{1}{3} \qquad (iv) \ \log_{1/27} \frac{1}{3}$$

One might question whether it is worth defining fractional powers and logarithms at all, given the severe restrictions on their existence in $\mathbb{Q}$. However, it is useful to have these concepts in place before moving on to the Real Numbers, where we will find that both fractional powers and logarithms do always exist as long as we stick to working with positive numbers.

6.5 Order in $\mathbb{Q}$

As when we defined order relations in $\mathbb{Z}$, we return to the formal definition of our present number system in terms of equivalence classes of ordered pairs of the previous class of numbers. The equivalence relation (Definition 6.1.1) can be written

$$(p, q) \equiv (t, s) \quad \text{if and only if} \quad p \times s = t \times q,$$

where $p, t \in \mathbb{Z}$ and $q, s \in \mathbb{Z}_+$, and we now define order relations $\oslash$ and $\oslash$:

Definition 6.5.1.
$$(p, q) \oslash (t, s) \quad \text{if and only if} \quad p \times s < t \times q.$$
$$(p, q) \oslash (t, s) \quad \text{if and only if} \quad p \times s > t \times q.$$

Notice how these definitions are analogous to Definition 4.4.1 for order in $\mathbb{Z}$. The relations $\oslash$ and $\oslash$ can easily be shown to be *well-defined*, i.e. that if (p, q) and (t, s) in the definitions were replaced by other ordered pairs from the same equivalence classes, the same order relation would still be valid. The definitions in terms of the trichotomous, transitive order relations on $\mathbb{Z}$ make it easy (but not trivial) to prove trichotomy and transitivity for our new relations on $\mathbb{Q}$.

Exercise 6.5.2. *Prove that the relation $\oslash$ in Definition 6.5.1 is transitive. You may assume any of the properties of order in* $\mathbb{Z}$.

Next, we note that a subset of the rationals, $\mathbb{Q}_Z := \{(p, q) \in \mathbb{Q} : q = 1\}$, has been shown to be isomorphic to $\mathbb{Z}$ with respect to addition and multiplication; we have already been treating the integers as a subset of the rationals (i.e. saying that $\mathbb{Z}$ is the same thing as $\mathbb{Q}_Z$) on this basis, but we should really only do this once we have verified that there is also an order isomorphism:

Theorem 6.5.3. *There is **order isomorphism** between the relations $\oslash, \oslash$ in $\mathbb{Q}_Z$ and the relations $<, >$ in $\mathbb{Z}$; i.e.*

$$(p, 1) \oslash (t, 1) \quad \text{if and only if} \quad p < t$$

and

$$(p, 1) \oslash (t, 1) \quad \text{if and only if} \quad p > t.$$

This follows immediately from putting $q = 1$ and $s = 1$ in Definition 6.5.1. Given this order isomorphism, we no longer need the new symbols $\ominus$ and $\oslash$; we can use the same symbols $<$ and $>$ and the same terminology, "less than" and "greater than", in $\mathbb{Q}$ as in $\mathbb{N}$ and $\mathbb{Z}$.

Now recall that we identify those rationals not in $\mathbb{Q}_Z$ as the quantities we commonly write as fractions. So Definition 6.5.1 can be written

$$\frac{p}{q} < \frac{t}{s} \quad \text{if and only if} \quad ps < tq; \qquad \frac{p}{q} > \frac{t}{s} \quad \text{if and only if} \quad ps > tq.$$

This is simply the common practice of putting fractions over a common denominator when it is required to compare their magnitude:

$$\frac{p}{q} = \frac{ps}{qs} \quad \text{and} \quad \frac{t}{s} = \frac{tq}{qs}.$$

Since a fraction m/n can be thought of as $m \times (1/n)$, our comparison of ps with tq is comparing the multiples of the quantity $1/qs$ represented by p/q and t/s.

Example 6.5.4. Determine order between $\frac{4}{5}$ and $\frac{13}{16}$.

Solution. The common denominator is $5 \times 16 = 80$.

$$\frac{4}{5} = \frac{4 \times 16}{5 \times 16} = \frac{64}{80} \quad \text{and} \quad \frac{13}{16} = \frac{13 \times 5}{16 \times 5} = \frac{65}{80}.$$

Since $65 > 64$, we have that $\frac{13}{16} > \frac{4}{5}$. $\qquad\square$

Exercise 6.5.5. *Determine order between: (i) $\frac{22}{7}$ and $\frac{355}{113}$; (ii) $-\frac{38}{27}$ and $-\frac{17}{12}$.*

Recall now that rationals were defined as *positive* or *negative* according to whether the numerator (first member of the ordered pair) is a positive or negative integer. The expected order relations with respect to zero are valid:

Theorem 6.5.6. *If $x \in \mathbb{Q}_+$, then $x > 0$. If $x \in \mathbb{Q}_-$, then $x < 0$.*

This can be verified by writing x in ordered-pair form, (p, q) where p is either a positive or negative integer, and using Definition 6.5.1 to compare with $(0, 1)$, the ordered-pair form of zero.

The remaining requirements for the rationals to be an ordered ring (see Definition 4.4.9) can now be verified.

Theorem 6.5.7. *For $x, y, z \in \mathbb{Q}$:*
(i) $x > y \Leftrightarrow x + z > y + z$.
(ii) If $x > 0$ and $y > 0$, then $xy > 0$.

Exercise 6.5.8. *Prove both parts of Theorem 6.5.7. For (i), write x, y, and z as ordered pairs of integers and use the definitions of addition and order in terms of these ordered pairs. For (ii), you should also use ordered-pair definitions of multiplication and order, noting by Definition 6.1.6 and Theorem 6.5.6 that with x represented by an ordered pair $(p, q) \in \mathbb{Z} \times \mathbb{Z}_+$, $x > 0$ if $p > 0$.*

The addition theorem has a corollary on subtraction, as in $\mathbb{Z}$:

Corollary 6.5.9. $x > y \Leftrightarrow x - z > y - z$.

All the remaining theorems on order from 4.4.10 to 4.4.20 follow from the ordered ring structure, so are valid in $\mathbb{Q}$ as in $\mathbb{Z}$; but the multiplication theorem is of such importance that it is worth restating here:

Theorem 6.5.10. *If $z > 0$ and $x > y$, then $xz > yz$. If $z < 0$ and $x > y$, then $xz < yz$.*

The methods that we used for solving algebraic inequalities in $\mathbb{Z}$ are based on theorems which follow from the ordered ring structure, so are also valid in $\mathbb{Q}$. In particular, Theorem 4.4.20 remains valid in $\mathbb{Q}$ and can be used to solve inequalities involving quadratic expressions. So, if we want to find all $x \in \mathbb{Q}$ satisfying $x^2 > 3x + 4$, the solution procedure in Example 4.4.25 remains valid and we obtain the solution set $\{x \in \mathbb{Q} : x > 4$ or $x < -1\}$, although the second form of solution set given in Example 4.4.25 has no counterpart in $\mathbb{Q}$.

The major difference between the Rationals and the Integers is that $\mathbb{Q}$ is not merely an ordered ring, but an **ordered field**, since multiplicative inverses and exact division are defined for all non-zero elements of $\mathbb{Q}$. Thus there are new theorems on how these affect order relations in $\mathbb{Q}$. Recalling that the multiplicative inverse of $x \in \mathbb{Q}$ may be written $\frac{1}{x}$, we have:

Theorem 6.5.11. *For $x, y \in \mathbb{Q}$:*

$$\text{if } x > y > 0, \quad \text{then } \frac{1}{y} > \frac{1}{x} > 0; \qquad \text{if } x < y < 0, \quad \text{then } \frac{1}{y} < \frac{1}{x} < 0.$$

Proof. It was pointed out in the proof of Theorem 6.2.6 that the multiplicative inverse of a positive number is positive; and Theorem 6.5.6 identifies "positive" with "greater than zero". Now, in both the cases $x > y > 0$ and $x < y < 0$, the ordered ring properties give $xy > 0$, so xy has a multiplicative inverse $\frac{1}{xy} > 0$. Hence by Theorem 6.5.10, multiplying all terms in the inequalities $x > y > 0$ and $x < y < 0$ by $\frac{1}{xy}$ preserves the order relations, yielding the required results.   □

Exercise 6.5.12. *Prove Theorem 6.5.11 using ordered-pair representations $x = (p, q)$ and $y = (t, s)$, noting Definition 6.5.1 for order in $\mathbb{Q}$ and the ordered-pair forms of multiplicative inverses in the proof of Theorem 6.2.6. [This will show how much easier it is to use our knowledge of ordered rings, as in the proof given above, than to work from basic definitions.]*

Example 6.5.13. $-3 < -2 < 0$; taking multiplicative inverses, $-\frac{1}{2} < -\frac{1}{3} < 0$.

Again using the fact that dividing by a rational number z is the same as multiplying by its multiplicative inverse $\frac{1}{z}$, Theorem 6.5.10 on multiplication and Theorem 6.5.11 on multiplicative inverses yield:

Corollary 6.5.14.

$$\text{If } z > 0 \text{ and } x > y, \quad \text{then } \frac{x}{z} > \frac{y}{z};$$

$$\text{If } z < 0 \text{ and } x > y, \quad \text{then } \frac{x}{z} < \frac{y}{z}.$$

So just as with multiplication, dividing both sides by a positive rational number preserves an order relation while dividing by a negative rational number reverses an order relation.

Example 6.5.15. We have already found that $\frac{13}{16} > \frac{4}{5}$. If we divide both sides by the negative number $-\frac{3}{4}$, we expect to find that $\frac{13}{16} \div \left(-\frac{3}{4}\right) < \frac{4}{5} \div \left(-\frac{3}{4}\right)$. Indeed,

$$\frac{13}{16} \div \left(-\frac{3}{4}\right) = -\left(\frac{13}{16} \times \frac{4}{3}\right) = -\frac{52}{48} = -\frac{13}{12},$$

$$\frac{4}{5} \div \left(-\frac{3}{4}\right) = -\left(\frac{4}{5} \times \frac{4}{3}\right) = -\frac{16}{15},$$

and

$$-\frac{13}{12} < -\frac{16}{15}.$$

To check the final order relation, we note that $-\frac{13}{12} = \frac{-13}{12}$ and $-\frac{16}{15} = \frac{-16}{15}$, and then use Definition 6.5.1: since $(-13) \times 15 = -195$, $(-16) \times 12 = -192$; and $-195 < -192$, we have that $\frac{-13}{12} < \frac{-16}{15}$.

We have not considered order in relation to exponentiation since Theorems 3.5.15 and 3.5.17 for $\mathbb{N}$. Even restricting attention to those rational powers of rational numbers that do exist in $\mathbb{Q}$, writing down theorems that would cover all possibilities for order in relation to exponentiation in $\mathbb{Q}$ would be long-winded. Fortunately such general theorems are not needed until we consider Real Numbers, but it is useful to have theorems on natural number powers of positive rationals:

Theorem 6.5.16. *For* $x, y \in \mathbb{Q}_+$ *and* $n \in \mathbb{N}$,

$$x > y \Leftrightarrow x^n > y^n.$$

Theorem 6.5.17. *For* $x \in \mathbb{Q}$ *with* $x > 1$ *and* $m, n \in \mathbb{N}$,

$$m > n \Leftrightarrow x^m > x^n.$$

Since we are only considering natural number powers, the proofs proceed as in Exercises 3.5.16 and 3.5.18.

Exercise 6.5.18. *Write down and prove a theorem similar to Theorem 6.5.17 but for the case where* $0 < x < 1$. *[Hint: Theorem 6.5.11 may be useful.]*

So far, the theory on order in $\mathbb{Q}$ has been entirely in line with the theory on order in $\mathbb{Z}$, at least when dealing with concepts that are defined in both number systems. We now come to a theorem which is in complete contrast to what we saw in $\mathbb{Z}$. Theorem 4.4.22 shows that there are "gaps" of length 1 between successive integers; but in the rationals, we have:

Theorem 6.5.19. *The rationals are **dense**: given* $a, b \in \mathbb{Q}$ *with* $a < b$, *there are countably infinitely many rationals* x_j *such that* $a < x_j < b$.[v]

Proof. There are many ways to construct an infinite set of rationals between a and b. A simple method is to let

$$x_j = a + \frac{b-a}{j+1} \quad \text{for each } j \in \mathbb{N}.$$

Starting from the number a, we have defined x_1 to be $\frac{1}{2}$ of the distance to b, x_2 to be $\frac{1}{3}$ of the distance to b, etc. Of course there are countably infinitely many other rationals between a and b (they must be countable, since the entire set of rational numbers is countable). $\square$

Theorem 6.5.19 says that we cannot find a gap between rational numbers; however close together two rationals are, we can always find more rationals between them. In particular, we can find rational numbers as close to zero as we want. Nevertheless, that does not prevent the rationals from having a property similar to Lemma 5.1.1 in the Integers:

---

[v] *Dense* is usually defined as meaning that there exists at least one rational x between any two other rationals a and b, but it is useful to observe that in fact one may find infinitely many rationals between a and b.

Theorem 6.5.20. *For any* $y \in \mathbb{Q}$ *and any* $x \in \mathbb{Q}_+$, $\exists\, k \in \mathbb{N}$ *such that* $kx > y$.

Proof. If $y \leq 0$, then $k = 1$ will suffice. If $y > 0$ we can write $x = \frac{m}{n}$ and $y = \frac{p}{q}$ for some $m, n, p, q \in \mathbb{Z}_+$, and letting $k = np + 1$ yields $kx > mp \geq \frac{p}{q}$ for all $m, p, q \in \mathbb{Z}_+$.   □

We can also prove a result similar to Lemma 5.1.2 in $\mathbb{Z}$, although this will first require a further lemma:

Lemma 6.5.21 (Bernoulli's Inequality). *For any* $n \in \mathbb{N}$ *and* $x > -1$,

$$(1 + x)^n \geq 1 + nx,$$

with equality only if $n = 1$ *or* $x = 0$.

If we remove the cases where equality applies, we have the **strict form of Bernoulli's inequality**:

$$\text{If } n \in \mathbb{N} \backslash \{1\} \text{ and } x > 0 \text{ or } -1 < x < 0, \text{ then } (1 + x)^n > 1 + nx. \qquad (6.1)$$

Proof. The cases of equality are easy to verify: if $n = 1$, then

$$(1 + x)^n = 1 + x \quad \text{and} \quad 1 + nx = 1 + x,$$

while if $x = 0$,

$$(1 + x)^n = 1 \quad \text{and} \quad 1 + nx = 1.$$

It remains to verify the strict form of the inequality, which we do by induction anchored at $n = 2$ (since $n = 1$ is excluded in the strict form): see comments below Theorem 3.7.23.
(a) For $n = 2$,

$$\begin{aligned}
(1 + x)^n &= (1 + x)^2 \\
&= 1 + 2x + x^2 \\
&> 1 + 2x \qquad \text{for } x \neq 0 \\
&= 1 + nx.
\end{aligned}$$

(b) Given the inductive hypothesis, $(1 + x)^n > 1 + nx$, we must show that $(1 + x)^{S(n)} > 1 + S(n)x$. Now,

$$\begin{aligned}
(1 + x)^{S(n)} &= (1 + x)(1 + x)^n \\
&> (1 + x)(1 + nx) \qquad [\textit{Inductive hypothesis \& Thm. 6.5.10, with } 1 + x > 0] \\
&= 1 + (n + 1)x + nx^2 \\
&> 1 + S(n)x. \qquad\qquad [nx^2 > 0 \textit{ when } n \in \mathbb{N} \textit{ and } x \neq 0] \qquad □
\end{aligned}$$

The property of the rationals that is similar to Lemma 5.1.2 in $\mathbb{Z}$ is:

Theorem 6.5.22. *For any* $y \in \mathbb{Q}$ *and* $x \in \mathbb{Q}$ *with* $x > 1$, $\exists\, k \in \mathbb{N}$ *such that* $x^k > y$.

Exercise 6.5.23. *Use Theorem 6.5.20 and the strict form of Bernoulli's inequality to prove Theorem 6.5.22.*

A further result, which will be useful when we consider powers of Real Numbers, is that n'th powers of rationals are dense within the positive rationals:

Lemma 6.5.24. *Given* $a, b \in \mathbb{Q}_+$ *with* $a < b$, *then for any* $n \in \mathbb{N}$ *there exists* $x \in \mathbb{Q}_+$ *with* $a < x^n < b$.

Proof. We first look for $m \in \mathbb{N}$ such that when the difference between two positive rational numbers is $1/m$, the difference between their n'th powers is less than $b - a$. So for $x \in \mathbb{Q}_+$ and $n \in \mathbb{N}$, we want

$$0 < \left(x + \frac{1}{m}\right)^n - x^n < b - a. \tag{6.2}$$

Now,

$$\left(x + \frac{1}{m}\right)^n - x^n = x^n \left(\left(1 + \frac{1}{mx}\right)^n - 1\right).$$

The next step is left as an exercise:

Exercise 6.5.25. *Prove by induction on n that*

$$\left(1 + \frac{1}{mx}\right)^n < 1 + \frac{2^n}{mx}.$$

if $mx \geq 2$.

Since also $(1 + 1/mx)^n > 1$ when $mx > 0$, we now have

$$0 < \left(x + \frac{1}{m}\right)^n - x^n < x^n \frac{2^n}{mx} = \frac{2^n x^{n-1}}{m}.$$

Thus to satisfy the inequalities (6.2) we need to choose m such that

$$m > \frac{2^n x^{n-1}}{b - a}, \tag{6.3}$$

as well as $m \geq 2/x$ (the condition required in Exercise 6.5.25); this is possible according to Theorem 6.5.20.

We want to fulfil these requirements with x satisfying $a < x^n < b$. We do not yet know that such x exists; but it is certainly possible to find $x_g \in \mathbb{Q}_+$ such that $x_g^n > b$ and $x_l \in \mathbb{Q}_+$ such that $x_l^n < a$, for any $a, b \in \mathbb{Q}_+$ and any $n \in \mathbb{N}$.[vi] We can then find m that satisfies (6.3) with $x = x_g$, and satisfies $m \geq 2/x$ with $x = x_l$; then from Theorems 6.5.16 and 6.5.11, this m will satisfy these conditions when $x^n < b$ (so that $x < x_g$) and $x^n > a$ (so that $x > x_l$).

Given this m, let p be the least natural number such that $(p/m)^n \geq b$, and let $x = (p-1)/m$; so

$$\left(x + \frac{1}{m}\right)^n \geq b \quad \text{and} \quad x^n < b.$$

Then when (6.2) applies,

$$x^n > \left(x + \frac{1}{m}\right)^n - (b - a) \geq b - (b - a) = a,$$

so x is a number whose n'th power is between a and b. □

We have found *one* rational number with n'th power between a and b; Theorem 6.5.19 then allows us to find *infinitely many* rational numbers with n'th powers between a and b. To do this, let $a_j = x_{j+1}$ and $b_j = x_j$, where x_j are the rationals between a and b as specified in the proof of Theorem 6.5.19, and then use Theorem 6.5.24 to find a rational whose n'th power lies between a_j and b_j for each $j \in \mathbb{N}$.

[vi] For the sake of brevity, we are not verifying this intuitively obvious fact here; but if you want verification, see Exercise 7.2.1.

6.6   Bounded Sets in $\mathbb{Q}$

The denseness of the rationals has profound implications for the theory of bounded sets in $\mathbb{Q}$, which is completely different from what we have seen in $\mathbb{N}$ and $\mathbb{Z}$. Theorem 6.5.19 immediately yields a contrast to Theorem 4.5.2 in $\mathbb{Z}$: whereas a set of integers that is bounded above and below must be finite, we have found an infinite set of rationals bounded below by a and above by b. Of course, the converse situation *is* the same in $\mathbb{Q}$ as in $\mathbb{Z}$: Theorem 3.7.19 implies that a finite subset of any ordered number system must be bounded above and below, since a greatest/least member is an upper/lower bound [*Definition 3.6.3*]. For subsets of $\mathbb{Z}$ we also had that the existence of an upper/lower bound implied the existence of a greatest/least member [*Theorem 4.5.1*], but this link is broken in $\mathbb{Q}$:

Theorem 6.6.1. *Not every subset of $\mathbb{Q}$ that is bounded above/below has a greatest/least member.*

A bounded subset of $\mathbb{Q}$ may or may not have a greatest/least member, and it is easy to find examples of both cases. Particularly simple examples are, for any given $a \in \mathbb{Q}$:

(i) the set $X_a := \{y \in \mathbb{Q} : y \geq a\}$ is bounded below and has the least member, $\min X_a = a$;

(ii) the set $Y_a := \{y \in \mathbb{Q} : y > a\}$ is bounded below but has no least member.

It is obvious from Definition 3.6.1 of "lower bound" and Definition 3.6.3 of "least member" that $\min X_a = a$. The case of Y_a is more interesting. We shall prove by contradiction that Y_a has no least member, by first supposing that a least member does exist, $l = \min Y_a$. Then $l \in Y_a$ so, by definition of Y_a, $l > a$. Theorem 6.5.19 says that there exist rational numbers between a and l, for example $(a + l)/2$. Now $(a + l)/2 \in Y_a$ because $(a + l)/2 > a$; but also $(a + l)/2 < l$, which contradicts l being the least member of Y_a. So the least member of Y_a cannot exist.

Now recall that in Section 3.6 we defined not only the greatest and least member of a set, but also the supremum and infimum of a set. We proved that if a greatest/least member existed, it would equal the supremum/infimum, but also mentioned that the converse was not true: a supremum/infimum may exist even if there is no greatest/least member. We now show that the set Y_a does have an infimum: $\inf Y_a = a$.

Recall that "infimum" means the greatest lower bound. The definition of "lower bound" shows immediately that a is a lower bound of Y_a. We show that it is the greatest lower bound by showing that a number $k > a$ cannot be a lower bound for Y_a, using a construction very similar to that used to prove that Y_a has no least member. Consider the number $(a + k)/2$: since $(a + k)/2 > a$ we have that $(a + k)/2 \in Y_a$; then since $(a + k)/2 < k$ the number k cannot be a lower bound for Y_a.

The construction of the set Y_a made it particularly easy to use Theorem 6.5.19 in the above proofs relating to least member and infimum (and of course it would be easy to construct a similar set with a supremum but no greatest member, using a similar proof). In some other examples, however, such proofs need to be tailored to the particular set in question.

Example 6.6.2. Show that the set

$$P := \{y \in \mathbb{Q} : y = 1 - \frac{1}{2^n}, n \in \mathbb{N}\}$$

has no greatest member, but that $\sup P = 1$.

Solution. If $\max P$ exists, then $\max P \in P$, so $\max P = 1 - 1/2^k$ for some $k \in \mathbb{N}$. Now, $1 - 1/2^{k+1} \in P$, since $k + 1 \in \mathbb{N}$ if $k \in \mathbb{N}$; and $2^{k+1} > 2^k$, so $1/2^{k+1} < 1/2^k$ [*Theorem 6.5.11*], so that $1 - 1/2^{k+1} > 1 - 1/2^k$ [*Theorem 4.4.16*]. This contradicts $1 - 1/2^k$ being the greatest member of P; so $\max P$ does not exist.

Since $1/2^n > 0$ for all $n \in \mathbb{N}$, we have that 1 is certainly an upper bound for P. Suppose there is an upper bound $u < 1$; then $u = 1 - y$ for some $y \in \mathbb{Q}_+$. Then $1/y > 0$ and from Theorem 6.5.22, $\exists k \in \mathbb{N}$ such that $2^k > 1/y$. Thus $0 < 1/2^k < y$ [*Theorem 6.5.11*] and so $1 - 1/2^k > 1 - y$, which contradicts $1 - y$ being an upper bound for P: there is no upper bound less than 1, so $\sup P = 1$.

$\square$

Exercise 6.6.3. *(a) For each of the following subsets of* $\mathbb{Q}$, *write down the greatest member and the least member if either or both exist. If either or both do not exist, prove the non-existence.*

$$(i) \left\{ x \in \mathbb{Q} : x = n + \frac{1}{n}, n \in \mathbb{N} \right\} \qquad (ii) \left\{ x \in \mathbb{Q} : x = 1 + \frac{1}{n}, n \in \mathbb{N} \right\}$$

$$(iii) \left\{ x \in \mathbb{Q} : x = 1 + \frac{1}{n}, n \in \mathbb{N}, n < 1000 \right\}$$

$$(iv) \left\{ x \in \mathbb{Q}; \ x = \frac{1}{k^3}, k \in \mathbb{N} \right\} \qquad (v) \left\{ x \in \mathbb{Q}; \ x = \frac{1}{k^3}, k \in \mathbb{Z} \backslash \{0\} \right\}$$

$$(vi) \left\{ x \in \mathbb{Q}; \ x = \frac{1}{k^3}, k \in \mathbb{Q}_+ \right\} \qquad (vii) \left\{ x \in \mathbb{Q}; \ x = \frac{1}{3^k}, k \in \mathbb{Z} \right\}$$

(b) For any case in (a) where either the greatest member or the least member does not exist, write down the supremum and/or infimum if they exist. In each such case, prove that your number is indeed the supremum/infimum.

A theme seems to be emerging: we don't need to worry if a subset of $\mathbb{Q}$ that is bounded above/below doesn't have a greatest/least member, because there will always be a supremum/infimum. Unfortunately the next theorem will end our complacency.

Theorem 6.6.4. *Not every subset of* $\mathbb{Q}$ *that is bounded above/below has a supremum/infimum.*

Proof. Let x be a rational number with $x > 1$ and $x \neq r^2$ for any $r \in \mathbb{Q}_+$, so that $\sqrt[2]{x}$ does not exist in $\mathbb{Q}$ according to Theorem 6.4.6. We shall show that the set

$$C_x := \{y \in \mathbb{Q} : y^2 < x\}$$

is bounded above by x, but has no supremum in $\mathbb{Q}$.[vii]

From Lemma 6.5.24, $\exists y \in \mathbb{Q}$ such that $1 < y^2 < x$, so C_x contains numbers greater than 1. For any $y > 1$, we have $y < y^2$; but $y^2 < x$ if $y \in C_x$, and so $y < x \ \forall y \in C_x$. So x is an upper bound for C_x.

[vii] We use a rather general specification for x, to show that a large class of sets have the property of being bounded above but having no supremum. If readers find the proof difficult, they may wish to rewrite it with a particular value of x inserted, e.g. consider the set $C_2 := \{y \in \mathbb{Q} : y^2 < 2\}$ and show that it is bounded above by 2 but has no supremum. Following the method in the main text, if t is a supposed supremum, define $u = (2t + 2)/(t + 2)$ so that $u - t = -(t^2 - 2)/(t + 2)$ and $u^2 - 2 = 2(t^2 - 2)/(t + 2)^2$, and find a contradiction to t being the supremum in each of the cases of $t^2 - 2$ being negative, positive or zero.

Let U_x be the set of upper bounds of C_x. We need to show that U_x has no least member, i.e. that C_x has no supremum. For contradiction, we suppose that U_x does have a least member, t. Clearly $t > 0$. Let

$$u = \frac{tx + x}{t + x},$$

so $u > 0$ also. Then

$$u - t = \frac{tx + x}{t + x} - t = -\frac{t^2 - x}{t + x} \tag{6.4}$$

and

$$u^2 - x = \frac{(tx + x)^2}{(t + x)^2} - x = \frac{(x^2 - x)(t^2 - x)}{(t + x)^2} \tag{6.5}$$

(where the reader can fill in the several lines of algebraic manipulation needed to obtain the right-hand formulae in (6.4) and (6.5)). Observe that $t + x > 0$ and $x^2 - x > 0$ (because $x > 1$), so whether $u - t$ or $u^2 - x$ are positive or negative depend entirely on whether $t^2 - x$ is positive or negative. Specifically:

(i) If $t^2 - x < 0$, then $u^2 - x < 0$ from (6.5) and $u - t > 0$ from (6.4). So $u^2 < x$ and $u > t$, meaning that u is a member of C_x greater than t, contradicting t being an upper bound for C_x.

(ii) If $t^2 - x > 0$, then $u - t < 0$ from (6.4) and $u^2 - x > 0$ from (6.5). So $u < t$ and $u^2 > x$, and from the definition of C_x, the latter inequality implies that $u^2 > y^2 \; \forall y \in C_x$; so from Theorem 4.4.29 we find $u > y \; \forall y \in C_x$, i.e. u is an upper bound for C_x. Since $u < t$, this contradicts t being the *least* upper bound for C_x.

(iii) By trichotomy, the only remaining possibility is $t^2 - x = 0$. But this is not possible since x was defined as having no square root in $\mathbb{Q}$.

So in all cases we have contradicted the existence of a least upper bound (supremum) of C_x in $\mathbb{Q}$. A similar argument can be used to show that C_x is bounded below by $-x$ but has no infimum. □

We can now say that the Rational Number system is *incomplete* in two senses. Firstly, exponentiation is not closed on $\mathbb{Q}$: roots and rational powers of rational numbers frequently cannot be found in $\mathbb{Q}$, or equivalently, equations of the form $x^n = c$ with $c \in \mathbb{Q}$ and $n \in \mathbb{N}$ frequently have no solution for x in $\mathbb{Q}$. Secondly, subsets of $\mathbb{Q}$ that are bounded above/below may have no supremum/infimum. These two issues are clearly related: our proof of Theorem 6.6.4 used the choice of a number x for which $\sqrt[2]{x} \notin \mathbb{Q}$ to show that the set C_x had no supremum, and one can certainly generalise to consider sets of the form $\{y \in \mathbb{Q} : y^n < x\}$ for $n \geq 2$ where $\sqrt[n]{x} \notin \mathbb{Q}$, which similarly have no supremum (although the proof is more difficult if $n > 2$). But one can also find sets whose definition is unrelated to the exponentiation issue, and which still have no supremum/infimum despite being bounded above/below.

This lack of a supremum/infimum may seem to be just an annoying technicality, but in fact it is of fundamental importance. The theory of differential and integral calculus developed in the 17th century was very successful in solving a wide variety of problems in physics and geometry, but during the 18th century it became clear that its logical foundations were not sound. Mathematicians began to seek a more rigorous theory, chief among them being Augustin-Louis Cauchy whose book *Cours d'Analyse*, published in 1821, set out many of the definitions and theorems which still form the basis of our understanding of limits, continuous functions, differentiation, and integration. This body of rigorous theory is now known as **Analysis**, and it includes as a fundamental requirement a **completeness** property. Readers can find several different "axioms of completeness" in different textbooks: they are all equivalent, in the sense that each one can be proved to be true if any of the others is true, but the simplest to state is that "every set that is bounded above/below has a supremum/infimum".[viii] So all the calculus that you may have learned at school is invalid if we only have the rational number system!

We shall define a number system which does satisfy an axiom of completeness in the next chapter, but first we shall see how the place-value system for writing numbers, which was introduced for integers in Section 5.2, can be extended to rationals. This will show where the familiar decimal representation of fractions comes from, but also turns out to have important theoretical ramifications.

6.7   Expressing Rational Numbers in Any Base

The place-value system for writing integers involves choosing a base $b \in \mathbb{N}\backslash\{1\}$, expressing an integer n as $n = r_K b^K + r_{K-1} b^{K-1} + \cdots + r_1 b + r_0$ where $r_k (k = 0, 1, 2, \ldots, K)$ are remainders found on successive divisions by b, and then writing the remainders as a string of digits where the position in the string indicates the power of b by which each r_k should be multiplied to reconstruct the integer n. We now know that negative integer powers of b are rational numbers, so the obvious way to extend the place-value system to rationals is to seek a representation in the form $r_K b^K + r_{K-1} b^{K-1} + \cdots + r_1 b + r_0 + r_{-1} b^{-1} + r_{-2} b^{-2} + \cdots$. As with integers, the procedure for doing this involves repeated use of the Division Theorem, but not in the same way. Also as with integers, we shall consider only positive rationals, with the base-b representation of a negative rational x being obtained by writing a minus sign in front of the representation of the positive rational $-x$.

Given a positive rational number written in fraction form, $\frac{m}{n}$ where $m \in \mathbb{Z}_+$ and $n \in \mathbb{Z}_+$, use the Division Theorem to find the unique quotient q_0 and remainder ρ_0, with $0 \leq \rho_0 < n$, such that

$$m = q_0 n + \rho_0.$$

Recalling that the fraction $\frac{m}{n}$ can be interpreted as the exact division of m by n, we can write the last equation as

$$\frac{m}{n} = q_0 + \frac{\rho_0}{n}. \tag{6.6}$$

We are now only interested in the case where $\frac{m}{n}$ is not an integer, so $\rho_0 > 0$, and we call q_0 the **integer part** of $\frac{m}{n}$.

Now multiply the remainder ρ_0 by the base b and divide the result by n according to the Division Theorem:

$$\rho_0 b = q_1 n + \rho_1. \tag{6.7}$$

Since $\rho_0 b > 0$, we must have $q_1 \geq 0$; also from (6.7) we obtain

$$q_1 = \frac{\rho_0 b}{n} - \frac{\rho_1}{n}$$

and since $\rho_0 < n$ and $\rho_1 \geq 0$, this yields $q_1 < b$. Equation (6.7) also gives

$$\frac{\rho_0}{n} = \frac{q_1}{b} + \frac{\rho_1}{bn},$$

[viii]Modern Analysis covers functions whose domains may be taken from a much wider class of mathematical objects than a number system, for which alternative statements of an axiom of completeness are more suitable. Indeed, one possible definition of Analysis is "the study of completeness"; although my preferred definition is that it is the theory of functions whose domain is an infinite set. If the domain is a finite set, everything about the function can be deduced from a finite list of ordered pairs of domain members and codomain members; but greater subtleties arise when the domain is infinite, and it is then that the importance of completeness of the codomain becomes apparent. The use in this book of a capital "A" for the branch of mathematics known as Analysis is to distinguish it from the many other uses of that word.

which we can substitute into (6.6) to obtain

$$\frac{m}{n} = q_0 + \frac{q_1}{b} + \frac{\rho_1}{bn}.$$

Now repeat the process of multiplying the remainder by the base and dividing by n; at the k'th iteration we obtain

$$\rho_{k-1} b = q_k n + \rho_k$$

with $0 \leq q_k < b$, so that

$$\frac{\rho_{k-1}}{b^{k-1} n} = \frac{q_k}{b^k} + \frac{\rho_k}{b^k n}$$

and

$$\frac{m}{n} = q_0 + \frac{q_1}{b} + \frac{q_2}{b^2} + \cdots + \frac{q_k}{b^k} + \frac{\rho_k}{b^k n} \tag{6.8}$$

which you can verify by induction on k.

Apart from the final term $\rho_k / b^k n$, we have achieved our objective: q_0 is an integer which can be represented in base b using positive powers of b, i.e. as $q_0 = r_K b^K + r_{K-1} b^{K-1} + \cdots + r_1 b + r_0$, and we then have a series of terms involving negative powers of b. We can write (6.8) as

$$\frac{m}{n} = r_K b^K + r_{K-1} b^{K-1} + \cdots + r_1 b + r_0 + q_1 b^{-1} + q_2 b^{-2} + \cdots + q_k b^{-k} + \cdots$$

in which the term $\rho_k / b^k n$ has been replaced with an ellipsis because we are yet to investigate what happens as our procedure continues to arbitrarily large k. We have previously represented integers as a string of digits, $r_K r_{K-1} \ldots r_1 r_0$ where $0 \leq r_i < b$ for $i = 0, 1, 2, \ldots, K$ and each digit is a symbol representing a number between 0 and $b - 1$; since we also have $0 \leq q_j < b$ for $j = 1, 2, \ldots, k$, we can extend the string-of-digits notation by writing the base-b representation of $\frac{m}{n}$ as

$$r_K r_{K-1} \ldots r_1 r_0 \cdot q_1 q_2 \ldots q_k \ldots , \tag{6.9}$$

in which the dot between r_0 and q_1 is known as a **decimal point** when the base is ten; for other bases we shall use the terminology, **base-b point**.

Readers should recognise that the rather long-winded derivation of (6.8) above is simply a detailed explanation of the standard procedure for division of integers to obtain a result in decimal form, but applicable to any base b. This can be seen if we set it out in "short division" style: we first write

$$n \overline{) m \cdot 0\, 0\, 0 \ldots} ,$$

recognising that $m \cdot 000 \ldots$ is a correct representation of the integer m because applying our procedure to the fraction $\frac{m}{1}$ would yield $\rho_0 = 0$ and then $q_1 = 0, \rho_1 = 0$ and all subsequent q's and ρ's would also be zero. The results of the calculations of quotients and remainders in the division of m by n would then be set out as

$$\begin{array}{c} q_0 \cdot\ \ q_1\ \ q_2\ \ q_3 \ldots \\ \hline n \overline{) m \cdot\ {}^{\rho_0}0\, {}^{\rho_1}0\, {}^{\rho_2}0 \ldots} \end{array}$$

(where we have omitted the representation of the integer part q_0 in base b); the quotients $q_1, q_2, q_3, \ldots$ are the digits after the base-b point, and the remainders ρ_0, ρ_1, ρ_2 are written in front of the zeroes to indicate the multiplication by b (since appending a zero at the end of the base-b representation of any integer multiplies it by b).

The generalisation to rational numbers of the rule about appending a zero is that shifting a base-b point one place to the right multiplies a number by b. Such a shift applied to (6.9) gives

$$r_K r_{K-1} \ldots r_1 r_0 q_1 \cdot q_2 \ldots q_k \ldots,$$

and since the point appears between the "units" digit and the digit multiplying b^{-1}, this is shorthand for

$$r_K b^{K+1} + r_{K-1} b^K + \cdots + r_1 b^2 + r_0 b + q_1 + q_2 b^{-1} + q_3 b^{-2} + \cdots + q_k b^{-k+1} + \cdots,$$

which is b times the number represented by (6.9). This is familiar in base ten: for example, we recognise that $10 \times 25 \cdot 03 = 250 \cdot 3$. Similarly, shifting the base-b point one place to the left divides a number by b.

We now investigate what happens as the iteration number k increases: does the process end, or can it continue indefinitely? These are the only two possibilities.

6.7.1 Terminating Base-b Representations

If $\rho_l = 0$ for some $l \in \mathbb{N}$ (with $\rho_k \neq 0$ for all $k < l$), then $q_{l+1} = 0$ and $\rho_{l+1} = 0$, and all subsequent quotients and remainders are also zero. Just as we don't usually need to write an integer m as $m \cdot 000 \ldots$, it is pointless to write a string of zeroes after the last non-zero digit q_l. So the base-b representation terminates l digits after the point.

Under what circumstances does a base-b representation of a fraction terminate? If we set $k = l$ and $\rho_l = 0$ in (6.8), we obtain

$$mb^l = n(q_0 b^l + q_1 b^{l-1} + \ldots + q_l),$$

so that $n \mid mb^l$. Assuming that $\frac{m}{n}$ is in lowest terms, Theorem 5.3.11 then yields

$$n \mid b^l$$

as a necessary and sufficient condition for the base-b representation to terminate l digits after the point. Since the prime factorisation of b^l will contain the same primes as that of b (but raised to higher powers), the base-b representation of $\frac{m}{n}$ will terminate if and only if the prime factorisation of b includes all the primes in the prime factorisation of the denominator n. When this condition is satisfied, we can determine l, the number of digits after the point, as follows. Suppose a prime p_i appears to the power α_i in the factorisation of b; then it will appear to the power $l\alpha_i$ in the factorisation of b^l. So to have $n \mid b^l$ we require the power of p_i in the prime factorisation of n to be no greater than $l\alpha_i$, for each prime p_i in the factorisation of b. The smallest value of l for which this criterion is satisfied for *all* of the primes will give the number of digits after the point.

Finding this value of l is useful when it comes to actually calculating a terminating representation. Noting that

$$\frac{m}{n} = \frac{m \times (b^l/n)}{b^l} \tag{6.10}$$

where b^l/n is an integer because $n \mid b^l$, we simply need to evaluate the integer $m \times (b^l/n)$, which we can think of as having a base-b point after the last digit, and then shift this point l places to the left for the division by b^l. This method is likely to be simpler than using short (or long) division, especially in bases other than ten for which the required multiplication tables are unfamiliar.

Example 6.7.1. Find the base-ten representation of $\frac{129}{80}$.

Solution. First show that the base-ten representation will terminate. Ten has the prime factorisation 2×5, while $80 = 2^4 \times 5$, so the base does include all the primes found in the denominator. For the number of digits after the point, we seek the least value of l such that $(2^4 \times 5) \mid (2 \times 5)^l$. This requires $2^4 \mid 2^l$ and $5 \mid 5^l$, so respectively $l \geq 4$ and $l \geq 1$. Thus $l = 4$.

Using formula (6.10), we then have that

$$\frac{129}{80} = \frac{129 \times (10^4/80)}{10^4} = \frac{129 \times 125}{10^4} = \frac{16125}{10^4} = 1 \cdot 6125. \quad \square$$

Example 6.7.2. Find the base-9 representation of $\left[\frac{38}{243}\right]_{\text{ten}}$.

Solution. The square bracket with subscript "ten", indicates that the numerator and denominator of the fraction are both represented in base ten. This notation is needed here for clarity because the example requires us to produce a representation in a different base from that in which the fraction is written. We can do our calculations in base ten, and only convert to base 9 near the end of the process.

The prime factorisations of the base and the denominator are respectively $9 = 3^2$ and $243 = 3^5$, so the base does include all prime factors in the denominator. Thus the base-9 representation will terminate with l digits after the point, where l is the least integer such that $3^5 \mid (3^2)^l$. This requires $2l \geq 5$, so $l = 3$. So using (6.10) we have (working in base ten)

$$\left[\frac{38}{243}\right]_{\text{ten}} = \left[\frac{38 \times (9^3/243)}{9^3}\right]_{\text{ten}} = \left[\frac{38 \times 3}{9^3}\right]_{\text{ten}} = \frac{114_{\text{ten}}}{b^3}.$$

If we can write the numerator in the required base (9), we then need to shift the point 3 places to the left for the division by b^3. Proceeding as in example 5.2.1,

$$114 = 12 \times 9 + 6$$
$$12 = 1 \times 9 + 3$$
$$1 = 0 \times 9 + 1;$$

so $114_{\text{ten}} = 136_9$ and

$$\left[\frac{38}{243}\right]_{\text{ten}} = \frac{136_9}{b^3} = 0 \cdot 136_9. \quad \square$$

To indicate that the integer part is zero, we have had to insert a zero before the point, even though there was no leading zero in the numerator 136. Zeroes may sometimes be needed between a base-b point and the first digit of the numerator in $\frac{m \times (b^l/n)}{b^l}$ if the numerator has fewer digits than l. For example, in finding a base-9 representation of $\left[\frac{38}{2187}\right]_{\text{ten}}$, where $2187 = 243 \times 9 = 3^7$, we would obtain

$$\left[\frac{38}{2187}\right]_{\text{ten}} = \frac{136_9}{b^4},$$

and shifting the base-9 point 4 places to the left in 136_9 would give $0 \cdot 0136_9$.

Exercise 6.7.3. *(i) Find the smallest base b in which $\left[\frac{31}{72}\right]_{\text{ten}}$ has a terminating base-b expansion, and find this expansion.*

(ii) List all the other bases less than seventy-two in which $\left[\frac{31}{72}\right]_{\text{ten}}$ would have a terminating expansion. For each such base, state the number of digits after the point in the expansion.

Exercise 6.7.4. *If a fraction (expressed in lowest terms) has a base-b expansion that termi-nates 2 digits after the point, what is/are the possible value(s) of the fraction's denominator for each of the following bases? (a) $b = 7$; (b) $b = 8$; (c) $b = $ twenty-four.*

Reconstructing a fraction from a terminating base-b representation is almost trivial. Reversing the procedure of shifting the decimal point, we have

$$r_K r_{K-1} \cdots r_1 r_0 \cdot q_1 q_2 \cdots q_l = \frac{r_K r_{K-1} \cdots r_1 r_0 q_1 q_2 \cdots q_l}{b^l}$$

in which the numerator is the base-b representation (with $K + 1 + l$ digits) of an integer. There may be common factors between the numerator and denominator, which can be cancelled to obtain the fraction in lowest terms.

Example 6.7.5. Find the fraction (with numerator and denominator represented in base ten) represented by: *(a)* $7 \cdot 244_{\text{ten}}$ *(b)* $7 \cdot 244_8$

Solution. (a) With all numbers represented in base ten,

$$7 \cdot 244 = \frac{7244}{10^3} = \frac{1811}{250}$$

where the final result in lowest terms was obtained by cancelling a common factor of 4.
(b)

$$7 \cdot 244_8 = \frac{7244_8}{b^3}$$

where $b = 8$. To obtain the numerator in a base-ten representation,

$$7244_8 = 7 \times 8^3 + 2 \times 8^2 + 4 \times 8 + 4 = 3748_{\text{ten}}.$$

The denominator is $8^3 = 512$. So

$$7 \cdot 244_8 = \left[\frac{3748}{512}\right]_{\text{ten}} = \left[\frac{937}{128}\right]_{\text{ten}} . \quad \square$$

Exercise 6.7.6. *Find the fraction (with numerator and denominator represented in base ten) represented by: (a)* $37 \cdot 575_{\text{ten}}$ *(b)* $37 \cdot 575_{\text{twelve}}$

6.7.2 Repeating Base-b Representations

If the denominator in $\frac{m}{n}$ has any prime factor(s) not found in the base b, then we can never have $n \mid b^k$ for any $k \in \mathbb{N}$, and so the remainder ρ_k in (6.8) can never be zero. This leaves $n - 1$ possible remainders in the repeated divisions by n. Starting with ρ_0 and generating remainders indefinitely, the Pigeonhole Principle then ensures that some value must appear more than once among the first n remainders, $\{\rho_0, \rho_1, \ldots, \rho_{n-1}\}$: so $\rho_{s+t} = \rho_s$ for some $s \geq 0$ and $s + t \leq n - 1$, where s is taken to be the index of the *first* appearance of a remainder value that subsequently reappears, and t to be the number of iterations until its *next* appearance. This t is called the **period** of the base-b representation: clearly $t \leq n - 1$. Now, the value of a remainder determines the values of all successive quotients and remainders in the process; so if a particular value of remainder is repeated, then the following sequence of quotients that provide the digits of the base-b representation will also be repeated; and this repetition of the same sequence of quotients will continue indefinitely. So after the base-b point, the representation will consist of s non-repeating digits, followed by an ordered set of t digits which is repeated indefinitely. To display such a repeating representation, we write the repeating digits only once but with an overbar, as in the following examples.

Example 6.7.7. Find the base-ten representation of $\frac{19}{66}$.

First observe that the denominator 66 contains prime factors (3 and 11) not found in the base. Working in base ten can easily be done in "short division" style:[ix]

$$
\begin{array}{r}
0\cdot\ \ 2\ \ \ 8\ \ \ 7 \\
66\ )\overline{19\cdot\ ^{19}0\ \ ^{58}0\ \ ^{52}0\ \ ^{58}0}
\end{array}
$$

where we have stopped the calculation because a remainder value, 58, has reappeared. So we know that the quotients since the first appearance of this remainder will form a repeating set, and the final base-ten representation is

$$
\frac{19}{66} = 0\cdot 2\overline{87},
$$

where the overbar indicates that we have an indefinitely repeating representation, $0\cdot 2878787\ldots$.[x] The period is $t = 2$, and the number of non-repeating digits between the point and the repeating set is $s = 1$.

Example 6.7.8. Find the base-7 representation of $\left[\frac{129}{80}\right]_{\text{ten}}$.

Solution. We have already found a base-ten representation of this fraction (in Example 6.7.1), but that is no help when finding a representation in a different base. Since short division in bases other than ten is unfamiliar, we set out the calculation to show the repeated use of the Division Theorem on "remainder $\times$ base", doing all the working in base ten.

$$
\begin{array}{rcl}
129 &=& 1 \times 80 + 49 \\
49 \times 7 = 343 &=& 4 \times 80 + 23 \\
23 \times 7 = 161 &=& 2 \times 80 + 1 \\
1 \times 7 = 7 &=& 0 \times 80 + 7 \\
7 \times 7 = 49 &=& 0 \times 80 + 49.
\end{array}
$$

The remainder value 49 has reappeared, having first appeared in the initial division which found the integer part to be 1. So all the quotients found after that initial division form a repeating set: we have

$$
\left[\frac{129}{80}\right]_{\text{ten}} = 1\cdot \overline{4200}_7
$$

with the overbar indicating that the representation continues indefinitely:

$$
1\cdot 420042004200\ldots_7.
$$

The period is 4 and there are no non-repeating digits after the point ($s = 0$).   □

[ix] In case readers are unclear about what is being done, here is the same calculation in longhand according to the procedure using the Division Theorem, as detailed above equation (6.8), with all working in base ten:

$$
\begin{array}{rcl}
19 &=& 0 \times 66 + 19 \\
19 \times 10 = 190 &=& 2 \times 66 + 58 \\
58 \times 10 = 580 &=& 8 \times 66 + 52 \\
52 \times 10 = 520 &=& 7 \times 66 + 58.
\end{array}
$$

[x] An alternative notation is to put a dot above the first and last digits in the repeating set: in our example we would write $0\cdot 2\dot{8}\dot{7}$.

Exercise 6.7.9. *For each of the following fractions and bases: (a) state whether the expansion of the fraction in the given base will terminate or repeat; (b) if it will terminate, first determine the number of digits after the point and then find the expansion; but if it will repeat, find the expansion and then state its period. Note that the numerators and denominators of the fractions are all written in base ten, and the stated bases are those in which you are required to find a representation.*

(i) $\dfrac{7917}{2500}$, *base ten* (ii) $\dfrac{31}{72}$, *base ten* (iii) $\dfrac{31}{72}$, *base 5* (iv) $\dfrac{3003}{1024}$, *base 8* (v) $\dfrac{248}{13}$, *base 9*

In both Examples 6.7.7 and 6.7.8, the value of $s+t$ was well below its maximum allowed value of $n-1$, where n is the denominator of the fraction. We may ask whether there is any way to predict the period t and the number s of non-repeating digits purely from the base b and the value of n. It turns out that we cannot definitely predict the period, but some simple number theory allows us to find s and place some restrictions on t.

First consider the case where b and n are coprime. In the procedure for generating the base-b representation of $\frac{m}{n}$ by equation (6.8), the remainders were given by

$$\begin{aligned} \rho_0 &= m - q_0 n \\ \rho_k &= \rho_{k-1}b - q_k n \qquad (k \in \mathbb{N}). \end{aligned}$$

If $\frac{m}{n}$ is in lowest terms, the first of these equations implies that ρ_0 is coprime with n, and induction on k using the second equation then implies that all remainders ρ_k are coprime with n when b is coprime with n. The second equation also yields

$$\rho_k \equiv \rho_{k-1}b \pmod{n}$$

from which

$$\rho_{l+t} \equiv \rho_l b^t \pmod{n} \quad \text{for any } l \in \mathbb{N} \cup \{0\} \quad \text{and } t \in \mathbb{N} \tag{6.11}$$

by induction on t. If t is the period of a base-b representation, then $\rho_{l+t} = \rho_l$ for some l, so we have

$$\rho_l \equiv \rho_l b^t \pmod{n}.$$

Since the remainders are coprime with n, Corollary 5.6.3 allows us to cancel ρ_l to obtain

$$b^t \equiv 1 \pmod{n}. \tag{6.12}$$

Substituting this into equation (6.11) with $l = 0$, we obtain

$$\rho_t \equiv \rho_0 \pmod{n}$$

which tells us that the repeating set of digits will start with q_1, the first digit after the point: there are no non-repeating digits if b and n are coprime, as was found in Example 6.7.8 where $b = 7$ and $n = 80$.

Suppose now that b and n are not coprime, and consider fractions of the form mb^s/n. For some $s \in \mathbb{N}$, all prime factors in n that are in common with b can be cancelled out, leaving the denominator coprime with b. So this fraction will have no non-repeating digits after the point. To recover the original fraction $\frac{m}{n}$ we need to divide by b^s, which shifts the point s places to the left, introducing s non-repeating digits. In Example 6.7.7, the denominator 66 shares a common factor of 2 with the base 10. We only need to multiply by the first power of the base to cancel this common factor from the denominator:

$$\frac{19 \times 10}{66} = \frac{19 \times 2 \times 5}{2 \times 33} = \frac{19 \times 5}{33}$$

in which the denominator has no common factors with the base. Hence there is one non-repeating digit after the point, as indeed was found.

Clearly we can only have non-repeating digits ($s \geq 1$) if the denominator n is composite; for if n is prime, it is either a factor of b, in which case we have a terminating expansion, or it shares no common factors with b. Furthermore, if n is prime and not a factor of b we can find the possible values of the period t using the following theorem due to Pierre de Fermat.[xi]

Theorem 6.7.10. *If p is prime and b is not a multiple of p, then*

$$b^{p-1} \equiv 1 \;(\mathrm{mod}\; p).$$

Proof. We use modular arithmetic in $\mathbb{Z}_p$: let b' be the remainder when b is divided by p, so that $[b'] \in \mathbb{Z}_p$ and b' is coprime with p (because p is prime). In proving Theorem 5.6.1 we showed that it was not possible to have $b'x_1 \equiv b'x_2 \;(\mathrm{mod}\; p)$ with $1 \leq x_2 < x_1 \leq p-1$ when b' is coprime with p. In other words, the function

$$f_b : \mathbb{Z}_p \backslash \{[0]\} \to \mathbb{Z}_p \backslash \{[0]\}$$

defined by

$$f_b : [x] \mapsto [b'] \times [x]$$

is an injection, with domain and codomain having the same cardinality, $p-1$. Hence by Theorem 3.7.18, it is a bijection: each element of $\mathbb{Z}_p \backslash \{[0]\}$ appears once in the set of $[x]$ values and once in the set of values of $[b'] \times [x]$. So the products of all members of each set are identical:

$$\prod_{x=1}^{p-1} [b'] \times [x] \;=\; \prod_{x=1}^{p-1} [x]$$

$$\therefore [b']^{p-1} \prod_{x=1}^{p-1} [x] \;=\; \prod_{x=1}^{p-1} [x].$$

Now, each x in the products is coprime with p (because $x < p$ and p is prime), so the product of all the x's is coprime with p; so we can use Corollary 5.6.3 to cancel the product from our last equation to yield $[b']^{p-1} = [1]$. Since by definition $b' \equiv b \;(\mathrm{mod}\; p)$, this implies $b^{p-1} \equiv 1 \;(\mathrm{mod}\; p)$. $\square$

Now, the condition $b^t \equiv 1 \;(\mathrm{mod}\; n)$ is *sufficient* for remainder values to repeat after t divisions, from equation (6.11); and if $b^t \equiv 1 \;(\mathrm{mod}\; n)$, then $b^{jt} \equiv 1 \;(\mathrm{mod}\; n)$ for any $j \in \mathbb{N}$. But if the denominator n of a fraction is prime and b is not a multiple of n, Theorem 6.7.10 tells us that $b^{n-1} \equiv 1 \;(\mathrm{mod}\; n)$. So the period t of the base-b expansion of the fraction must satisfy $jt = n - 1$ for some $j \in \mathbb{N}$: the period is a divisor of $n - 1$. In cases where $j = 1$, the period is $n - 1$, the maximum value allowed by the Pigeonhole Principle; but there is no way of determining for which prime denominators n and which bases b this occurs, without actually finding the base-b representation. Indeed, the study of periods of base-b representations is a fascinating area of Number Theory.

Example 6.7.11. How many non-repeating digits are there after the point, and what is the period, in the base-twelve representation of $\left[\frac{7}{480}\right]_{\mathrm{ten}}$?

[xi]This theorem is commonly known as *Fermat's Little Theorem*, which seems a rather inappropriate designation: although the statement and proof of the theorem are quite simple, it is of great importance in Number Theory.

Solution. Writing all numbers in base ten, we have $b = 12 = 2^2 \times 3$ and $n = 480 = 2^5 \times 3 \times 5$. The base-twelve representation will be repeating, since 480 contains a prime factor, 5, that is not found in 12. But 480 is not coprime with 12, so we seek s such that the denominator in $7 \times 12^s/480$ (when reduced to lowest terms) *is* coprime with 12. From the prime factorisations, this means that $(2^5 \times 3)$ must be a divisor of $(2^2 \times 3)^s$. Considering powers of 2, this requires $s = 3$ (the powers of 3 would only require $s = 1$). There are 3 non-repeating digits.

The repeating part of the expansion is determined by the factor of the denominator that has not been cancelled, which is 5. Because this is prime, we know that the period must be a divisor of $5 - 1$: possible values of period are 4, 2, and 1.

You may like to check that in fact $\left[\frac{7}{480}\right]_{\text{ten}} = 0 \cdot 021\overline{2497}_{\text{twelve}}$, so $s = 3$ and $t = 4$ in accord with our deductions. $\square$

Exercise 6.7.12. *Without calculating the representation, determine the number of non-repeating digits and the possible values of period for the base-ten representation of* $\frac{83}{1750}$*.*

6.7.3   Fractions from Repeating Base-b Representations

Given a repeating base-b representation, can you reconstruct the fraction that it represents? A representation with s non-repeating digits and period t is written

$$q_0 \cdot q_1 \cdots q_s \overline{q_{s+1} \cdots q_{s+t}}$$

(where we haven't written out the integer part q_0 in base b). This is shorthand for

$$q_0 + \frac{q_1}{b} + \cdots + \frac{q_s}{b^s} + \left(\frac{q_{s+1}}{b^{s+1}} + \cdots + \frac{q_{s+t}}{b^{s+t}}\right) + \left(\frac{q_{s+1}}{b^{s+t+1}} + \cdots + \frac{q_{s+t}}{b^{s+2t}}\right) + \cdots$$

where the groups of terms in parentheses contain the same set of repeating digits $q_{s+1} \cdots q_{s+t}$ but with the denominator increasing by a factor of b^t from one group to the next; so it can be written more concisely as

$$q_0 + \frac{q_1}{b} + \cdots + \frac{q_s}{b^s} + \left(\frac{q_{s+1}}{b^{s+1}} + \cdots + \frac{q_{s+t}}{b^{s+t}}\right)\left(1 + \frac{1}{b^t} + \frac{1}{b^{2t}} + \cdots\right). \tag{6.13}$$

Now multiply the last factor by $(b^t - 1)$:

$$(b^t - 1)\left(1 + \frac{1}{b^t} + \frac{1}{b^{2t}} + \cdots\right) = \left(b^t + 1 + \frac{1}{b^t} + \cdots\right) - \left(1 + \frac{1}{b^t} + \frac{1}{b^{2t}} + \cdots\right)$$
$$= b^t$$

in which the terms 1 and $1/b^t$ certainly cancel between the two parentheses on the right, and we trust that because the ellipsis indicates that the pattern of terms of the form $1/b^{jt}$ continues in both parentheses, all subsequent terms also cancel. But in mathematics we should not take anything on trust; indeed, it is not at all clear what it means to add a series of terms that continues indefinitely. This will be investigated in Section 6.8; for now, you are asked to accept the validity of the calculation presented here, and await its proof below. The last equation suggests that we can write

$$1 + \frac{1}{b^t} + \frac{1}{b^{2t}} + \cdots = \frac{b^t}{b^t - 1},$$

so that our base-b representation can be written

$$q_0 + \frac{q_1}{b} + \cdots + \frac{q_s}{b^s} + \left(\frac{q_{s+1}}{b^{s+1}} + \cdots + \frac{q_{s+t}}{b^{s+t}}\right)\frac{b^t}{b^t - 1}. \tag{6.14}$$

This is certainly a rational number: what we have shown is:

Theorem 6.7.13. *Every rational number has a terminating or repeating representation in any base, and every terminating or repeating representation in any base corresponds to a rational number.*

Expression (6.14) could be treated as a formula for the rational number represented by $q_0 \cdot q_1 \cdots q_s \overline{q_{s+1} \cdots q_{s+t}}$, but is rather cumbersome. To reconstruct the rational number, the essential idea is the multiplication by $(b^t - 1)$. So first multiply $q_0 \cdot q_1 \cdots q_s \overline{q_{s+1} \cdots q_{s+t}}$ by b^t, which shifts the base-b point t places to the right, and then subtract $q_0 \cdot q_1 \cdots q_s \overline{q_{s+1} \cdots q_{s+t}}$. Since the shift of t places leaves repeating digits in the same position, the subtraction will cancel out the indefinitely repeating digits, as in the following example.

Example 6.7.14. Find the fraction in base ten which is represented in base 7 by $0 \cdot 3\overline{354}_7$.

Solution. The period is $t = 3$, with 1 non-repeating digit after the point. Multiplying by b^3 (where $b = 7$) we have $335 \cdot \overline{4354}_7$, which still has the repeating digits in the second to fourth positions after the point. The subtraction gives[xii]

$$(b^3 - 1) \times 0 \cdot 3\overline{354}_7 = 335 \cdot 4\overline{354}_7 - 0 \cdot 3\overline{354}_7$$
$$= 335 \cdot 1_7.$$

To convert to base ten, first note that

$$335 \cdot 1_7 = \frac{3351_7}{7}$$

and then that

$$3351_7 = 3 \times 7^3 + 3 \times 7^2 + 5 \times 7 + 1 = 1212_{\text{ten}}.$$

Finally recall that $\frac{3351_7}{7}$ is $(b^t - 1)$ times the required number, where $b^t - 1 = 7^3 - 1 = 342_{\text{ten}}$. So

$$0 \cdot 3\overline{354}_7 = \frac{1212_{\text{ten}}}{7 \times 342_{\text{ten}}} = \left[\frac{202}{399}\right]_{\text{ten}}$$

in lowest terms.   □

Exercise 6.7.15. *Find the fractions in lowest terms in base ten that are represented by the following representations in various bases:*
   (i) $2 \cdot 8\overline{37}_{\text{ten}}$ *(ii)* $2 \cdot 8\overline{37}_{\text{twelve}}$ *(iii)* $0 \cdot 01\overline{011000}_2$ *(iv)* $106 \cdot 2\overline{43}_7$

There is a special case of repeating base-b representations which deserves attention. This is where the period is 1 and the single repeating digit is $b - 1$; so in base ten the repeating digit would be 9. Setting $t = 1$ and $q_{s+1} = b - 1$ in our base-b representation (6.14), it becomes

$$q_0 + \frac{q_1}{b} + \cdots + \frac{q_s}{b^s} + \frac{b-1}{b^{s+1}} \times \frac{b}{b-1},$$

[xii]In this example the subtraction is straightforward even in base 7, since there is no carrying to the next column to the left, *cf.* Example 5.2.4. More generally, one does need to be aware of the base when doing this step.

which simplifies to

$$q_0 + \frac{q_1}{b} + \cdots + \frac{q_s + 1}{b^s},$$

This can be written as a single fraction with denominator b^s, which in turn has a *terminating* base-b representation with s digits after the point, the last digit being $(q_s + 1)$.

Example 6.7.16. In base ten, consider $44 \cdot 51\overline{9}$, which has $s = 2$, $t = 1$, and the repeating digit $q_{s+1} = 9 = b - 1$. Noting also that the last non-repeating digit is $q_s = 1$, so that $q_s + 1 = 2$, the rational number represented by $44 \cdot 51\overline{9}$ is

$$44 + \frac{5}{10} + \frac{2}{10^2},$$

which has the terminating representation $44 \cdot 52$.

Note that if the repeating digit $(b - 1)$ follows the base-b point immediately (with no non-repeating digits), then the digit $(q_s + 1)$ in the terminating representation becomes $q_0 + 1$, i.e. 1 is added to the integer part. For example, in base ten $299 \cdot \overline{9}$ is a representation of the number $299 + 1 = 300$.

We could view this "special case" as indicating that every rational number, including integers, has a repeating representation in any base; indeed, if it has a terminating representation, then it also has *two* repeating representations, since we terminated an expansion when all subsequent quotients (representing digits) were zero. So in the base-ten examples above, 44.52 can also be written $44 \cdot 52\overline{0}$ or $44 \cdot 51\overline{9}$, and the integer 300 can be written $300 \cdot \overline{0}$ or $299 \cdot \overline{9}$. To understand why these various representations of the same number are possible, and indeed to properly understand what an indefinitely repeating expansion means at all, we need to introduce some Analysis, specifically the concepts of Sequences and Series.

6.8 Sequences and Series

We have suggested that Analysis can be defined as the study of functions whose domain is an infinite set. The simplest kind of infinite set is $\mathbb{N}$:[xiii] so the following definition makes a **sequence** the simplest concept in Analysis.

Definition 6.8.1. A **sequence** is a function whose domain is $\mathbb{N}$.

In this book the codomain $\mathbb{S}$ of a sequence $f : \mathbb{N} \to \mathbb{S}$ will always be a number system, for example, $\mathbb{Q}$ in which case we refer to "a sequence of rational numbers"; but more generally one may have sequences of mathematical objects from other kinds of codomain sets. In any case, the values of $f(n)$ for each $n \in \mathbb{N}$ are called the **terms** of the sequence.

Note on notation. *Having defined a sequence as a type of function, we have used the standard notation for functions. This seems the obvious thing to do, but in fact it is usual practice to write the terms of a sequence f as f_n (with the independent variable as a subscript rather than in parentheses), and to denote the sequence as a whole by (f_n). This notation has some advantages, but we shall stick with the standard function notation.*

If we have a sequence of numbers from some number system $\mathbb{S}$, we can add those numbers:

[xiii]Firstly, $\mathbb{N}$ is the smallest infinite set, in the sense of Theorem 3.7.25. Furthermore, subsets of $\mathbb{N}$ are infinite if and only if they have no upper bound (Theorem 3.7.20). So we could say that $\mathbb{N}$ is only infinite "in the upwards direction"; this is in contrast to $\mathbb{Q}$ which, although its cardinality is the same as $\mathbb{N}$, could be described as "internally infinite", in that we can find an infinite set of rational numbers in between any two given rationals (Theorem 6.5.19).

Definition 6.8.2. If $f : \mathbb{N} \to \mathbb{S}$ is a sequence, then for each $n \in \mathbb{N}$ we define the n'th **partial sum** of the sequence to be

$$\sum_{k=1}^{n} f(k).$$

Since each partial sum is a number in $\mathbb{S}$, we have just defined a new sequence:

Definition 6.8.3. A sequence of partial sums is called a **series**. Given a sequence $f : \mathbb{N} \to \mathbb{S}$, the corresponding series is the function $\sigma_f : \mathbb{N} \to \mathbb{S}$ defined by

$$\sigma_f : n \mapsto \sum_{k=1}^{n} f(k).$$

The relevance of these definitions to repeating base-b representations should now be becoming apparent. We wrote such representations in the form (6.13) which involved the quantity

$$1 + \frac{1}{b^t} + \frac{1}{b^{2t}} + \cdots,$$

in which the ellipsis indicated that we could add terms of the form $1/b^{jt}$ (with $j \in \mathbb{N}$) indefinitely. With a little sleight of hand, we suggested that this sum of countably infinitely many terms was in fact equal to $b^t/(b^t - 1)$. But we can now recognise that what we have is a series: in the notation of Definition 6.8.3, it is the series $\sigma_f : \mathbb{N} \to \mathbb{Q}$ consisting of partial sums of the sequence with $f(n) = 1/b^{(n-1)t}$ (where we have written the power of b as $(n-1)t$ so that the first term (with $n = 1$) is $1/b^0 = 1$ as required). So to give our sleight of hand some rigorous underpinning, we need some theory of series.

The problem is that we don't know what it means to add infinitely many terms; this is not merely a matter of practicality, that we don't have an infinite amount of time to do infinitely many additions; it a conceptual difficulty. Since a series is just a particular kind of sequence (i.e. of partial sums of some other sequence), our problem is equivalent to defining the "infinity'th term of a sequence". So we are looking at the behaviour of $f(n)$ as n becomes larger without bound. In some sequences, $f(n)$ appears to approach a particular numerical value; for example, in the sequence $f : \mathbb{N} \to \mathbb{Q}$ defined by $f(n) = \frac{1}{n}$, the value of $f(n)$ appears to approach 0 as n becomes larger; but $\frac{1}{n}$ is not actually equal to 0 for any $n \in \mathbb{N}$. The sense in which $\frac{1}{n}$ "appears to approach" 0 is that the difference between $\frac{1}{n}$ and 0 becomes smaller as n becomes larger; indeed, smaller than any desired "margin of error". Suppose we set our maximum allowed margin of error to be some positive rational number ϵ; so $\epsilon = \frac{p}{q}$ for some $p, q \in \mathbb{Z}_+$. Then if $n > q$, we have

$$\frac{1}{n} < \frac{1}{q} \qquad\qquad [\textit{Theorem 6.5.11}]$$
$$\leq \frac{p}{q} = \epsilon;$$

and since $\frac{1}{n} > 0 > -\epsilon$, we can write

$$-\epsilon < \frac{1}{n} - 0 < \epsilon \quad \text{whenever } n > q, \qquad\qquad (6.15)$$

to show that the margin of error has not been exceeded on either the positive or negative side. We can write (6.15) more concisely using modulus signs, as

$$\left| \frac{1}{n} - 0 \right| < \epsilon \quad \text{whenever } n > q.$$

If we set $\epsilon = \frac{1}{100}$, the difference between $\frac{1}{n}$ and 0 would be less than ϵ when $n > 100$; if we set $\epsilon = \frac{1}{1000}$ we would need $n > 1000$ in order to ensure that the difference between $\frac{1}{n}$ and 0 is less than ϵ; but the important thing to note is that, however small we set ϵ, we can always make the difference between $\frac{1}{n}$ and 0 smaller by ensuring that n is large enough. This motivates the following definition:

Definition 6.8.4. A sequence $f : \mathbb{N} \to \mathbb{S}$ **converges** to a **limit** $L \in \mathbb{S}$ if, for any $\epsilon > 0$, there exists $N \in \mathbb{N}$ such that

$$|f(n) - L| < \epsilon \quad \text{for all} \quad n > N.$$

In such a case we write

$$f(n) \to L \text{ as } n \to \infty$$

(or simply $f(n) \to L$), or

$$\lim_{n \to \infty} f(n) = L.^{\text{xiv}}$$

So we are allowed to choose the margin of error ϵ as small as we like, and convergence requires the difference (positive or negative) between sequence values $f(n)$ and the supposed limit L to be smaller than our chosen ϵ if we take n to be large enough.[xv] We have already seen this to be true for the case where $f(n) = \frac{1}{n}$ and $L = 0$. But our immediate objective is to put our understanding of repeating base-b representations of rational numbers on a rigorous footing; so we would like to show that the series

$$1 + \frac{1}{b^t} + \frac{1}{b^{2t}} + \cdots$$

converges to $b^t/(b^t - 1)$, because the concept of convergence is the only way to give meaning to the addition of infinitely many terms.

Now, according to Definition 6.8.3, convergence of the *series* means that the *sequence of partial sums,*

$$\sigma_f(n) = \sum_{k=1}^{n} \frac{1}{b^{(k-1)t}}$$

converges. To determine that this convergence does happen, it is useful to have an explicit formula for the partial sums. Now, multiplication of the entire series by $(b^t - 1)$ gave us the formula $b^t/(1 - b^t)$ for what we expect the series to converge to, but actually that process was not valid with the sum of infinitely many terms; whereas we can multiply a *partial sum* $\sum_{k=1}^{n} 1/b^{(k-1)t}$ (the sum of finitely many terms) by $(b^t - 1)$ to obtain $b^t - 1/b^{(n-1)t}$ (with all other terms cancelling), so that

$$\sum_{k=1}^{n} \frac{1}{b^{(k-1)t}} = \frac{b^t - 1/b^{(n-1)t}}{b^t - 1},$$

which you should verify by induction on n. So by Definition 6.8.4, convergence of the

---

[xiv] The symbol ∞ ("infinity") used here is simply a part of the notation for convergence of a sequence. ∞ is not a number, and should never be treated as such.

[xv] The modulus in the definition embodies the idea of "difference (positive or negative)". This idea works when the codomain of f is a number system that is ordered and also dense, in the sense of Theorem 6.5.19. For more general codomains, we replace the idea of "difference" with a *metric*, which defines a concept of "distance" between two elements in the codomain. The metric is a mapping from pairs of elements in the codomain to an ordered, dense number system.

sequence of partial sums to $b^t/(1 - b^t)$ requires that, for any $\epsilon > 0$, there exists $N \in \mathbb{N}$ such that

$$\left| \frac{b^t - 1/b^{(n-1)t}}{b^t - 1} - \frac{b^t}{b^t - 1} \right| < \epsilon$$

whenever $n > N$, which simplifies to the requirement

$$\left| \frac{1}{b^{(n-1)t}(b^t - 1)} \right| < \epsilon. \tag{6.16}$$

Now, since $b \geq 2$ and $t \geq 1$, we have $b^t \geq 2$, so that $b^t - 1 \geq 1$. Also, since $b^{(n-1)t} = (b^t)^{n-1}$, Theorem 6.5.22 with $b^t \geq 2$ yields that there exists $N \in \mathbb{N}$ such that $b^{(N-1)t} > 1/\epsilon$. Then whenever $n > N$,

$$b^{(n-1)t}(b^t - 1) > b^{(N-1)t}(b^t - 1)$$

$$> \frac{1}{\epsilon} \times 1$$

so that

$$0 < \frac{1}{b^{(n-1)t}(b^t - 1)} < \epsilon;$$

thus our requirement (6.16) is fulfilled.

We have shown that a repeating base-b representation of a rational number is actually a series that converges to that rational number. Our proof is rather untypical of proofs of convergence of series; in general, we may not know what limit a series is supposed to converge to, and we may not have a convenient formula for partial sums of the series. There is nevertheless a large body of theory that enables one to prove (or disprove) convergence of a wide variety of sequences and series. We shall derive a few useful theorems as and when they are needed in later chapters, but for a thorough treatment of sequences and series you should refer to Analysis textbooks.

Investigations

1. This investigation considers bijections from $\mathbb{N} \times \mathbb{N}$ to $\mathbb{N}$ and more generally from $\mathbb{N}^k$ to $\mathbb{N}$, where $\mathbb{N}^k$ is the Cartesian product of k copies of $\mathbb{N}$. Theorem 6.3.2 gave an *injection* (with $A_1 = A_2 = \cdots = A_k = \mathbb{N}$), whereas we now seek explicit *bijections*.

   (a) First consider the algorithm used to biject $\mathbb{Q}$ to $\mathbb{N}$ at the end of Section 6.3. Without the complications involved with $\mathbb{Q}$ (the need for coprimeness of two integers, and the inclusion of negative rationals), it is possible to find an explicit formula for a bijection $f : \mathbb{N} \times \mathbb{N} \to \mathbb{N}$.

      (i) Show that, for any $q \in \mathbb{N} \setminus \{1\}$ there are $\frac{1}{2}q(q - 1)$ pairs $(m, n) \in \mathbb{N} \times \mathbb{N}$ such that $m + n \leq q$.

      (ii) Hence obtain a formula for the natural number obtained for a given pair (m, n) by the algorithm in Section 6.3 (without the consideration of coprimeness or inclusion of negative m). [Hint: first obtain a formula in terms of q and n, and then substitute $q = m + n$.]

(b) Now consider a method based on unique prime factorisation.

   (i) Explain why the function $f_2 : \mathbb{N} \times \mathbb{N} \to \mathbb{N}$ defined by

$$f_2 : (a, b) \mapsto 2^{a-1}(2b - 1)$$

   is a bijection.

   (ii) Find a function $h_3(c)$ that will make

$$g_3 : \mathbb{N} \times \mathbb{N} \times \mathbb{N} \to \mathbb{N}, \quad (a, b, c) \mapsto 2^{a-1}3^{b-1}h_3(c)$$

   a bijection. Your $h_3(c)$ should be an increasing function, i.e. $h_3(c_2) > h_3(c_1)$ if $c_2 > c_1$, and may have a multi-part definition, i.e. of the form

$$h(c) = \begin{cases} \text{[formula 1]} & \text{if} \quad \text{[condition 1]} \\ \text{[formula 2]} & \text{if} \quad \text{[condition 2]} \\ \quad \cdots & \cdots & \cdots \end{cases}$$

[5]

   (iii) Using a similar method, construct a bijection

$$f_4 : \mathbb{N} \times \mathbb{N} \times \mathbb{N} \times \mathbb{N} \to \mathbb{N}, \quad (a, b, c, d) \mapsto 2^{a-1}3^{b-1}5^{c-1}h_4(d)$$

   with $h_4(d)$ as an increasing function to be found. Find the values of a, b, c, d such that $f_4(a, b, c, d) = 7236$. Also find values of a, b, c, d such that $f_4(a, b, c, d) = 833$.

   (iv) Write down the principle for constructing bijections $f_k : \mathbb{N}^k \to \mathbb{N}$ by this method. You may wish to refer to the sets $\mathbb{Z}_d^{\phi}$ defined in Investigation 13 of Chapter 5.

2. (a) Find the base-ten (decimal) representations of $1/7, 2/7, 3/7, 4/7, 5/7$, and $6/7$. Describe and explain any symmetries you observe within and between the six periodic decimal representations.

   (b) Find the base-7 representations of $1/5, 2/5, 3/5$, and $4/5$. Do these representations display similar symmetries to those found in part (a)?

   (c) Find the decimal representations of $m/13$ for $m = 1, 2, \ldots, 12$. What are the similarities and differences from the base-ten representations of $m/7$ in part (a)?

   (d) Find the decimal representations of the following fractions (with prime denominators), and hence find their periods:

$$\frac{1}{41}, \quad \frac{1}{73}, \quad \frac{1}{137}, \quad \frac{1}{239}, \quad \frac{1}{271}, \quad \frac{1}{4649}$$

   Verify that in each case the denominator is a divisor of $(10^t - 1)$ where t is the period. How does this follow from Theorem 6.7.10?

   (e) Find the base-5 expansions of the following fractions (with prime denominators), and hence find their periods:

$$\left[\frac{1}{13} \right]_{\text{ten}}, \quad \left[\frac{1}{31} \right]_{\text{ten}}, \quad \left[\frac{1}{71} \right]_{\text{ten}}$$

   What does Theorem 6.7.10 imply that the denominators should be divisors of? Verify this in each case.

3. Theorem 6.7.10 determines the possible periods of base-b representations of $\frac{m}{n}$ when n is prime; but what about when n is composite? If n has any prime factors in common with b, they only produce non-repeating digits; so now consider the case of composite n which is coprime with b.

   (a) Find the periods of the base-ten representations of $1/21$, $1/33$, $1/91$, $1/123$, $1/407$, $1/451$. Write down the prime factorisations of each the denominators, and find the periods of the base-ten representations of fractions with each of these primes as denominators (some of these were found in the previous question). Compare the periods for the composite denominators with those for the denominators with the prime factors of those composites. Note that in the above cases, the composite denominator has each of its prime factors only to the first power: formulate a hypothesis for the period of the base-b representation in this case, and explain why this hypothesis is true.

   (b) Now consider cases where the denominator is the square of a prime. Some of the periods in this case are quite long, so it would help if you use computer software that can find decimal representations to unlimited (or at least large) numbers of digits.

   Find the periods of the base-ten representations of $1/49$, $1/121$, and $1/169$, and compare with the periods of the base-ten representations of $1/7$, $1/11$, and $1/13$. Formulate a hypothesis for the case where the denominator is the square of a prime. Why does $1/9$ not satisfy this hypothesis?

Chapter 7

Real Numbers, $\mathbb{R}$

7.1 The Requirements for Our Next Number System

We have identified two defects of the Rational Numbers. Firstly, exponentiation is not closed in $\mathbb{Q}$: for most choices of $a \in \mathbb{Q}\backslash\{0\}$ and $b \in \mathbb{Q}\backslash\mathbb{Z}$ there does not exist a rational number a^b if exponentiation is defined to have properties consistent with those in $\mathbb{N}$. An alternative statement of this issue is that most **algebraic equations** of the form

$$x^n = c$$

have no solution for x in $\mathbb{Q}$ when $c \in \mathbb{Q}\backslash\{0\}$ and $n \in \mathbb{N}\backslash\{1\}$; for if $b = \frac{m}{n}$ where $m \in \mathbb{Z}$ and $n \in \mathbb{Z}_+$ (as in our definition of rational numbers), the number $x = a^b$ would be a solution of $x^n = a^m$, where $a^m \in \mathbb{Q}\backslash\{0\}$ if $a \in \mathbb{Q}\backslash\{0\}$ and $m \in \mathbb{Z}$. However, a complete resolution of the problem of solving algebraic equations, and hence of being able to always exponentiate non-zero numbers, will await a further number system, the Complex Numbers. Our attention here is focused on resolving the second defect identified in $\mathbb{Q}$, that sets could be bounded above/below without having a supremum/infimum.

Thus we seek to define a number system, to be denoted $\mathbb{R}$, that retains the "good" properties of $\mathbb{Q}$ but also satisfies the **Axiom of Completeness**:

Every subset of $\mathbb{R}$ that is bounded above/below has a supremum/infimum.

The good properties of $\mathbb{Q}$ are in relation to arithmetic and order (clearly we need order relations so that we can define supremum and infimum):

- There must be addition and multiplication operations satisfying the requirements of a field. These were set down in Definition 5.6.12, which refers back to Definition 4.3.5 and in turn to Definition 4.3.1, but it is worth reiterating the full set of requirements here. Addition and multiplication must both be associative and commutative; there must be an identity element for each operation; each element must have both an additive and a multiplicative inverse element, except that the identity element for addition has no multiplicative inverse element; and multiplication must distribute over addition.

- There must be order relations, *greater than* and *less than*, satisfying trichotomy and transitivity, with $a < b$ if and only if $b > a$. The order relations must be preserved when the same number is added to both sides. Numbers should be partitioned into positive and negative subsets and zero, where zero is the identity element for addition and the positive and negative numbers are respectively greater and less than zero. The product of two positive numbers must be positive.

When defining $\mathbb{Z}$ and $\mathbb{Q}$ we not only retained useful properties of arithmetical operations and order from the respective previous number systems $\mathbb{N}$ and $\mathbb{Z}$, but we retained the *members* of the previous number system as a subset of the new system – or, more precisely, the new number system contained a subset which was *isomorphic* to the previous system with respect to addition, multiplication, and order. We shall similarly require that $\mathbb{R}$ contains a subset $\mathbb{R}_Q$ with addition, multiplication, and order isomorphisms with $\mathbb{Q}$.

The number system that we shall define to meet these requirements could best be described as the **Continuous Numbers**. But that is not what they are called; instead, the terminology, **Real Numbers** has become established. Calling one set of numbers "real" suggests that certain other numbers are not real; and many readers will have encountered the term, **Imaginary Numbers**. But according to Dedekind's notion that "Numbers are free creations of the human mind", all numbers are imaginary: they are products of the human imagination. Nevertheless, the name *Real Numbers* is used universally, so we shall stick with it.

The definition of $\mathbb{R}$ that we shall adopt was devised by Dedekind, although the idea can be traced back to the writing of Eudoxus in ancient Greece. Most modern textbooks prefer a definition in terms of **Cauchy sequences** (to be mentioned briefly in Section 7.7); the two definitions can be shown to be equivalent. We prefer Dedekind's definition because it only involves elementary concepts, whereas Cauchy sequences require ideas from Analysis; but the disadvantage of Dedekind's definition will become apparent when we define arithmetic operations in $\mathbb{R}$, where both the definitions and the proofs of theorems become somewhat cumbersome.

7.2 Dedekind Cuts

Suppose we have a positive rational c and a natural number n such that the equation $x^n = c$ has no solution in $\mathbb{Q}$. Observe that we *can* find positive rational numbers x_l and x_g such that

$$x_l^n < c \quad \text{and} \quad x_g^n > c.$$

Exercise 7.2.1. *Using theorems on order for rational numbers, prove that if $c \in \mathbb{Q}_+$ and $n \in \mathbb{N}$, then there exist $x_l, x_g \in \mathbb{Q}_+$ such that $x_l^n < c$ and $x_g^n > c$. [Hints: consider the cases $c > 1$ and $0 < c < 1$ separately. If you can satisfy the requirements when $n = 1$, can you satisfy them for any $n \in \mathbb{N}$?]*

So we can partition the positive rationals into two non-empty sets, those whose n'th power is greater than c and those whose n'th power is less than c. Now, Theorem 6.5.16 shows that if $x_l^n < c < x_g^n$, then $x_l < x_g$; so our partition based on the magnitudes of n'th powers of numbers is also a partition based on the magnitudes of the numbers themselves. Thus we have separated the positive rational numbers into a greater set and a lesser set as a way of specifying a number, the n'th root of c, that may not itself be rational. This is the essence of the **Dedekind cut**.

Definition 7.2.2. A **Dedekind cut in the rationals** is a set α with the following properties:

 (a) $\alpha \subset \mathbb{Q}$ with $\alpha \neq \emptyset$;

 (b) If $r \in \alpha$ and $s \in \mathbb{Q}\backslash\alpha$, then $r < s$;

 (c) α has no greatest member.

Property (a) simply says that a Dedekind cut must contain some, but not all, rational numbers; so both α and $\mathbb{Q}\backslash\alpha$ are non-empty. Property (b) separates the rationals into a greater and a lesser set, with all rationals in α being less than all rationals that are not in α. We have defined the Dedekind cut as the lesser set, whereas some authors prefer to define it as the greater set. Similarly, some authors prefer to specify that the greater set should have no least member, rather than our property (c).

Applying trichotomy to property (b), we immediately have:

Corollary 7.2.3. *Let α be a Dedekind cut. If $r \in \alpha$ and $q < r$, then $q \in \alpha$. If $s \in \mathbb{Q}\backslash\alpha$ and $t > s$, then $t \in \mathbb{Q}\backslash\alpha$.*

Some authors frame the definition of Dedekind cuts in the form of this corollary rather than our property (b).

It is important to observe that there is no gap between the lesser and greater sets, a property which is formally stated in the following lemma.

Lemma 7.2.4. *If α is a Dedekind cut, then for any $\epsilon \in \mathbb{Q}_+$ we can find $r \in \alpha$ such that $r + \epsilon \in \mathbb{Q}\backslash\alpha$.*

Proof. If α is a Dedekind cut, there exist $r \in \alpha$ and $s \in \mathbb{Q}\backslash\alpha$ and, given $\epsilon \in \mathbb{Q}_+$ we can find $k \in \mathbb{N}$ such that $k\epsilon > s - r$ [*Theorem 6.5.20*]. If the lemma is false for the given ϵ, then $r + \epsilon \in \alpha$ and, by induction, $r + k\epsilon \in \alpha$. But $r + k\epsilon > s$, so this would contradict requirement (b) in Definition 7.2.2. □

We can also make a useful observation about our greater set:

Theorem 7.2.5. *If α is a Dedekind cut, then $\mathbb{Q}\backslash\alpha$ is the set of all upper bounds of α.*

Proof. For any $s \in \mathbb{Q}\backslash\alpha$, we have $r < s$ for all $r \in \alpha$, so that s is an upper bound of α [*Definition 3.6.1(i)*]. If a member of α was an upper bound of α, it would be the greatest member of α [*Definition 3.6.3(i)*], contradicting part (c) of Definition 7.2.2. So every member of $\mathbb{Q}\backslash\alpha$, and no rational number not in $\mathbb{Q}\backslash\alpha$, is an upper bound for α. □

While a Dedekind cut does not have a greatest member, it may or may not have a supremum in $\mathbb{Q}$.

Definition 7.2.6. Let α be a Dedekind cut in the rationals.
(i) If $\sup \alpha$ exists in $\mathbb{Q}$, then α is a **rational cut**;
(ii) If α has no supremum in $\mathbb{Q}$, then α is an **irrational cut**.

We can think of the rationals set out on a number line, with greater numbers to the right of lesser numbers, and imagine cutting the line at some point, with numbers to the left of the cutting point constituting the set α. There may be a rational number at the cutting point, in which case it is allocated to $\mathbb{Q}\backslash\alpha$, and is the least member of that set, i.e. the supremum of α [*Theorem 7.2.5*]. But the cutting procedure is still valid when there is no rational number at the cutting point, in which case we have an irrational cut. We consider each cut, whether rational or irrational, to define a number in our new number system:

Definition 7.2.7. A **Real Number** is a Dedekind cut in the rationals.

We can now start talking about **irrational numbers**, a phrase that was *not* used in the previous chapter when we found that most roots and logarithms of rational numbers did not exist in $\mathbb{Q}$: an irrational number is simply an irrational cut, as defined in Definition 7.2.6. Whereas Theorem 6.5.19 showed that there are no "gaps" between rational numbers, Theorem 6.6.4 revealed the possibility of cutting the rational number line and finding a "hole" with no number at the cutting point. The Real Number system fills in those "holes" with irrational numbers, giving us a **continuum** of numbers so that we find a real number wherever we choose to cut the number line.

The following examples will show (informally) how rational numbers and roots of rational numbers can be represented as Dedekind cuts, and therefore fit into $\mathbb{R}$; formal verification that our representations of roots are correct can only be made once we have defined arithmetic operations in $\mathbb{R}$. But readers should bear in mind that $\mathbb{R}$ may include numbers that cannot be constructed from rationals by arithmetic operations or by solution of algebraic equations with rational coefficients; and later we shall see that in fact $\mathbb{R}$ *must* include such numbers.

Example 7.2.8. For any $x \in \mathbb{Q}$, the set $\{r \in \mathbb{Q} : r \leq x\}$ is *not* a Dedekind cut: it has a greatest member, x, contravening requirement (c) of Definition 7.2.2.

Example 7.2.9. For any $x \in \mathbb{Q}$, the set $\alpha_x := \{r \in \mathbb{Q} : r < x\}$ is a Dedekind cut.

Proof. We need to verify that each of the requirements in Definition 7.2.2 is satisfied.
(a) $x - 1 \in \alpha_x$ and $x \notin \alpha_x$, so both α_x and $\mathbb{Q}\backslash\alpha_x$ are non-empty.
(b) By trichotomy, $\mathbb{Q}\backslash\alpha_x = \{s \in \mathbb{Q} : s \geq x\}$. So, for all $r \in \alpha_x$ and $s \in \mathbb{Q}\backslash\alpha_x$, we have $r < x \leq s$ and so, by transitivity, $r < s$.
(c) See the verification following Theorem 6.6.1 that the set Y_a has no least member. A similar procedure will verify that α_x has no greatest member.   □

Observe that for each $x \in \mathbb{Q}$ we can define a Dedekind cut (i.e. a member of $\mathbb{R}$) of the form α_x in Example 7.2.9, and for each such $\alpha_x \in \mathbb{R}$ there corresponds a rational number x. So if we let
$$\mathbb{R}_Q := \{\alpha_x \in \mathbb{R} : x \in \mathbb{Q}\},$$
we have a bijection between $\mathbb{R}_Q$ and $\mathbb{Q}$. $\mathbb{R}_Q$ is the set of all rational cuts; having established the bijection, we have taken the first step towards finding isomorphisms which will allow us to treat rational numbers as a subset of the real numbers.

Example 7.2.10. For any $x \in \mathbb{Q}$, the set $\beta_x := \{r \in \mathbb{Q} : r^2 < x\}$ is *not* a Dedekind cut. If $x \in \mathbb{Q}_-$ or $x = 0$, then $\beta_x = \emptyset$, contravening requirement (a) of Definition 7.2.2. But if $x \in \mathbb{Q}_+$, then $0 \in \beta_x$ while $-x - 1 \in \mathbb{Q}\backslash\beta_x$ with $-x - 1 < 0$, contravening requirement (b) of Definition 7.2.2. To see that $-x - 1 \in \mathbb{Q}\backslash\beta_x$, note that $\mathbb{Q}\backslash\beta_x = \{s \in \mathbb{Q} : s^2 \geq x\}$ (by trichotomy); since $(-x-1)^2 = x^2 + 1 + 2x$, in which $x^2 + 1 > 0$ and $2x > x$ if $x \in \mathbb{Q}_+$, we have $(-x-1)^2 > x$ so $(-x-1) \in \mathbb{Q}\backslash\beta$.

The difficulty in this example arose from the fact that the square of a negative number is positive, and would arise similarly with other even powers: the set $\{r \in \mathbb{Q} : r^{2m} < x\}$ would not be a Dedekind cut for any $m \in \mathbb{N}$. But this is easy to remedy:

Example 7.2.11. For any $x \in \mathbb{Q}_+$, the set $\tilde{\beta}_x := \{r \in \mathbb{Q} : r^2 < x\} \cup \mathbb{Q}_- = \beta_x \cup \mathbb{Q}_-$ is a Dedekind cut.

Proof. (a) $0 \in \tilde{\beta}_x$ and $x + 1 \notin \tilde{\beta}_x$ (verified similarly to $-x - 1 \in \mathbb{Q}\backslash\beta_x$ in the previous example), so both $\tilde{\beta}_x$ and $\mathbb{Q}\backslash\tilde{\beta}_x$ are non-empty.
(b) Including all negative rationals in $\tilde{\beta}_x$ means that $s^2 \geq x$ and $s \geq 0$ for any $s \in \mathbb{Q}\backslash\tilde{\beta}_x$, while if $r \in \tilde{\beta}_x$, then either $r^2 < x$ or $r < 0$ (or both). In the former case, we have $r^2 < x \leq s^2$, and since $s \geq 0$, Theorem 4.4.29 yields $r < s$. In the latter case, $r < 0 \leq s$ so $r < s$.
(c) A greatest member for $\tilde{\beta}_x$ certainly cannot be found in $\mathbb{Q}_-$, since $0 \in \beta_x$. If $\sqrt[3]{x}$ does not exist in $\mathbb{Q}$, then the proof of Theorem 6.6.4 shows that β_x, and hence $\tilde{\beta}_x$, has no greatest

member. On the other hand, if $\sqrt[2]{x} = y \in \mathbb{Q}$, Theorem 4.4.29 implies that $r^2 < x$ if and only if $r < y$ (for $r \geq 0$), so $\tilde{\beta}_x$ is actually the rational cut α_y (see Example 7.2.9) which has no greatest member. $\square$

Note on notation. *We shall use the symbols α and β for general Dedekind cuts, except where these symbols carry a subscript representing a rational number, as in $\alpha_y, \beta_x, \tilde{\beta}_x$, when they will represent the specific types of cuts in the above examples.*

Clearly we can generalise the idea in Example 7.2.11 to define cuts of the form $\{r \in \mathbb{Q} : r^{2m} < x\} \cup \mathbb{Q}_-$ for any $x \in \mathbb{Q}_+$ and $m \in \mathbb{N}$, whereas when working with odd powers, $\{r \in \mathbb{Q} : r^{2m+1} < x\}$ is a cut for any $x \in \mathbb{Q}$ and $m \in \mathbb{N}$. In the latter case we can allow x to be negative and we do not need "$\cup \mathbb{Q}_-$" in the definition of the cut, because with odd powers we have $x < y \Leftrightarrow x^{2m+1} < y^{2m+1}$ for all $x, y \in \mathbb{Q}$, not merely for positive rationals as in Theorem 6.5.16.

Exercise 7.2.12. *Given Theorem 6.5.16 for positive rationals x and y and other theorems on order in $\mathbb{Q}$, together with the rule for odd powers that $(-x)^{2m+1} = -x^{2m+1}$, show that*

$$x < y \Leftrightarrow x^{2m+1} < y^{2m+1} \quad \text{for all} \quad x, y \in \mathbb{Q}.$$

Exercise 7.2.13. *For each of the following subsets of $\mathbb{Q}$, determine whether the set is a cut; if it is a cut, state whether it is a rational cut or an irrational cut; if it is a rational cut, write down the rational number that it corresponds to under the bijection between $\mathbb{R}_Q$ and $\mathbb{Q}$; if it is not a cut at all, explain why it is not a cut:*

$$(i) \; \{a \in \mathbb{Q} : a < -1\} \quad (ii) \; \{a \in \mathbb{Q} : a^2 < 1\} \quad (iii) \; \{a \in \mathbb{Q} : a^2 < 1\} \cup \mathbb{Q}_-$$

$$(iv) \; \{a \in \mathbb{Q} : a^2 < 7\} \cup \mathbb{Q}_- \quad (v) \; \{a \in \mathbb{Q}_- : a^2 > 1\} \quad (vi) \; \{a \in \mathbb{Q} : a^3 < -1\}$$

$$(vii) \; \{a \in \mathbb{Q} : a^2 < -1\} \quad (viii) \; \left\{a \in \mathbb{Q} : a = \frac{1}{q}, q \in \mathbb{Q}_-\right\} \quad (ix) \; \{a \in \mathbb{Q} : a = -q^2, q \in \mathbb{Q}\}$$

Our intuition is that Dedekind cuts of the forms $\{r \in \mathbb{Q} : r^{2m} < x\} \cup \mathbb{Q}_-$ and $\{r \in \mathbb{Q} : r^{2m+1} < x\}$ are representations of roots of x, respectively $\sqrt[2m]{x}$ (or $x^{1/2m}$) for positive rational x and $\sqrt[2m+1]{x}$ (or $x^{1/(2m+1)}$) for any $x \in \mathbb{Q}$. This is because the n'th-root operation is the inverse of the n'th-power (exponentiation) operation, and the cut partitions the rationals into subsets whose n'th power (with $n = 2m$ or $n = 2m + 1$) is less or greater than x. But we have no proof that defining $\sqrt[n]{x}$ by these cuts actually produces a real number whose n'th power is a real number corresponding to the rational x. Such a proof will require us first to define arithmetic operations in $\mathbb{R}$, but for now we shall accept that the representation is correct.

Exercise 7.2.14.   *(a) Explain why $-\sqrt[2m]{x}$ (with $m \in \mathbb{N}$ and $x \in \mathbb{Q}_+$) can be represented by the Dedekind cut $\{r \in \mathbb{Q}_- : r^{2m} > x\}$.*

    (b) Write down Dedekind cuts to represent each of the following irrational numbers:

$$(i) \; \sqrt[5]{18} \quad (ii) \; \sqrt[6]{7} \quad (iii) \; -\sqrt[6]{7} \quad (iv) \; \sqrt[7]{-6}$$

It will be useful at this stage to define what is meant by **positive, negative,** and **zero** in the Real Numbers. The zero cut is simply the cut corresponding to the rational number 0 under the bijection between $\mathbb{R}_Q$ and $\mathbb{Q}$, i.e.

$$\alpha_0 := \{r \in \mathbb{Q} : r < 0\}.$$

Introducing the obvious notation $\mathbb{R}_+$ and $\mathbb{R}_-$ for the sets of positive and negative real numbers, respectively, and recalling that all Real Numbers α are Dedekind cuts, which are subsets of $\mathbb{Q}$:

Definition 7.2.15. $\alpha \in \mathbb{R}_+$ if $0 \in \alpha$; $\alpha \in \mathbb{R}_-$ if $0 \notin \alpha$ and $\alpha \neq \alpha_0$.

We have defined Real Numbers as Dedekind cuts, but have not yet verified that they fulfil our objective of satisfying the Axiom of Completeness. To do this, we require order relations to be defined, so we shall consider order before arithmetic. We shall then consider addition, the first of the arithmetic operations: the definition of this operation in $\mathbb{R}$ will appear quite simple, but proving that this definition does actually yield a Dedekind cut, and then that it satisfies all the standard properties of addition, is quite a cumbersome business. The proofs will be given in full, and do not involve any difficult concepts although some are quite long. We shall then move on to multiplication, for which the definition is complicated due to the need to treat positive and negative numbers separately; but we shall leave most of the proofs as exercises for the reader since they are along similar lines to those for addition (at least for multiplication of positive numbers). Finally, we shall consider exponentiation, first justifying rigorously our intuitive representations of roots of rational numbers; this will allow us to define all rational powers and logarithms of positive rational numbers in $\mathbb{R}$, but because these operations usually yield irrational numbers a new approach will be required to define exponentiation of general real numbers. While studying the arithmetic operations, we will also establish their effects on order relations, as well as establishing isomorphisms between the operations on the rational cuts $\mathbb{R}_Q$ and on the rational numbers $\mathbb{Q}$.

7.3 Order and Bounded Sets in $\mathbb{R}$

Recalling that a Dedekind cut is a set of rationals satisfying certain requirements, order relations can be defined in a particularly simple way:

Definition 7.3.1.

$$\alpha \olessthan \beta \quad \text{if} \quad \alpha \subset \beta;$$

$$\alpha \ogreaterthan \beta \quad \text{if} \quad \beta \subset \alpha.$$

We can immediately establish order isomorphism between the relations $\olessthan$ and $\ogreaterthan$ on the rational cuts $\mathbb{R}_Q$ and the relations $<$ and $>$ on the rational numbers $\mathbb{Q}$.

Theorem 7.3.2. *Let α_x and α_y be the rational cuts corresponding to rational numbers x and y. Then $\alpha_x \olessthan \alpha_y$ if and only if $x < y$, and $\alpha_x \ogreaterthan \alpha_y$ if and only if $x > y$.*

Proof. The rational cuts are defined by $\alpha_x := \{r \in \mathbb{Q} : r < x\}$ and $\alpha_y := \{r \in \mathbb{Q} : r < y\}$.

If $x < y$, $\exists s \in \mathbb{Q}$ such that $x < s < y$ [*Theorem 6.5.19*], so $s \notin \alpha_x$ but $s \in \alpha_y$. Now suppose $r \in \alpha_x$, so that $r < x$. Since $x < s$, transitivity implies $r < s$; and since $s \in \alpha_y$ and α_y is a cut, this implies that $r \in \alpha_y$ [*Corollary 7.2.3*]. So every $r \in \alpha_x$ is also a member of α_y, while $s \in \alpha_y$ is not a member of α_x; i.e. $\alpha_x \subset \alpha_y$ or, according to Definition 7.3.1, $\alpha_x \olessthan \alpha_y$.

Conversely, if $\alpha_x \olessthan \alpha_y$, i.e. if $\alpha_x \subset \alpha_y$, there exists $s \in \mathbb{Q}$ such that $s \notin \alpha_x$ but $s \in \alpha_y$. Now, $s \notin \alpha_x$ means that $s \geq x$ (from the definition of α_x, using trichotomy); while $s \in \alpha_y$ means that $s < y$. So $x \leq s < y$, and transitivity then yields $x < y$.

The proof for the relations $\ogreaterthan$ and $>$ is similar. $\qquad\square$

This order isomorphism means that we can use the symbols $<$ and $>$, and the terms "less than" and "greater than", for the relations in Definition 7.3.1; but we will still need to use this definition, which says that a cut is less than or greater than another cut according to whether one of the cuts is a proper subset of the other, to verify that the usual properties of order relations are true for all Real Numbers. To emphasise this, we shall retain the notations $\oslash$ and $\ominus$ in some of the proofs below, to show the distinction between the relations in ℝ and in ℚ.

Firstly, it is immediately obvious from the definition that $\alpha < \beta$ if and only if $\beta > \alpha$. Next, we shall verify the essential properties of trichotomy and transitivity.

Theorem 7.3.3. *The order relations in* ℝ *are trichotomous: for any* $\alpha, \beta \in$ ℝ*, one and only one of* $\alpha = \beta$ *or* $\alpha \oslash \beta$ *or* $\alpha \ominus \beta$ *is true.*

Proof. By the definitions of *proper subset* and *equal* for sets, the relations $\subset$ and $=$ are mutually exclusive. So Definition 7.3.1 shows that if $\alpha = \beta$ then it is not the case that either $\alpha \oslash \beta$ or $\alpha \ominus \beta$. It then remains to show that if $\alpha \neq \beta$ then one and only one of the options $\alpha \oslash \beta$ or $\alpha \ominus \beta$ is true. From Definition 7.3.1 this amounts to showing that if $\alpha \not\subset \beta$, then $\beta \subset \alpha$.

Now if $\alpha \not\subset \beta$, there exists $x \in \alpha$ such that $x \notin \beta$, i.e. $x \in$ ℚ$\backslash\beta$. Then for any $y \in \beta$ we have $y < x$ since β is a cut [*Definition 7.2.2, part (b)*]. But $y < x$ with $x \in \alpha$ implies that $y \in \alpha$, since α is also a cut [*Corollary 7.2.3*]. We have shown that every $y \in \beta$ is also a member of α, so that $\beta \subseteq \alpha$; and since $\beta \neq \alpha$, we have $\beta \subset \alpha$ as required.   □

Theorem 7.3.4. *The order relations on* ℝ *are transitive.*

Proof. This follows immediately from Definition 7.3.1 and the fact that the proper subset relation is transitive: for any sets A, B, C such that $A \subset B$ and $B \subset C$, a member of A will also be a member of B and then of C.   □

Next, we verify that positive and negative real numbers (as defined in Definition 7.2.15) are respectively greater and less than the zero cut.

Theorem 7.3.5.

$$(i) \quad \alpha \in ℝ_- \text{ if and only if } \alpha \oslash \alpha_0;$$

$$(ii) \quad \alpha \in ℝ_+ \text{ if and only if } \alpha \ominus \alpha_0.$$

Proof of (i). If $\alpha \in ℝ_-$, then $0 \in$ ℚ$\backslash\alpha$ from Definition 7.2.15; so since α is a cut, each $r \in \alpha$ has $r < 0$ and hence $r \in \alpha_0$ (by definition of α_0). Thus $\alpha \subseteq \alpha_0$; but since $\alpha \neq \alpha_0$ from Definition 7.2.15, we have $\alpha \subset \alpha_0$, i.e. $\alpha \oslash \alpha_0$.

Conversely, if $\alpha \oslash \alpha_0$, i.e. $\alpha \subset \alpha_0$, then $\alpha \neq \alpha_0$ and since $0 \notin \alpha_0$, we have $0 \notin \alpha$.   □

Exercise 7.3.6. *Prove part (ii) of Theorem 7.3.5.*

We can now establish that Dedekind cuts fulfil the requirement stated at the beginning of this chapter for the Real Numbers:

Theorem 7.3.7. *Dedekind cuts satisfy the Axiom of Completeness: every set of Dedekind cuts that is bounded above has a supremum which is a Dedekind cut.*

Proof. It is worth recalling first what the terminology in the statement of the theorem means. Let A be a set of Dedekind cuts (so A is a set of sets of rational numbers), and let α denote a general member of A. An upper bound for A is a cut ψ such that $\alpha \leq \psi$ for each $\alpha \in A$, so that

$$\alpha \subseteq \psi \ \forall \alpha \in A \tag{7.1}$$

according to Definition 7.3.1. Let U_A be the set of all such upper bounds for A: we need to show that if U_A is non-empty, then it has a least member. We shall show that this least member of U_A, the supremum of A, is in fact

$$\lambda := \bigcup_{\alpha \in A} \alpha.$$

Here λ has been defined as the *union* of all the members of A, so it is a set of rationals consisting of every rational contained in every cut in A. We need to show first that λ itself is a cut, and then that it is the least upper bound of A.

(a) λ is non-empty since all the cuts $\alpha \in A$ are non-empty. With the premise that A is bounded above, there does exist $\psi \in \mathbb{R}$ satisfying (7.1), so from the definition of λ we have $\lambda \subseteq \psi$. Since $\mathbb{Q} \backslash \psi$ is non-empty (because ψ is a cut), we deduce from Theorem 2.1.7 that $\mathbb{Q} \backslash \lambda \neq \emptyset$.

(b) Take any $r \in \lambda$ and $s \in \mathbb{Q} \backslash \lambda$. Because $r \in \lambda$, we have $r \in \alpha$ for some $\alpha \in A$. Since $s \notin \lambda$, then $s \notin \alpha$ for any $\alpha \in A$. Because each $\alpha \in A$ is a cut, we have $r < s$.

(c) For any $r \in \lambda$, we have $r \in \alpha$ for some $\alpha \in A$. Since α is a cut, with no greatest member, we have $q > r$ for some $q \in \alpha$. So $q \in \lambda$, and we have shown that no $r \in \lambda$ is its greatest member.

From the definition of λ, we have $\alpha \subseteq \lambda$, and hence $\alpha \leq \lambda$, for each $\alpha \in A$; so λ is an upper bound of A. We have already noted that any upper bound ψ satisfies $\lambda \subseteq \psi$, and hence $\lambda \leq \psi$; so λ is the *least* upper bound of A. □

We may regard Theorem 7.3.7 as confirming that Real Numbers form a continuum. When we cut the Rational Number line according to the rules in Definition 7.2.2, we found that the cutting point only corresponded to a previously existing (i.e. rational) number if the lesser set had a supremum (from Definition 7.2.6 and the order isomorphism between rational cuts and rational numbers); otherwise, a new number was defined at the cutting point. If we now cut the *Real Number line* according to the same rules, no new numbers can be generated; the lesser set always has a supremum.

7.4　Addition in $\mathbb{R}$

As usual we symbolise the new addition operation by $\oplus$ until we have established an isomorphism with the addition operation on the previous number system.

Definition 7.4.1. If α and β are Dedekind cuts,

$$\alpha \oplus \beta := \{t \in \mathbb{Q} : t = r + q \text{ where } r \in \alpha \text{ and } q \in \beta\}.$$

We add each rational in the cut α to each rational in β, and the set of resulting rationals is defined to be $\alpha \oplus \beta$; we now need to verify that this set is a cut.

(a) Since α and β are non-empty, take any $r \in \alpha$ and $q \in \beta$ to yield $t = r + q \in \alpha \oplus \beta$ so that $\alpha \oplus \beta$ is non-empty. To show that $\mathbb{Q} \backslash (\alpha \oplus \beta)$ is non-empty, take $s \in \mathbb{Q} \backslash \alpha$ and $p \in \mathbb{Q} \backslash \beta$; if $s + p \in \alpha \oplus \beta$, there would exist $r \in \alpha$ and $q \in \beta$ such that $s + p = r + q$; but because α and β are Dedekind cuts, we have $s > r$ and $p > q$. Thus $s + p > r + q$, contradicting the requirement $s + p = r + q$, so that $s + p \notin \alpha \oplus \beta$.

(b) Suppose $t \in \alpha \oplus \beta$ and $u \in \mathbb{Q} \backslash (\alpha \oplus \beta)$; we need to show that $t < u$. Suppose to the contrary that $t \geq u$. Then since $t = r + q$ with $r \in \alpha$ and $q \in \beta$, we have $r = t - q \geq u - q$. So the rational number $u - q$ is not greater than r, so not in $\mathbb{Q} \backslash \alpha$; thus $u - q \in \alpha$. Since

$u = (u - q) + q$ where $u - q \in \alpha$ and $q \in \beta$, we have $u \in \alpha \oplus \beta$, contradicting the definition of u. Hence we cannot have $t \geq u$.

(c) Suppose that $\alpha \oplus \beta$ has a greatest member, $\max(\alpha \oplus \beta) = r_g + q_g$ for some $r_g \in \alpha$ and $q_g \in \beta$. Since α and β have no greatest members, there exist $r_h \in \alpha$ with $r_h > r_g$ and $q_h \in \beta$ with $q_h > q_g$. But then $r_h + q_h \in \alpha \oplus \beta$ with $r_h + q_h > r_g + q_g$, contradicting $r_g + q_g$ being the greatest member of $\alpha \oplus \beta$.

We want our addition operation in $\mathbb{R}$ to have the same arithmetic properties as addition in $\mathbb{Q}$. This is covered by:

Theorem 7.4.2. $\langle \mathbb{R}, \oplus \rangle$ *is an Abelian group, with α_0 as the additive identity element.*

Proof. The commutative and associative properties follow immediately from the same properties for addition in $\mathbb{Q}$, given Definition 7.4.1 for $\oplus$ in terms of addition in $\mathbb{Q}$.

Next we need to show that $\alpha_0 := \{r \in \mathbb{Q} : r < 0\}$ is an additive identity element for $\mathbb{R}$. Now if $t \in \alpha_0 \oplus \beta$, then $t = r + q$ with $r < 0$ and $q \in \beta$, so $t < q$ and hence $t \in \beta$. Conversely, if $t \in \beta$ then since β has no greatest member there exists $q \in \beta$ with $t < q$ so that $t = r + q$ for some $r < 0$. We have shown that $t \in \alpha_0 \oplus \beta$ if and only if $t \in \beta$, i.e. that $\alpha_0 \oplus \beta = \beta$. The commutative property then yields that $\beta \oplus \alpha_0 = \beta$, which establishes α_0 as an identity element for $\oplus$.

Finally we need to establish the existence of an additive inverse $-\beta$ for each $\beta \in \mathbb{R}$, such that $(-\beta) \oplus \beta = \beta \oplus (-\beta) = \alpha_0$. If we think of cutting the rational number line at a point β, we would intuitively want $-\beta$ to be a cut the same distance on the other side of zero from β. The numbers to the left of the $-\beta$ cut would be a "mirror image" (reflected in zero) of the numbers to the right of the β cut. This motivates the definition,

$$-\beta := \{r \in \mathbb{Q} : -r \in \mathbb{Q} \backslash \beta\},$$

which we shall show to be valid when β is an irrational cut; but if β is rational we must specify

$$-\beta := \{r \in \mathbb{Q} : -r \in \mathbb{Q} \backslash \beta, r \neq -\min(\mathbb{Q} \backslash \beta)\}$$

to prevent $-\beta$ having a greatest member, since a rational cut β has a supremum which is the least member of $\mathbb{Q} \backslash \beta$, so that $-\min(\mathbb{Q} \backslash \beta)$ would be the greatest member of $\{r \in \mathbb{Q} : -r \in \mathbb{Q} \backslash \beta\}$. This deals with part (c) of the definition of Dedekind cut. For part (a), there exist rational numbers both within and outside $\mathbb{Q} \backslash \beta$, so taking their additive inverses (in $\mathbb{Q}$) provides rationals within and outside $-\beta$. For part (b), if $r \in (-\beta)$ and $s \in \mathbb{Q} \backslash (-\beta)$, then $-r \in \mathbb{Q} \backslash \beta$ and $-s \notin \mathbb{Q} \backslash \beta$ (for irrational β), i.e. $-s \in \beta$. But since β is a Dedekind cut, this means $-s < -r$, so that $r < s$ as required. For rational β there is the possibility that $s = -\min(\mathbb{Q} \backslash \beta)$, but since $-r \in \mathbb{Q} \backslash \beta$ and $r \neq -\min(\mathbb{Q} \backslash \beta)$ we have $-r > -\min(\mathbb{Q} \backslash \beta) = -s$ so that $r < s$ again.

We have completed the verification that $-\beta$ is a Dedekind cut; we now need to show that $\beta \oplus (-\beta) = \alpha_0$ (and it then follows from the commutative property that $(-\beta) \oplus \beta = \alpha_0$). This means showing that $q + r < 0$ (i.e. $q + r \in \alpha_0$) if and only if $q \in \beta$ and $r \in (-\beta)$, for any given $\beta \in \mathbb{R}$. Now if $q \in \beta$ and $r \in (-\beta)$ so that $-r \in \mathbb{Q} \backslash \beta$, we have $q < -r$ and hence $q + r < 0$. Conversely, for any $t < 0$ we must find $q \in \beta$ and $r \in (-\beta)$ such that $q + r = t$, i.e. that $-r = q - t \in \mathbb{Q} \backslash \beta$; but since $-t > 0$, Lemma 7.2.4 does confirm the existence of $q \in \beta$ such that $q - t \in \mathbb{Q} \backslash \beta$. $\qquad\square$

Exercise 7.4.3. *Referring to Definition 7.2.15, verify that if $\beta \in \mathbb{R}_+$, then $(-\beta) \in \mathbb{R}_-$.*

We can now establish the isomorphism between $\mathbb{R}_\mathbb{Q}$ and $\mathbb{Q}$ with their respective addition operations. We have already found a bijection, with a rational cut α_x corresponding to each rational number x (using the notation introduced in Example 7.2.9), so it remains to show:

Theorem 7.4.4. *If $x + y = z$, then $\alpha_x \oplus \alpha_y = \alpha_z$.*

Proof. If $r \in \alpha_x$ and $q \in \alpha_y$ then $r < x$ and $q < y$ so that $r + q < x + y = z$, i.e. $r + q \in \alpha_z$. Conversely, if $t \in \alpha_z$ so that $t < z$, let $\epsilon = z - t > 0$ and then let $r = x - \epsilon/2 < x$ and $q = y - \epsilon/2 < y$ so that $r + q = z - \epsilon = t$ with $r \in \alpha_x$ and $q \in \alpha_y$.   □

This isomorphism allows us to use the usual symbol $+$ for the addition operation in $\mathbb{R}$; we no longer need the symbol $\oplus$. We can also introduce subtraction in $\mathbb{R}$ (with its usual symbol): according to Theorem 4.2.13, we can do this inverse operation of addition because the additive inverse element exists for each $\beta \in \mathbb{R}$ (as shown in Theorem 7.4.2):

Definition 7.4.5.
$$\alpha - \beta := \alpha + (-\beta).$$

Having covered the arithmetic properties of addition, we can also show that addition or subtraction of a real number to both sides of an order relation preserves that relation, as in $\mathbb{Z}$ and $\mathbb{Q}$:

Theorem 7.4.6. *Let α, β, γ be real numbers. Then*

$$\alpha < \beta \Leftrightarrow \alpha + \gamma < \beta + \gamma;$$

$$\alpha < \beta \Leftrightarrow \alpha - \gamma < \beta - \gamma.$$

Proof. We first show that if $\alpha < \beta$, then $\alpha + \gamma < \beta + \gamma$. From Definition 7.3.1, we need to suppose that $\alpha \subset \beta$ and show that it follows that $(\alpha + \gamma) \subset (\beta + \gamma)$. Given any $t \in (\alpha + \gamma)$, we have $t = r + q$ where $r \in \alpha$ and $q \in \gamma$ [*Definition 7.4.1*]. Because $\alpha \subset \beta$, having $r \in \alpha$ implies that $r \in \beta$, and hence $r + q \in (\beta + \gamma)$. So $t \in (\beta + \gamma)$ for any $t \in (\alpha + \gamma)$, i.e. $(\alpha + \gamma) \subseteq (\beta + \gamma)$. But if $(\alpha + \gamma) = (\beta + \gamma)$, that would imply that $\alpha = \beta$ (by subtraction of γ in the equality), so we can only have $(\alpha + \gamma) \subset (\beta + \gamma)$

The similar theorem for subtraction then follows immediately from Definition 7.4.5, that subtraction is simply the addition of an additive inverse number. The cancellation law, that if $\alpha + \gamma < \beta + \gamma$ then $\alpha < \beta$, is then implied by the fact that cancellation of the addition of γ is the same as subtraction of γ; and similarly, cancellation of a subtraction is the same as an addition.   □

Further theorems on order can then be stated, in particular those involving additive inverses:

Theorem 7.4.7.

(i) If $\alpha < \beta$, then $-\alpha > -\beta$.

(ii) If $\alpha > \beta$, then $-\alpha < -\beta$.

Theorem 7.4.8. *If $\alpha > 0$, then $-\alpha < 0$. If $\alpha < 0$, then $-\alpha > 0$.*

These are simply Theorems 4.4.10 and 4.4.15, rewritten in our notation for real numbers, and they follow from Theorem 7.4.6 in the same way that the integer theorems follow from Theorem 4.4.6.

Real numbers have been classified as rational cuts or irrational cuts, with the former being isomorphic to rational numbers under addition; so since $\mathbb{Q}$ is closed under addition, the addition of two rational cuts definitely produces a rational cut. But what happens when we add a rational to an irrational, or add two irrationals?

Theorem 7.4.9. *The sum of a rational and an irrational is always irrational.*

Proof. Given $\alpha \in \mathbb{R}_Q$ (rational) and $\beta \in \mathbb{R} \backslash \mathbb{R}_Q$ (irrational), suppose to the contrary that $\alpha + \beta$ is rational. Since $\beta = (\alpha + \beta) - \alpha$ and the subtraction of two rationals yields a rational, this contradicts β being irrational. Hence $\alpha + \beta$ cannot be rational. □

When two irrationals are added, the result may be either rational or irrational. There is no general procedure for determining which, but a few cases are easy to deal with. A particularly simple example is that if α is irrational, $\alpha + (-\alpha) = 0$ which is rational; from Theorem 7.4.9, $-\alpha$ cannot then be rational, and we have added two irrationals to obtain a *rational*. On the other hand, the simplest cases where the sum of two irrationals is itself *irrational* are where both irrationals are roots of rationals; but the verification of irrationality involves multiplication and exponentiation, so examples will be left until later.

7.5 Multiplication in $\mathbb{R}$

The definition of Dedekind cuts involves order in $\mathbb{Q}$; and the effect of multiplication on an order relation (in $\mathbb{Q}$) depends on whether it is being multiplied by a positive or negative number (or zero); so to define multiplication of cuts we will need to take into account whether the cuts being multiplied are positive, negative or zero. Defining addition of cuts was simple because order is preserved by addition in $\mathbb{Q}$; and the simplest case for multiplication is where order is preserved, which is when we are multiplying positive numbers. However, we can start with a definition of multiplication which includes zero as well as positive numbers, i.e. for all *non-negative* real numbers. For this purpose we introduce the notation α^+ for the set of non-negative rationals in the cut α,

$$\alpha^+ := \{r \in \alpha : r \geq 0\}. \tag{7.2}$$

Note that $\alpha^+ = \emptyset$ if α is the zero cut as well as when $\alpha \in \mathbb{R}_-$

Definition 7.5.1. If $\alpha, \beta \in \mathbb{R}_+ \cup \{\alpha_0\}$ (i.e. α and β are non-negative Dedekind cuts),

$$\alpha \otimes \beta := \{t \in \mathbb{Q} : t = r \times q \text{ where } r \in \alpha^+ \text{ and } q \in \beta^+\} \cup \mathbb{Q}_-.$$

This follows the general idea in Definition 7.4.1 of addition, but removes negative rationals from α and β because the products of all negative rationals would be all positive rationals, and we certainly don't want *all* positive rationals in the cut $\alpha \otimes \beta$. On the other hand, we do need all negative rationals in $\alpha \otimes \beta$, so we finally include $\mathbb{Q}_-$ in our definition. Note in particular that when multiplying the zero cut, because $\alpha_0^+ = \emptyset$ we find $\alpha_0 \otimes \beta = \mathbb{Q}_- = \alpha_0$ for any $\beta \in \mathbb{R}$, as expected.

Exercise 7.5.2. *Show that $\alpha \otimes \beta$ as given in Definition 7.5.1 is a Dedekind cut.*

Because order is preserved on multiplication by a positive rational, this exercise is no more difficult than our proof that $\alpha \oplus \beta$ is a Dedekind cut.

We can then show that multiplication of positive real numbers has the usual properties:

Exercise 7.5.3. *Show that $\langle \mathbb{R}_+, \otimes \rangle$ is an Abelian group, with α_1 as the multiplicative identity element. [Hints: As with addition, verifying the commutative and associative laws is trivial. With α_1 being the rational cut corresponding to the rational number 1, proving that it fulfils the role of a multiplicative identity element is not difficult. To work out how to* *specify the multiplicative inverse of a positive cut β, note that if $0 < r < x$, then $\frac{1}{r} > \frac{1}{x}$, with the reversal of the order relation indicating a similar strategy to that used when defining the additive inverse, $-\beta$.]*

We can now define multiplications in which one or both of α and β are negative by simply using additive inverses in combination with Definition 7.5.1 for multiplication of the positive cuts $(-\alpha)$ and/or $(-\beta)$, to give the expected results:

Definition 7.5.4.

If $\alpha \in \mathbb{R}_-$ and $\beta \in \mathbb{R}_+$, then $\alpha \otimes \beta = -((-\alpha) \otimes \beta)$.

If $\alpha \in \mathbb{R}_+$ and $\beta \in \mathbb{R}_-$, then $\alpha \otimes \beta = -(\alpha \otimes (-\beta))$.

If $\alpha \in \mathbb{R}_-$ and $\beta \in \mathbb{R}_-$, then $\alpha \otimes \beta = (-\alpha) \otimes (-\beta)$.

These definitions yield cuts because taking an additive inverse yields a cut and multiplication of positive cuts yields a cut; and the required properties (commutative, associative, and existence of multiplicative inverses) are also easily verified for these cases once we know that the properties apply to multiplication of positive cuts. There is one further property required in order to establish that $\mathbb{R}$ has all the same arithmetic properties as $\mathbb{Q}$: that is the distributive law. This follows immediately from the definitions of $\oplus$ and $\otimes$ in terms of the arithmetic operations in $\mathbb{Q}$, in which multiplication distributes over addition. Thus we have:

Theorem 7.5.5. $\langle \mathbb{R}, \oplus, \otimes \rangle$ *is a field.*

We already have an isomorphism between the addition operations in $\mathbb{R}_Q$ and $\mathbb{Q}$, and a similar isomorphism exists between the multiplication operations.

Exercise 7.5.6. *Show that there is isomorphism between $\otimes$ in $\mathbb{R}_Q$ and $\times$ in $\mathbb{Q}$.*

This isomorphism allows us to dispense with the symbol $\otimes$ and start using the usual symbol $\times$ for multiplication in $\mathbb{R}$. The inverse operation of multiplication, exact division (except by zero), can also be done in $\mathbb{R}$ as in $\mathbb{Q}$ because a multiplicative inverse $\frac{1}{\beta}$ exists for every number $\beta \in \mathbb{R} \backslash \{\alpha_0\}$:

$$\alpha \div \beta := \alpha \times \frac{1}{\beta}.$$

The field structure then ensures that any further formulae involving addition, subtraction, multiplication, division, and additive and multiplicative inverses are the same in $\mathbb{R}$ as in $\mathbb{Q}$.

We now have isomorphisms between the arithmetic operations in $\mathbb{R}_Q$ and $\mathbb{Q}$ and also between order in $\mathbb{R}_Q$ and $\mathbb{Q}$. Thus the rational cuts $\alpha_x \in \mathbb{R}_Q$ behave in every way identically to the corresponding rational numbers $x \in \mathbb{Q}$, so we will no longer need to distinguish between them; we can consider $\mathbb{Q}$ to be a subset of $\mathbb{R}$, writing x rather than α_x, 1 rather than α_1, etc., although we will persist with the Greek letter notation even for rational cuts when we want to emphasise their definition as cuts.

Determining whether the products of rational and irrational numbers are rational or irrational follows similar principles to those applying to sums:

Exercise 7.5.7.

(i) Prove that if α is rational and non-zero, and β is irrational, then $\alpha \times \beta$ is irrational.

(ii) If α is irrational, find an irrational β (in terms of α) such that $\alpha \times \beta$ is rational.

The requirement for a field to be an **ordered field** is that the order relations should have the same five properties as in an ordered ring, see Definition 4.4.9. We have already obtained all of these except the last, the multiplication property:

Theorem 7.5.8. *If* $\alpha > 0$ *and* $\beta > 0$, *then* $\alpha\beta > 0$.

Proof. Theorem 7.3.5 identifies being positive (i.e. membership of $\mathbb{R}_+$) with being greater than zero. Referring to Definition 7.5.1, if α and β are positive, there exist $r \in \alpha^+$ and $q \in \beta^+$ with $r > 0$ and $q > 0$. Hence there exists $t \in (\alpha \otimes \beta)$ such that $t = r \times q > 0$, so $\alpha \otimes \beta > 0$.   □

Having now established that the Real Numbers are an ordered field, all the order properties that are consequent on this structure (including those for an ordered ring) can be stated without further proof. We restate some of the most important properties here, in the notation we have been using for $\mathbb{R}$.

Theorem 7.5.9. *Let* α, β, γ *be real numbers. Then:*

$$\text{If } \gamma > 0 \text{ and } \alpha < \beta, \text{ then } \alpha \times \gamma < \beta \times \gamma;$$

$$\text{If } \gamma < 0 \text{ and } \alpha < \beta, \text{ then } \alpha \times \gamma > \beta \times \gamma;$$

$$\text{If } \gamma > 0 \text{ and } \alpha < \beta, \text{ then } \alpha \div \gamma < \beta \div \gamma;$$

$$\text{If } \gamma < 0 \text{ and } \alpha < \beta, \text{ then } \alpha \div \gamma > \beta \div \gamma.$$

Theorem 7.5.10. *(i) If* $\gamma > 0$, *then* $\frac{1}{\gamma} > 0$. *If* $\gamma < 0$, *then* $\frac{1}{\gamma} < 0$.
(ii) If $0 < \alpha < \beta$, *then* $\frac{1}{\alpha} > \frac{1}{\beta}$. *If* $0 > \alpha > \beta$, *then* $\frac{1}{\alpha} < \frac{1}{\beta}$.

Since the effect of multiplication on an order relation is the same in $\mathbb{R}$ as in $\mathbb{Z}$ and $\mathbb{Q}$, we could write down a theorem of the same form as Theorem 4.4.20 for $\mathbb{R}$, and algebraic inequalities in $\mathbb{R}$ can then be solved in the same way as was done in the Examples in $\mathbb{Z}$ following Theorem 4.4.20.

In $\mathbb{Z}$ and $\mathbb{Q}$ it was quite easy to prove theorems of the form, "For any $y \in \mathbb{S}$ and any $x \in \mathbb{S}_+$, $\exists k \in \mathbb{N}$ such that $kx > y$", where $\mathbb{S}$ is the relevant number system and multiplication by a natural number is permissible since $\mathbb{N}$ is considered to be a subset of our other number systems. Given the continuum nature of $\mathbb{R}$, it is not so obvious that a similar theorem applies in $\mathbb{R}$: might there exist a positive real number α so close to zero, and another real number β so large, that no natural number multiple k is sufficiently large to yield $k\alpha > \beta$? The next theorem shows that no such α and β exist in $\mathbb{R}$.

Theorem 7.5.11 (The Archimedean property of $\mathbb{R}$**).** *For any* $\beta \in \mathbb{R}$ *and any* $\alpha \in \mathbb{R}_+$, $\exists k \in \mathbb{N}$ *such that* $k\alpha > \beta$.

Proof. Suppose to the contrary that there does exist $\beta \in \mathbb{R}$ such that $k\alpha \leq \beta$ for every $k \in \mathbb{N}$. Then β is an upper bound for the set

$$B := \{\zeta \in \mathbb{R} : \zeta = k\alpha, k \in \mathbb{N}\}.$$

So B has a supremum in $\mathbb{R}$ [*Theorem 7.3.7*]; let $\sigma = \sup B$. Since $\alpha > 0$, we have $\sigma - \alpha < \sigma$, so $\sigma - \alpha$ is *not* an upper bound of B (it is less than the least upper bound). Thus $\exists m \in \mathbb{N}$ with $m\alpha > \sigma - \alpha$, so that $\sigma < (m+1)\alpha$. But $(m+1) \in \mathbb{N}$, so $(m+1)\alpha \in B$, and we have contradicted σ being an upper bound for B. We conclude that there does *not* exist $\beta \in \mathbb{R}$ such that $k\alpha \leq \beta$ for every $k \in \mathbb{N}$.   □

The Archimedean property is key to showing **denseness** properties of $\mathbb{R}$, similar to Theorem 6.5.19.

Theorem 7.5.12. *Both the rational numbers and the irrational numbers are dense in* $\mathbb{R}$: *given* $\alpha, \beta \in \mathbb{R}$ *with* $\alpha < \beta$, *there exist countably infinitely many rational numbers* x_n *and countably infinitely many irrational numbers[i]* γ_n *such that*

$$\alpha < x_n < \beta \quad and \quad \alpha < \gamma_n < \beta.$$

Proof. We first find a single rational x such that $\alpha < x < \beta$. If $\alpha < 0 < \beta$ we can take $x = 0$. Otherwise, since $\alpha < \beta$ we have $\beta - \alpha > 0$, and so from Theorem 7.5.11 there exists $k \in \mathbb{N}$ such that $k(\beta - \alpha) > 1$,

$$\therefore \frac{1}{k} < \beta - \alpha. \tag{7.3}$$

Now consider the case where $\alpha \geq 0$, so $\beta > 0$. Given some $k \in \mathbb{N}$ satisfying (7.3), let m be the least multiple of $\frac{1}{k}$ which is greater than α:

$$m := \min\left\{n \in \mathbb{N} : \frac{n}{k} > \alpha\right\},$$

where such m exists because of Theorem 7.5.11 and the Well-Ordering property of $\mathbb{N}$. So $\frac{m}{k} > \alpha$ but $\frac{m-1}{k} \leq \alpha$, and hence

$$\frac{m}{k} = \frac{m-1}{k} + \frac{1}{k} < \alpha + (\beta - \alpha) = \beta,$$

where we have used (7.3) to obtain the inequality. We have found a rational number $x = \frac{m}{k}$ satisfying $\alpha < x < \beta$.

Having found x between α and β, we can use the same method to find a rational number y between x and β; then Theorem 6.5.19 gives us countably infinitely many rationals x_n between x and y, and hence between α and β. Finally take any irrational number γ such that $0 < \gamma < 1$ (for example $\gamma = 1/\sqrt[2]{2}$), and let $\gamma_n = x_n - \gamma(x_n - x_{S(n)})$ for each $n \in \mathbb{N}$ (noting that the designation of x_n in the proof of Theorem 6.5.19 yields $x_n > x_{S(n)}$). This yields one irrational number between each adjacent pair of rationals x_n and $x_{S(n)}$, and so provides the countably infinitely many irrationals between α and β.

For the case where $\alpha < \beta \leq 0$, the proof above provides rationals and irrationals between the non-negative numbers $-\beta$ and $-\alpha$, and the required x_n and γ_n are just the negatives of these rationals and irrationals. $\square$

7.6 Exponentiation in $\mathbb{R}$

We start in the same way that we defined exponentiation in $\mathbb{Q}$ in Section 6.4. Natural number powers, α^n where $\alpha \in \mathbb{R}$ and $n \in \mathbb{N}$, are defined inductively as in Definition 3.4.1; the inductive definition is in terms of multiplication, so from Definition 7.5.1 and using the inductively defined product notation (Definition 3.7.8), the n'th power of a positive cut α is

$$\alpha^n = \{t \in \mathbb{Q} : t = \prod_{k=1}^{n} r_k \text{ where } r_k \in \alpha^+ (k = 1, 2, \ldots, n)\} \cup \mathbb{Q}_-, \tag{7.4}$$

where α^+ is defined in equation (7.2). The inductive definition implies that the Laws of Indices (the theorems on exponentiation in Section 3.4) must be satisfied for natural number

---

[i]Later we shall show that there are more than countably infinitely many irrational numbers between any α and β in $\mathbb{R}$.

powers, so we then define $\alpha^0 := 1$, $\alpha^{-n} := \frac{1}{\alpha^n}$ and $\alpha^{1/n} := \sqrt[n]{\alpha}$ for $n \in \mathbb{N}$ and $\alpha \neq 0$ to ensure that these Laws of Indices remain valid for all rational powers. As in $\mathbb{Q}$, we can be certain that all integer powers of real non-zero numbers exist in $\mathbb{R}$, since the definition of α^{-n} involves division which is closed on $\mathbb{R}\backslash\{0\}$. Where we depart from the theory in $\mathbb{Q}$ is in determining the range of values of $\alpha \in \mathbb{R}$ for which $\sqrt[n]{\alpha}$, and hence any fractional powers of α, exist in $\mathbb{R}$.

We can restrict attention to positive α, because Definition 7.5.4 for multiplication of negative real numbers ensures that we retain the usual rules, $(-\alpha)^{2m} = \alpha^{2m}$ and $(-\alpha)^{2m+1} = -\alpha^{2m+1}$; so for $\alpha \in \mathbb{R}_-$, even roots $\sqrt[2m]{\alpha}$ certainly do not exist in $\mathbb{R}$, while odd roots are given by $\sqrt[2m+1]{\alpha} = -\sqrt[2m+1]{-\alpha}$ if they do exist. We have already posed the representation of the n'th root of a positive rational x as a Dedekind cut,

$$\sqrt[n]{x} = \{r \in \mathbb{Q} : r^n < x\} \cup \mathbb{Q}_- \tag{7.5}$$

(which applies for both even and odd n when $x > 0$, although "$\cup\mathbb{Q}_-$" is redundant for odd n). We know that this is a Dedekind cut (by generalising the proof given for Example 7.2.11), so to show that it is a true representation of the n'th root of x, it remains to verify that its n'th power is the rational cut α_x corresponding to the rational number x:

Theorem 7.6.1. *Let*

$$\gamma_x := \{r \in \mathbb{Q} : r^n < x\} \cup \mathbb{Q}_-$$

for some $x \in \mathbb{Q}_+$. *Then*

$$(\gamma_x)^n = \alpha_x := \{r \in \mathbb{Q} : r < x\}.$$

Proof. As above, we adopt the notation γ_x^+ for the set of non-negative rationals in the cut γ_x; all cuts involved in the proof include the whole of $\mathbb{Q}_-$, so we only have to be concerned with the non-negative rationals.

Now from (7.4) we have

$$(\gamma_x)^n = \{t \in \mathbb{Q} : t = \prod_{k=1}^n r_k \text{ where } r_k \in \gamma_x^+ (k = 1, 2, \dots, n)\} \cup \mathbb{Q}_-,$$

and $r_k \in \gamma_x^+$ means that $0 \leq r_k^n < x$. This implies that

$$0 \leq \prod_{k=1}^n r_k^n < x^n$$

$$\therefore 0 \leq \left(\prod_{k=1}^n r_k\right)^n < x^n$$

$$\therefore 0 \leq \prod_{k=1}^n r_k < x \qquad [\textit{Theorem 6.5.16}]$$

so that $\prod_{k=1}^n r_k \in \alpha_x$. Conversely, if $t \in \alpha_x^+$ so that $0 \leq t < x$, we need to show that there exist $r_k \in \gamma_x^+ (k = 1, 2, \dots, n)$ such that $t = \prod_{k=1}^n r_k$. To do this, let r_1 be a rational number satisfying $t < r_1^n < x$ (which is possible according to Lemma 6.5.24), let $r_k = r_1$ for $k = 2, \dots, n-1$, and let $r_n = t/r_1^{n-1} < r_1$. Then $t = \prod_{k=1}^n r_k$ and $r_k^n < x$ for each $k = 1, 2, \dots, n$, so that $r_k \in \gamma_x^+$. $\square$

Having defined roots of positive rationals as cuts, we can proceed to define x^y for any $x \in \mathbb{Q}_+$ and $y \in \mathbb{Q}$, by writing y in fraction form, $y = \frac{m}{n} = \frac{1}{n} \times m$ where $m \in \mathbb{Z}$ and $n \in \mathbb{Z}_+$: to satisfy the laws of indices, we have

$$x^y = x^{m/n} := \sqrt[n]{x^m}, \tag{7.6}$$

which can then be written directly as a cut,

$$x^{m/n} := \{t \in \mathbb{Q} : t^n < x^m\} \cup \mathbb{Q}_-. \tag{7.7}$$

Thus all rational powers of positive rational numbers exist in $\mathbb{R}$, as do integer powers of all real numbers, and these all satisfy the laws of indices. We still need to define what it means to take an irrational power of any number, but before doing that we shall develop further theory on rational powers.

The laws of indices (and basic properties of multiplication) are useful in proving the irrationality of numbers formed as sums and products of roots and rational powers:

Example 7.6.2. Show that $\sqrt[2]{2} + \sqrt[2]{3}$ is irrational.

Solution. We know from Theorem 6.4.6 that both $\sqrt[2]{2}$ and $\sqrt[2]{3}$ are irrational. Squaring our sum of square roots:

$$(\sqrt[2]{2} + \sqrt[2]{3})^2 = 2 + 3 + 2\sqrt[2]{2}\sqrt[2]{3}$$
$$= 5 + 2\sqrt[2]{6},$$
$$\therefore \frac{1}{2}\left((\sqrt[2]{2} + \sqrt[2]{3})^2 - 5\right) = \sqrt[2]{6}. \tag{7.8}$$

Now if $\sqrt[2]{2} + \sqrt[2]{3}$ was rational, then so would be $\frac{1}{2}\left((\sqrt[2]{2} + \sqrt[2]{3})^2 - 5\right)$, and hence equation (7.8) would be equating a rational number to the irrational $\sqrt[2]{6}$, which is impossible. Hence $\sqrt[2]{2} + \sqrt[2]{3}$ must be irrational. □

Exercise 7.6.3. *Show that each of the following numbers are irrational:*
(i) $\sqrt[2]{2} + \sqrt[2]{3} + \sqrt[2]{5}$
     [Hint: Let $r = \sqrt[2]{2} + \sqrt[2]{3} + \sqrt[2]{5}$ and square the equation $r - \sqrt[2]{5} = \sqrt[2]{2} + \sqrt[2]{3}$.]
(ii) $\sqrt[2]{2} + \sqrt[3]{3}$
(iii) $\sqrt[2]{2} + \sqrt[3]{3}$ *[Hint: Let $r = \sqrt[2]{2} + \sqrt[3]{3}$ and cube the equation $r - \sqrt[2]{2} = \sqrt[3]{3}$.]*
(iv) $\sqrt[2]{p} + \sqrt[2]{q}$ *where p, q are coprime positive integers with both $\sqrt[2]{p}$ and $\sqrt[2]{q}$ irrational.*

Next we consider exponentiation of order relations, to the extent that exponentiation has been defined so far in $\mathbb{R}$. As in $\mathbb{Q}$, we start with theorems on natural number powers of positive real numbers:

Theorem 7.6.4. *For $\alpha, \beta \in \mathbb{R}_+$ and $n \in \mathbb{N}$,*

$$\alpha > \beta \Leftrightarrow \alpha^n > \beta^n.$$

Theorem 7.6.5. *For $\alpha \in \mathbb{R}$ with $\alpha > 1$ and $m, n \in \mathbb{N}$,*

$$m > n \Leftrightarrow \alpha^m > \alpha^n.$$

We are still using the inductive definition of natural number powers, with multiplication of positive real numbers having the same order properties as for natural numbers, so the proofs of these theorems in $\mathbb{N}$ remain valid here. However, we now know that roots of positive rational numbers exist in $\mathbb{R}$, so by considering the case where $\alpha^n = x \in \mathbb{Q}_+$ and $\beta^n = y \in \mathbb{Q}_+$, Theorem 7.6.4 yields:

Corollary 7.6.6. *For $x, y \in \mathbb{Q}_+$ and $n \in \mathbb{N}$,*

$$x > y \Leftrightarrow \sqrt[n]{x} > \sqrt[n]{y}.$$

Then for any positive rational exponent $a = \frac{m}{n}$, since $x^a = \sqrt[n]{x^m}$ we can combine Theorem 7.6.4 and Corollary 7.6.6 to yield:

Corollary 7.6.7. *For* $x, y, a \in \mathbb{Q}_+$,

$$x > y \Leftrightarrow x^a > y^a.$$

Furthermore, if $a, b \in \mathbb{Q}_+$, we can write $a = \frac{m}{n}$ and $b = \frac{p}{q}$ where $m, n, p, q \in \mathbb{Z}_+$. Also, let $\gamma = x^{1/nq}$ so that $x^a = \gamma^{mq}$ and $x^b = \gamma^{pn}$. Then since $a > b \Leftrightarrow mq > pn$, Theorem 7.6.5 yields:

Corollary 7.6.8. *For* $x \in \mathbb{Q}$ *with* $x > 1$ *and* $a, b \in \mathbb{Q}_+$,

$$a > b \Leftrightarrow x^a > x^b.$$

If $0 < x < 1$, then $1/x > 1$ so with $a > b$, Corollary 7.6.8 gives $(1/x)^a > (1/x)^b$ and hence $1/x^a > 1/x^b$. Theorem 7.5.10 (ii) then gives us:

Corollary 7.6.9. *For* $x \in \mathbb{Q}$ *with* $0 < x < 1$ *and* $a, b \in \mathbb{Q}_+$,

$$a > b \Leftrightarrow x^b > x^a.$$

A theorem of the same form as Theorem 6.5.22 in $\mathbb{Q}$ is valid in $\mathbb{R}$:

Theorem 7.6.10. *For any* $\beta \in \mathbb{R}$ *and* $\alpha \in \mathbb{R}$ *with* $\alpha > 1$, $\exists k \in \mathbb{N}$ *such that* $\alpha^k > \beta$.

To prove this, note that Bernoulli's inequality (Lemma 6.5.21), and particularly its strict form (6.1), remains valid for $x \in \mathbb{R}$ (possibly irrational), since the proof is by induction on the exponent n; combining this with the Archimedean property (Theorem 7.5.11) allows a proof of Theorem 7.6.10 by the same method as in $\mathbb{Q}$. Related to this theorem is a useful theorem on roots: informally, we can say that n'th roots of positive rationals become closer to 1 as n increases:

Exercise 7.6.11. *For any* $x > 1$, *let* R_x *be the set of* n'*th roots of* x,

$$R_x := \{\gamma \in \mathbb{R} : \gamma = x^{1/n}, n \in \mathbb{N}\}.$$

Use Theorem 7.6.10 to show that $\inf(R_x) = 1$.[ii]

We now consider the exponentiation of a general positive real number by a real number. Because exponentiation is not closed in $\mathbb{Q}$, a definition along the lines of Definitions 7.4.1 and 7.5.1, involving $\{t \in \mathbb{Q} : t = r^q \ldots\}$, would not make sense since for most choices of $r, q \in \mathbb{Q}$ the number r^q would not exist in $\mathbb{Q}$. Even if we ignored this, and just considered all the values of r^q that were rational, the proof that the resulting set is a Dedekind cut could not be done in the same way as for addition and multiplication, because those proofs involved the inverse operations which cannot be done for exponentiation in $\mathbb{Q}$. However, the Axiom of Completeness will resolve this difficulty.

As when defining multiplication in $\mathbb{R}$, we need to consider several classes of numbers, starting with the case where the operation preserves order when applied to rational numbers. For the non-commutative operation of exponentiation, this is where the base is greater than 1 and the exponent is positive (the conditions for Corollary 7.6.8). So we now propose the definition:

Definition 7.6.12. *If* $\alpha, \beta \in \mathbb{R}$ *with* $\alpha > 1$ *and* $\beta > 0$, *then*

$$\alpha^\beta := \sup\{t \in \mathbb{R} : t = r^q \text{ where } r \in \alpha^+ \text{ and } q \in \beta^+ \backslash \{0\}\}.$$

[ii]If $0 < x < 1$, we have $\sup(R_x) = 1$ similarly. In both cases, $x > 1$ and $0 < x < 1$, the theorems are true when $x \in \mathbb{Q}$ is replaced with $\alpha \in \mathbb{R}$, but we haven't yet formally defined non-integer powers of real numbers.

Here, the strictly positive rational powers q (in $\beta^+ \backslash \{0\}$)[iii] of non-negative rational bases r (in α^+) are known to exist in $\mathbb{R}$; so according to the Axiom of Completeness (Theorem 7.3.7), to verify that α^β has been defined as a real number we only need to verify that the set in the definition is bounded above. Take $s \in \mathbb{Q}\backslash\alpha$ and $p \in \mathbb{Q}\backslash\beta$, so that $r < s$ for every $r \in \alpha$ and $q < p$ for every $q \in \beta$. Then $r^q < r^p < s^p$ [*Corollaries 7.6.7 and 7.6.8*]. So s^p is an upper bound of $\{t \in \mathbb{R} : t = r^q$ where $r \in \alpha^+$ and $q \in \beta^+\backslash\{0\}\} \cup \mathbb{Q}_-$, as required.

Thus we have verified that α^β is a Dedekind cut, i.e. a set of rationals, even though in Definition 7.6.12 it is the supremum of a set of (mostly irrational) numbers formed as rational powers. But each of those rational powers r^q is itself a cut: see equation (7.7); and the rational numbers in the cut α^β are simply those in the union of all the cuts r^q of which α^β is the supremum. This allows us to check that the Laws of Indices are obeyed by real powers. We state these laws as a theorem, and we provide a proof of one of them: similar reasoning may be used to prove the others.

Theorem 7.6.13 (Laws of indices in $\mathbb{R}$). *For $\alpha, \beta, \gamma, \delta \in \mathbb{R}_+$ with $\alpha > 1$ and $\beta > 1$:*

(a) $\alpha^{\gamma+\delta} = \alpha^\gamma \times \alpha^\delta$

(b) $(\alpha \times \beta)^\gamma = \alpha^\gamma \times \beta^\gamma$

(c) $\alpha^{\gamma \times \delta} = (\alpha^\gamma)^\delta$

Proof of (a). Given $\alpha > 1$ and $\gamma, \delta \in \mathbb{R}_+$, the multiplication $\alpha^\gamma \times \alpha^\delta$ yields a Dedekind cut which, according to Definition 7.5.1, consists of products of rationals in the cuts α^γ and α^δ. But these rational products can now be identified as products of all pairs of rationals from cuts r^q and r^p of which α^γ and α^δ are the respective suprema according to Definition 7.6.12. Now, the rational powers are known to obey the law $r^q \times r^p = r^{q+p}$ (by *definition* for $p, q \in \mathbb{Q}\backslash\mathbb{N}$, see following equation (7.4)); and the cut $\gamma + \delta$ consists of all sums $q + p$ of rationals q and p in the cuts γ and δ. So the rationals in the cut $\alpha^\gamma \times \alpha^\delta$ are the rationals in all the cuts representing the rational powers r^{q+p}, which are the rationals in the cut $\alpha^{\gamma+\delta}$. $\qquad\square$

We can then ensure that the Laws of Indices are obeyed for exponentiation with all real exponents and all positive real bases by *defining* α^β such that the laws are satisfied in those cases not covered by Definition 7.6.12 and Theorem 7.6.13:

Definition 7.6.14. For $\alpha, \beta \in \mathbb{R}$:

$1^\beta := 1$

If $\beta > 0$, then $0^\beta := 0$;

If $\alpha > 0$, then $\alpha^0 := 1$;

If $0 < \alpha < 1$ and $\beta > 0$, then $\alpha^\beta := \dfrac{1}{(1/\alpha)^\beta}$;

If $\alpha > 0$ and $\beta < 0$, then $\alpha^\beta := \dfrac{1}{\alpha^{-\beta}}$.

In the last two stages of this definition, we are defining new cases in terms of what has been previously defined: when $0 < \alpha < 1$, $1/\alpha > 1$ so is covered by Definition 7.6.12; similarly when $\beta < 0$, we have $-\beta > 0$ which is covered by that definition.

The laws governing exponentiation in order relations take the same form for real powers as those for rational powers in Corollaries 7.6.7–7.6.9. We state these laws as a theorem, and they can be proved by similar reasoning to that used above for the proof of Theorem 7.6.13(a).

---

iii The exclusion of zero from β^+ is to avoid including 0^0 in our set.

Theorem 7.6.15 (Exponentiation in order relations in $\mathbb{R}$). *For any* $\alpha, \beta, \gamma, \delta \in \mathbb{R}_+$:

> *(a)* $\alpha > \beta \Leftrightarrow \alpha^\gamma > \beta^\gamma$;
>
> *(b)* $\gamma > \delta \Leftrightarrow \alpha^\gamma > \alpha^\delta$ *for* $\alpha > 1$;
>
> *(c)* $\gamma > \delta \Leftrightarrow \alpha^\gamma < \alpha^\delta$ *for* $0 < \alpha < 1$.

Relations involving negative powers then follow from $\alpha^\gamma = 1/\alpha^{-\gamma}$ *and:*

> *(d)* *If* $0 < \alpha < \beta$, *then* $\dfrac{1}{\alpha} > \dfrac{1}{\beta}$.

Although the laws governing exponentiation take the same form in $\mathbb{R}$ as in $\mathbb{Q}$, we cannot find any kind of isomorphism between exponentiation in $\mathbb{R}_\mathbb{Q}$ and in $\mathbb{Q}$ because the operation is not closed in $\mathbb{Q}$. However, we do expect our definitions of exponentiation for general real numbers to be consistent with the earlier definitions for rational powers of rational numbers in the case where α and β are rational cuts. Recalling that the rational cut corresponding to the rational number x is simply $\alpha_x = \{r \in \mathbb{Q} : r < x\}$, Definition 7.6.12 gives

$$\alpha_x^{\alpha_y} = \sup\{t \in \mathbb{R} : t = r^q \text{ where } 0 \le r < x \text{ and } 0 < q < y\},$$

and preservation of order according to Corollaries 7.6.7 and 7.6.8 would then yield this supremum as x^y when $x > 1$ and $y > 0$. Definition 7.6.14 then covers the cases where $0 \le x \le 1$ and/or $y \le 0$ by defining the exponentiation of real numbers to be consistent with the properties of rational powers of rational numbers.

It should also be clear that it is hopeless to try to extend exponentiation of real numbers to the case where the base is negative. Definition 7.6.12 works with $\alpha > 1$ because of Corollary 7.6.8, which states that positive rational powers x^a and x^b are ordered in the same way as the exponents a and b, provided $x > 1$. If $0 < x < 1$, the ordering of the rational powers is simply reversed from the ordering of the exponents. But if $x < 0$, with $a = \frac{m}{n}$ and $b = \frac{p}{q}$ where m, n, q are all odd integers but p is even, then (7.6) would give x^a as negative while x^b is positive; and if n was even, x^a would not be defined at all. So ordering of exponents a and b gives no information about ordering of the rational powers x^a and x^b, and there is no sensible way to define α^β as a cut when α is negative.[iv]

Similarly, we only seek to define logarithms to positive bases of positive real numbers. With the logarithm defined as the second inverse of exponentiation, we require $\log_\alpha \beta = \gamma$ when $\alpha^\gamma = \beta$. It is not immediately clear that such $\gamma \in \mathbb{R}$ exists for any given $\alpha, \beta \in \mathbb{R}_+$, but the Axiom of Completeness assures us that if we set

$$\log_\alpha \beta := \sup\{t \in \mathbb{R} : \alpha^t < \beta\} \tag{7.9}$$

for $\alpha > 1$ and $\beta > 0$, that will be a Real Number. The restriction $\alpha > 1$ ensures that the set in (7.9) is bounded above, so that the Axiom of Completeness applies; and the definition clearly satisfies the "second inverse" requirement in a similar way to the Dedekind cuts used to represent roots ("first inverses") in Section 7.2. For $0 < \alpha < 1$, Corollary 6.4.17 shows that we must define

$$\log_\alpha \beta := -\log_{1/\alpha} \beta,$$

in terms of the logarithm to base $1/\alpha$ where $1/\alpha > 1$.

[iv] In the early days of electronic calculators, an error message would appear if one attempted to take a cube root (or other odd root) of a negative number, because the calculator would be using an algorithm which treated the power $1/3$ as a real number and approximated it to some level of precision.

In summary, exponentiation of a positive real number by a real exponent has now been defined to always yield a positive real number, and logarithms to a positive real base of a positive real number always yield a real number. Exponentiation was originally introduced in $\mathbb{N}$ as a binary operation, i.e. closed in that number system. We were unable to define exponentiation as closed in $\mathbb{Z}$ or $\mathbb{Q}$, but we now have the slightly less unsatisfactory situation that it is closed on $\mathbb{R}_+$ but not on $\mathbb{R}$. Further resolution of this issue awaits our definition of the Complex Numbers.

With the binary operations of addition and multiplication, we found that combining a rational number with an irrational number by these operations always yielded an irrational number. This was a simple consequence of the inverse operations, subtraction and division, being closed in the rationals (except for zero). In contrast, neither exponentiation nor its inverse operations (roots and logarithms) are closed in $\mathbb{Q}$, and we now show that it is possible for an exponentiation of any combination of rational and/or irrational numbers to yield either a rational or an irrational result. First, Corollary 6.4.7 shows that a rational power of a rational number may be either rational or irrational. Next, consider rational powers of the irrational number $2^{1/2}$: the cases $2^{1/2} \wedge 1 = 2^{1/2}$ and $2^{1/2} \wedge 2 = 2$ (where for clarity we use the $\wedge$ notation from Section 3.4 for exponentiation) show that rational powers of irrational numbers can be either irrational or rational. It remains to consider irrational powers of either rational or irrational numbers. Since the equation $x \wedge \alpha = z$ is equivalent to $\log_x z = \alpha$, Corollary 6.4.22 shows that an irrational power of a rational number may be rational. The remaining possibilities to be verified are:

(a) "rational $\wedge$ irrational = irrational";

(b) "irrational $\wedge$ irrational = rational";

(c) "irrational $\wedge$ irrational = irrational".

For this purpose, consider the number,

$$
\begin{aligned}
\theta \quad &:= \quad 2^{1/2} \wedge 2^{1/2} & &\text{(7.10)} \\
&= \quad 2 \wedge \left(\frac{1}{2} \times 2^{1/2}\right) & &[\textit{Theorem 7.6.13(c)}] \\
&= \quad 2 \wedge 2^{-1/2}. & &\text{(7.11)}
\end{aligned}
$$

Here θ has been expressed both as an irrational power of an irrational number and as an irrational power of a rational number; but we do not know whether θ itself is rational or irrational.[v] However, now consider the numbers

$$
\theta \wedge 2^{1/2} = \left(2^{1/2} \wedge 2^{1/2}\right) \wedge 2^{1/2} = 2^{1/2} \wedge \left(2^{1/2} \times 2^{1/2}\right) = 2^{1/2} \wedge 2 = 2 \qquad \text{(7.12)}
$$

and

$$
\theta \wedge 2^{-1/2} = \left(2^{1/2} \wedge 2^{1/2}\right) \wedge 2^{-1/2} = 2^{1/2} \wedge \left(2^{1/2} \times 2^{-1/2}\right) = 2^{1/2} \wedge 1 = 2^{1/2}. \qquad \text{(7.13)}
$$

If θ is irrational, then possibilities (a), (b), (c), respectively are exemplified by equations (7.11), (7.12), (7.10). But if θ is rational, possibilities (a) and (b) are exemplified by (7.13) and (7.10), while to exemplify (c) we can show that

$$
\left(\theta \times 2^{1/2}\right) \wedge \left(2^{1/2} - 1\right) = 2^{1/2}. \qquad \text{(7.14)}
$$

---

[v]Nothing in the theory that we have covered can reveal whether θ is rational or irrational, but later we shall mention the *Gelfond-Schneider Theorem*, which shows that it is irrational.

Exercise 7.6.16.

(a) Supposing that θ is rational, why do we know that both $\theta \times 2^{1/2}$ and $2^{1/2} - 1$ are irrational?

(b) Noting the definition of θ in (7.10), use the laws of indices to show that equation (7.14) is true.

7.7 Expressing Real Numbers in Any Base

In this section we shall start by seeking representations of Real Numbers in any base $b \in \mathbb{N}\backslash\{1\}$, and will then find that this involves an understanding of sequences (as it did when finding base-b representations of Rational Numbers); this will lead us to a digression concerning an alternative way to define Real Numbers, before we return to the business of finding base-b representations. Our interest is mainly in irrational numbers; from Section 6.7 we already know how to express a rational number in any base, and we know that a number has a terminating or repeating representation in any base if and only if it is in $\mathbb{Q}$ [*Theorem 6.7.13*]. Thus a representation of an irrational number in any base must have infinitely many digits after the point, and cannot repeat indefinitely. This does not exclude the possibility of some pattern in the digits that is not indefinitely repeating but does allow one to find the k'th digit for any $k \in \mathbb{N}$ by a finite calculation; but in general we must expect that there is no such procedure to find the complete representation of an irrational number in any base. The best we can hope for is to find representations in some base which *approximate* a given irrational number to any desired precision. In particular, if $b \in \mathbb{N}\backslash\{1\}$ is a chosen base, we can approximate a positive real number[vi] α to within a precision of $1/b^k$ for some $k \in \mathbb{N}$ if we can find an integer q_k such that

$$\frac{q_k}{b^k} \le \alpha < \frac{q_k + 1}{b^k}. \tag{7.15}$$

Then the rational number q_k/b^k will have a terminating expansion with k digits after the point in base b, which provides an approximate representation of α in base b. It is not difficult to confirm that this is possible, for any $k \in \mathbb{N}$:

Theorem 7.7.1. *Given $\alpha \in \mathbb{R}_+$, $b \in \mathbb{N}\backslash\{1\}$ and $k \in \mathbb{N}$, there exists a unique $q_k \in \mathbb{Z}$ satisfying (7.15).*

Proof. From the Archimedean property of $\mathbb{R}$, there exists $m \in \mathbb{N}$ such that $m > \alpha b^k$; so the set

$$M := \{m \in \mathbb{N} : m > \alpha b^k\}$$

is non-empty, and hence has a least member (by the Well-ordering of $\mathbb{N}$). Let $q_k = \min(M) - 1$; then $q_k + 1 \in M$, meaning that

$$q_k + 1 > \alpha b^k;$$

but $q_k \notin M$ (since q_k is less than the least member of M), so by trichotomy applied to the definition of M we have

$$q_k \le \alpha b^k.$$

[vi]As in $\mathbb{Q}$, representations of negative real numbers $\alpha < 0$ are found by considering the positive number $-\alpha$ and then placing a minus sign in front of the representation.

Dividing our inequalities for $q_k + 1$ and q_k by b^k, we obtain (7.15).

Uniqueness is easy to see: any integer $\widetilde{q_k} < q_k$ will not satisfy $\alpha < (\widetilde{q_k} + 1)/b^k$, while any integer $\widetilde{q_k} > q_k$ will not satisfy $\widetilde{q_k}/b^k \leq \alpha$. $\square$

There is one integer q_k satisfying (7.15) for each $k \in \mathbb{N}$, so we have a countably infinite set of approximations to the real number α. The next two theorems give a clearer understanding of how these approximations represent α.

Theorem 7.7.2. *Given* $\alpha \in \mathbb{R}$ *with* q_k *satisfying (7.15) for each* $k \in \mathbb{N}$,

$$\alpha = \sup\left\{\frac{q_k}{b^k} : k \in \mathbb{N}\right\}.$$

Proof. From (7.15), $q_k/b^k \leq \alpha$ for each $k \in \mathbb{N}$, so α is an upper bound for the set of approximations q_k/b^k. To show that it is the supremum, we need to verify that no other upper bound is less than α.

Suppose to the contrary that β is an upper bound and $\beta < \alpha$. Then $\alpha - \beta > 0$, so $1/(\alpha - \beta) > 0$, and $\exists K \in \mathbb{N}$ such that $b^K > 1/(\alpha - \beta)$ [*Theorem 7.6.10*]. Thus (since $\alpha - \beta > 0$), we have

$$(\alpha - \beta)b^K > 1$$
$$\therefore \beta b^K < \alpha b^K - 1$$
$$\therefore \beta < \alpha - \frac{1}{b^K}$$
$$< \frac{q_K}{b^K}. \qquad\qquad [\textit{Inequalities (7.15)}]$$

The fact that $\beta < q_K/b^K$ for some $K \in \mathbb{N}$ contradicts β being an upper bound for the set of approximations. Hence α is the *least* upper bound. $\square$

The countably infinite set of approximations constitute a *sequence* according to Definition 6.8.1, since for each $k \in \mathbb{N}$ there is an approximation q_k/b^k. Formally, for any given $\alpha \in \mathbb{R}_+$ and $b \in \mathbb{N}\backslash\{1\}$, the sequence is a function $Q_{\alpha,b} : \mathbb{N} \to \mathbb{R}$ defined by

$$Q_{\alpha,b} : k \mapsto \frac{q_k}{b^k} \quad \text{where } q_k \text{ satisfies (7.15).} \tag{7.16}$$

In Section 6.8 we interpreted repeating representations of rational numbers as series (which are sequences of partial sums), *converging* to the rational numbers being represented. For real numbers, we now have the following theorem on our sequences of approximations:

Theorem 7.7.3. *The sequence* $Q_{\alpha,b}(k)$ *defined in (7.16) converges to* α.

Proof. Given any $\epsilon > 0$, $\exists N \in \mathbb{N}$ such that $b^N > \frac{1}{\epsilon}$ [*Theorem 7.6.10*], so $1/b^N < \epsilon$ [*Theorem 7.5.10 (ii)*]. Also, $1/b^k < 1/b^N$ for all $k > N$. From (7.15),

$$0 \leq \alpha - \frac{q_k}{b^k} < \frac{1}{b^k}$$

for each $k \in \mathbb{N}$; so when $k > N$ we have

$$0 \leq \alpha - Q_{\alpha,b}(k) < \frac{1}{b^k} < \epsilon$$

so that

$$|Q_{\alpha,b}(k) - \alpha| < \epsilon.$$

Thus the sequence $Q_{\alpha,b}(k)$ of base-b approximations converges to α according to Definition 6.8.4 of convergence. $\square$

The last two theorems highlight the contrast between $\mathbb{Q}$ and $\mathbb{R}$. If α is irrational, Theorem 7.7.2 finds an irrational supremum in $\mathbb{R}$ for a set of rational numbers (the base-b approximations to α), where no supremum exists within $\mathbb{Q}$; and Theorem 7.7.3 finds a sequence of rational numbers converging to an irrational limit in $\mathbb{R}$, where that sequence does not converge at all in $\mathbb{Q}$. How do we know that it does not converge at all in $\mathbb{Q}$? If it did, it would have some rational limit, L, which must differ from its irrational limit α in $\mathbb{R}$.[vii] But a sequence cannot converge to two different limits: for suppose $f(n)$ converged to both L_1 and L_2 where $L_2 > L_1$ (without loss of generality). Let $\epsilon = (L_2 - L_1)/3$; we can find N such that both $-\epsilon < f(n) - L_1 < \epsilon$ and $-\epsilon < f(n) - L_2 < \epsilon$ for all $n > N$ (since $f(n)$ converges to both L_1 and L_2). Inserting our chosen value of ϵ, you can easily verify that $f(n) - L_1 < \epsilon$ contradicts $-\epsilon < f(n) - L_2$.

There is an intuitive notion that a sequence $f(n)$ converges to a limit L if the terms $f(n)$ become ever closer to L as n increases. This is fulfilled by the sequence $Q_{\alpha,b}(k)$ in $\mathbb{R}$, but not in $\mathbb{Q}$, where the values of $Q_{\alpha,b}(k)$ may appear to approach some limiting value, and yet the sequence does not converge. But what we shall prove to be true in $\mathbb{Q}$ as in $\mathbb{R}$ is that the terms become closer *to each other* as k increases. This property is formalised in the following definition:-

Definition 7.7.4. A sequence $f : \mathbb{N} \to \mathbb{S}$ is a **Cauchy sequence** if, for any $\epsilon > 0$, there exists $N \in \mathbb{N}$ such that
$$|f(m) - f(n)| < \epsilon$$
whenever $m > N$ and $n > N$.[viii]

We shall prove the following theorem, which carries the implication that our sequence of base-b approximations is Cauchy[ix] but will also be useful in achieving the original objective of this section, to develop representations of real numbers in any base.

Theorem 7.7.5. *If integers q_k satisfy (7.15) for all $k \in \mathbb{N}$ for some given $\alpha \in \mathbb{R}$ and $b \in \mathbb{N}\backslash\{1\}$, then*
$$b^{m-n}q_n \le q_m < b^{m-n}(q_n + 1)$$
for any $m, n \in \mathbb{N}$ with $m > n$.

Proof. Replacing k with m and with n in (7.15), and multiplying by b^m in each case, we obtain
$$q_m \le \alpha b^m < q_m + 1$$
and
$$b^{m-n}q_n \le \alpha b^m < b^{m-n}(q_n + 1).$$
Using transitivity among these inequalities, we find $b^{m-n}q_n < q_m + 1$, so that $b^{m-n}q_n \le q_m$ [*Theorem 4.4.22*], and also $q_m < b^{m-n}(q_n + 1)$. The last two inequalities are those in Theorem 7.7.5. □

[vii]We are being rather loose by talking about "some rational limit, L". Is L a rational number in $\mathbb{Q}$ or a rational cut in $\mathbb{R}$? Strictly, we should prove that a sequence of rational numbers converges in $\mathbb{Q}$ if and only if the sequence of corresponding rational cuts converges in $\mathbb{R}$. Pedantic readers may wish to complete this proof, which uses the order isomorphism and the denseness properties, so that a real ϵ' can always be found between 0 and a rational ϵ, and *vice versa*.

[viii]As for Definition 6.8.4, the modulus in this definition can be replaced with a metric in more general codomains.

[ix]A mark of Augustin-Louis Cauchy's greatness as the father of Analysis is the common use of his surname as an adjective in this way.

To see that this theorem implies that $Q_{\alpha,b}(k)$ is Cauchy, take any $\epsilon > 0$ and find $N \in \mathbb{N}$ such that $1/b^N < \epsilon$ (as was done when proving Theorem 7.7.3). Divide the inequalities in Theorem 7.7.5 by b^m to obtain $q_n/b^n \leq q_m/b^m < (q_n + 1)/b^n$, so that

$$0 \leq \frac{q_m}{b^m} - \frac{q_n}{b^n} < \frac{1}{b^n},$$

and since $1/b^n < 1/b^N$ when $n > N$, we have shown that the difference between the m'th and n'th terms of the sequence is less than $1/b^N$, hence less than ϵ.

Just as a set that is bounded above must have a supremum in $\mathbb{R}$ but may not have one in $\mathbb{Q}$, a Cauchy sequence must converge to a limit in $\mathbb{R}$ but need not do so in $\mathbb{Q}$. So, just as we constructed $\mathbb{R}$ from $\mathbb{Q}$ by taking Dedekind cuts in the Rational Number line, we can construct $\mathbb{R}$ from $\mathbb{Q}$ by finding Cauchy sequences of Rational Numbers. If a Dedekind cut has a supremum in $\mathbb{Q}$, or if a Cauchy sequence converges to a limit in $\mathbb{Q}$, it represents a member of $\mathbb{R}_Q$, the subset of Real Numbers that biject to Rational Numbers; whereas if a Dedekind cut has no supremum in $\mathbb{Q}$, or a Cauchy sequence does not converge in $\mathbb{Q}$, it represents an irrational number, i.e. a Real Number which does not correspond to any Rational Number.[x] The Axiom of Completeness for Dedekind cuts is that every set that is bounded above has a supremum; an alternative Axiom of Completeness is that every Cauchy sequence converges. The proof that the two constructions of $\mathbb{R}$, and hence the two Axioms of Completeness, are equivalent requires showing that if we define $\mathbb{R}$ by means of Dedekind cuts, then every Cauchy sequence will converge to a limit in $\mathbb{R}$, and that if we define $\mathbb{R}$ by means of Cauchy sequences, then every set that is bounded above has a supremum. Readers are referred to Analysis textbooks if they wish to pursue this further; we shall content ourselves with having one definition of Real numbers, by Dedekind cuts.

Returning now to finding base-b approximations to a real number α, Theorem 7.7.5 assures us that our sequence of approximations satisfying (7.15) has the property that if $m > n$, the first n digits (after the point) of the m'th approximation will simply be the digits of the n'th approximation. For example, by the procedure presented below, you may verify that the irrational number $\sqrt[2]{2}$ has the following approximations in base ten: $q_1/b = 14/10 = 1 \cdot 4$, $q_2/b^2 = 141/10^2 = 1 \cdot 41$, $q_3/b^3 = 1414/10^3 = 1 \cdot 414, \ldots$, in which the first digit after the point (4) in the first approximation is retained in the second and third approximations, the second digit (1) in the second approximation is retained in the third (and all subsequent) approximations, etc. We are so used to this that it seems obvious, but it is not clear from (7.15). However, consider the integers in the inequalities in the statement of Theorem 7.7.5: the last $m - n$ digits of $b^{m-n}q_n$ and of $b^{m-n}(q_n + 1)$ are zeroes; the digit immediately before these zeroes is greater by 1 in $b^{m-n}(q_n + 1)$ than in $b^{m-n}q_n$;[xi] all digits before that are the same in the two integers; and these digits must also be the same in any number between those two integers, in particular q_m. The "strictly less than" relation in the right-hand inequality in Theorem 7.7.5 then ensures that the n'th digit in q_m is the same as that in $b^{m-n}q_n$ rather than the greater digit in $b^{m-n}(q_n + 1)$. Returning to the example of approximations to $\sqrt[2]{2}$ in base ten, with $n = 1$ and $m = 3$ we have $b^{3-1}q_1 = 10^2 \times 14 = 1400$, $b^{3-1}(q_1 + 1) = 10^2 \times 15 = 1500$, and Theorem 7.7.5 then requires that $1400 \leq q_3 < 1500$. So q_3 must have digits $14\ldots$, and is in fact 1414.

This property means that our sequence of approximations to a real number can just be written as a string of digits, which we can regard as the base-b representation of the

---

[x] We actually need to be more careful with Cauchy sequences: different Cauchy sequences may represent the same Real number, and it is necessary to define an equivalence relation, with each equivalence class of Cauchy sequences representing a unique Real number.

[xi] In the case where that digit in $b^{m-n}q_n$ is $b - 1$, the corresponding digit in $b^{m-n}(q_n + 1)$ will be zero, and there will also be a difference of 1 in the previous digit (and possibly earlier digits, until a digit not equal to $b - 1$ is encountered); but the argument here remains valid.

real number. If that number is irrational, we will have an infinite, non-repeating string of digits, so we will only ever be able to write down an approximation consisting of finitely many digits. How to find that approximation in practice depends on how the irrational number arises; in many cases of practical importance, that may be as the solution of an algebraic equation, or as the limit of a sequence. A particularly simple case is that of roots of positive rational numbers (which are solutions of algebraic equations of the form $\alpha^n = c$). An approximation to the number $\sqrt[n]{c}$ with k digits after the point in base b satisfies

$$\frac{q_k}{b^k} \leq \sqrt[n]{c} < \frac{q_k + 1}{b^k};$$

since all numbers in these inequalities are positive, multiplying through by b^k and taking n'th powers yields

$$q_k^n \leq cb^{kn} < (q_k + 1)^n$$

in which cb^{kn} is rational, so that finding q_k satisfying the inequalities requires only rational arithmetic and order calculations (or in fact integer arithmetic and order, once any denominator of a non-integer c is multiplied out), although in practice, an electronic calculator or computer is likely to be needed for most such calculations.

Example 7.7.6. Find approximations to $\sqrt[2]{6}$, (a) to 2 digits after the point in base ten, and (b) to 3 digits after the point in base 7.

Solution. (a) Pressing a few buttons on a calculator will give us an approximation to $\sqrt[2]{6}$ to more than 2 digits after the point, but we shall nevertheless use this as our first example to illustrate the theory. With $b = 10$ and $k = 2$, we require q_2 such that

$$\frac{q_2}{10^2} \leq \sqrt[2]{6} < \frac{q_2 + 1}{10^2}.$$

Multiplying through by 10^2 and squaring throughout, we obtain

$$q_2^2 \leq 6 \times 10^4 < (q_2 + 1)^2.$$

The integer $q_2 = 244$ satisfies this, since $244^2 = 59,536 \leq 60,000$ and $245^2 = 60,025 > 60,000$.[xii] Since q_k/b^k is written by placing a point k places from the end of the base-b representation of the integer q_k, this yields $2 \cdot 44_{\text{ten}}$ as our required approximation to $\sqrt[2]{6}$.
(b) We can do the calculation in base ten, only converting to base 7 at the last stage. With $b = 7$ and $k = 3$, we require q_3 such that

$$\frac{q_3}{7^3} \leq \sqrt[2]{6} < \frac{q_3 + 1}{7^3}.$$

Multiplying through by 7^3 and squaring throughout, we obtain

$$q_3^2 \leq 6 \times 7^6 < (q_3 + 1)^2.$$

Noting that $6 \times 7^6 = 705,894$, we obtain (by finding the square root of 705,894 using a calculator, and taking the integer part) that $q_3 = 840_{\text{ten}}$; the required inequalities are satisfied since $840^2 = 705,600 \leq 705,894$ and $841^2 = 707,281 > 705,894$. Converting 840_{ten} to base 7 by the procedure in Example 5.2.1, we obtain $840_{\text{ten}} = 2310_7$, so our required base-7 approximation to $\sqrt[2]{6}$ is $2 \cdot 310_7$. □

[xii]But how did we find this? We used a calculator to find the square root of 60,000, which has integer part 244.

Exercise 7.7.7. *Find approximations to the following irrational numbers, with the given numbers of digits after the point in the given bases:-*

(a) $\sqrt[2]{3}$; *2 digits in base eleven;*   (b) $\sqrt[2]{3}$; *6 digits in base 3;*

(c) $\sqrt[3]{11}$; *3 digits in base 7;*   (d) $\left(\frac{3}{4}\right)^{4/3}$; *4 digits in base 6;*

(e) $\left(\frac{3}{2}\right)^{3/5}$; *2 digits in base 4.*

Exercise 7.7.8. *Referring to your calculation for part (d) of the previous exercise, find a rational number x and an irrational number α such that*

$$\frac{49}{72} < x < \alpha < \left(\frac{3}{4}\right)^{4/3}.$$

7.8   Cardinality of $\mathbb{R}$

Cardinality is defined by bijections. This was originally done for finite sets [Definition 3.7.4], but was then extended to infinite sets; in particular, we denoted the cardinality of $\mathbb{N}$ as $\aleph_0$. Every other infinite set that we have encountered has had a bijection to $\mathbb{N}$, so has the same cardinality $\aleph_0$. This has included (maybe rather surprisingly) the number systems $\mathbb{Z}$ and $\mathbb{Q}$. But what about $\mathbb{R}$?

Theorem 7.8.1. $\mathbb{R}$ *is uncountable.*

We shall show that $\mathbb{R}$ does *not* have a bijection to $\mathbb{N}$, by the **diagonal argument** devised by Georg Cantor. The proof uses the theory of base-b representations of Real Numbers.

Proof. Consider just the real numbers between 0 and 1: to be precise, the set $\mathbb{R}_1 := \{\alpha \in \mathbb{R} : 0 \le \alpha < 1\}$. Every $\alpha \in \mathbb{R}_1$ can be represented in any base b by an infinite string of digits after the point (with just a zero before the point); in the case of a rational number with a terminating base-b representation, there is an infinite string of zeroes after the last non-zero digit, and we do not allow the alternative representation which has an infinite string of the digit $b-1$. Conversely, any string of digits $0 \cdot \rho_1\rho_2\rho_3 \ldots$ with $0 \le \rho_j < b\,(j = 1, 2, 3, \ldots)$ does represent a real number in $\mathbb{R}_1$: it can be interpreted as an infinite set of rational approximations q_k/b^k obtained by truncating the string after the k'th digit for $k = 1, 2, 3, \ldots$, and the supremum of the set of all these rationals is a real number: see Theorem 7.7.2.

Now suppose the real numbers in $\mathbb{R}_1$ to be countable: then they can be labelled with natural numbers, as $\alpha_1, \alpha_2, \alpha_3 \ldots$. So we can set out the list of all real numbers in $\mathbb{R}_1$ with their base-b representations:

$$\alpha_1 : 0 \cdot r_{11}r_{12}r_{13} \ldots$$
$$\alpha_2 : 0 \cdot r_{21}r_{22}r_{23} \ldots$$
$$\alpha_3 : 0 \cdot r_{31}r_{32}r_{33} \ldots$$
$$\vdots$$

in which r_{ij} is the j'th digit after the point in the representation of the i'th real number.

Now write down a string of digits,

$$0 \cdot \rho_1\rho_2\rho_3 \ldots$$

in which the only requirement is that $\rho_1 \neq r_{11}, \rho_2 \neq r_{22}, \rho_3 \neq r_{33}, \ldots$: each of the numbers $\alpha_1, \alpha_2, \alpha_3, \ldots$ has its digit on the *diagonal* of the above array of base-b representations differing from the corresponding digit in the string $0 \cdot \rho_1 \rho_2 \rho_3 \ldots$; so this new string is certainly not included in that list of base-b representations. But it *does* represent a real number in $\mathbb{R}_1$, contradicting the supposition that all such real numbers were included in the list. So there is no bijection between $\mathbb{R}_1$ and $\mathbb{N}$: we can say that $\mathbb{R}_1$ has a greater cardinality than $\mathbb{N}$, and since $\mathbb{R}_1 \subset \mathbb{R}$, $\mathbb{R}$ also has a greater cardinality than $\mathbb{N}$. $\square$

Exercise 7.8.2.

(i) Working in base 2, where the only possible digits are 0 and 1, each ρ_j in the string of digits $0 \cdot \rho_1 \rho_2 \rho_3 \ldots$ in the above proof is fully determined by the diagonal digits r_{jj}, for each $j \in \mathbb{N}$: if $r_{jj} = 0$, then $\rho_j = 1$, and vice versa. So it might be thought that the real numbers α_j could be arranged in such an order that the string $0 \cdot \rho_1 \rho_2 \rho_3 \ldots$ would end with an infinite string of 1's, which is not allowed (so the proof would fail). Why can this not happen?

(ii) Suppose that an attempt is made to use Cantor's diagonal argument to show that $\mathbb{Q}$ is uncountable, by listing the base-b expansions of all the rationals and trying to construct a base-b expansion of a rational number that is not in the list. At what point does the attempted proof break down?

In $\mathbb{Q}$ we found that there were countably infinitely many rational numbers between any two rationals a and b, so that the cardinality of the set $\{x \in \mathbb{Q} : a < x < b\}$ is the same as the cardinality of the whole of $\mathbb{Q}$: they both have a bijection to $\mathbb{N}$. A similar situation pertains to $\mathbb{R}$: given any two real numbers y and z, the set of reals between y and z has the same cardinality as the whole of $\mathbb{R}$. Before we prove this, it is useful to introduce some terminology and notation:

Definition 7.8.3. Given $y, z \in \mathbb{R}$ with $y < z$, we define:

The **closed interval** $[y, z] := \{\alpha \in \mathbb{R} : y \leq \alpha \leq z\}$.

The **open interval** $(y, z) := \{\alpha \in \mathbb{R} : y < \alpha < z\}$.[xiii]

The **semi-open intervals** $[y, z) := \{\alpha \in \mathbb{R} : y \leq \alpha < z\}$ and $(y, z] := \{\alpha \in \mathbb{R} : y < \alpha \leq z\}$.

The **semi-infinite intervals** $(-\infty, z] := \{\alpha \in \mathbb{R} : \alpha \leq z\}$, $\quad (-\infty, z) := \{\alpha \in \mathbb{R} : \alpha < z\}$, $[y, \infty) := \{\alpha \in \mathbb{R} : \alpha \geq y\}$, $\quad (y, \infty) := \{\alpha \in \mathbb{R} : \alpha > y\}$.[xiv]

Note in particular that the set $\mathbb{R}_1$ used in the proof of Theorem 7.8.1 is the interval $[0, 1)$. All closed, open, and semi-open intervals are **bounded**: they have an upper and a lower bound.

Theorem 7.8.4. *Every interval has the same cardinality as $\mathbb{R}$.*

Proof. Recall first the definition that two sets have the same cardinality if there is a bijection between them.

We start with the trickiest part of the proof, which is to show that open, semi-open, and closed intervals can all have the same cardinality. A bijection f from the semi-open interval

---

[xiii]Although the notation (y, z) is the same as for an ordered pair, in practice it should always be clear from the context whether an open interval or an ordered pair is intended.

[xiv]Note that the "interval" terminology and notation should only be used in $\mathbb{R}$. For example, if m, n are integers with $m < n$, do *not* write $[m, n]$ for the set $\{k \in \mathbb{Z} : m \leq k \leq n\}$. Note also that ∞ and $-\infty$ should not be treated as numbers; they are simply symbols in the notation for certain kinds of sets of real numbers.

$[0, 1)$ to the open interval $(0, 1)$ can be defined as

$$f : 0 \mapsto \frac{1}{2}$$

$$f : \frac{1}{n} \mapsto \frac{1}{n+1} \quad \text{for each } n \in \mathbb{N} \backslash \{1\}$$

$$f : \alpha \mapsto \alpha \quad \text{when } \alpha \neq 0 \text{ and } \alpha \neq \frac{1}{n} \text{ for any } n \in \mathbb{N}.$$

Exercise 7.8.5. *Write down the inverse bijection f^{-1} from $(0, 1)$ to $[0, 1)$.*

Bijections from $[0, 1]$ to $[0, 1)$ and from $(0, 1]$ to $(0, 1)$ can be defined using the same trick, except for mapping 1 to $\frac{1}{2}$ rather than 0 to $\frac{1}{2}$. Recalling that the relation of having a bijection is symmetric and transitive, we now have a bijection between any two of the four types of interval (closed, open, two types of semi-open) consisting of numbers between 0 and 1.

Next we find a bijection g between $(0, 1)$ and (y, z) for arbitrary $y, z \in \mathbb{R}$ with $y < z$. Such a bijection is defined by

$$g : \alpha \mapsto y + (z - y)\alpha.$$

To see that this is a bijection, note that the inverse function is

$$g^{-1} : \beta \mapsto \frac{\beta - y}{z - y}$$

and that if $y < \beta < z$, then $0 < \frac{\beta - y}{z - y} < 1$. The same function will biject from the closed and semi-open intervals $[0, 1]$, $[0, 1)$ and $(0, 1]$ to the respective closed and semi-open intervals $[y, z]$, $[y, z)$ and $(y, z]$. Then symmetry and transitivity will give us a bijection between any two bounded intervals (closed, open or semi-open); for example, a bijection from $[y_1, z_1]$ to (y_2, z_2) is made by bijecting first from $[y_1, z_1]$ to $[0, 1]$, then to $(0, 1)$, and finally to (y_2, z_2) using the bijections specified above.

Bijections between bounded and semi-infinite intervals may be devised as follows. From $[0, 1)$ to $[y, \infty)$ (for arbitrary $y \in \mathbb{R}$), the function h_+ defined by

$$h_+ : \alpha \mapsto y + \frac{\alpha}{1 - \alpha}$$

is a bijection, with its inverse function

$$h_+^{-1} : \beta \mapsto \frac{\beta - y}{1 + \beta - y}$$

yielding $0 \leq \frac{\beta - y}{1 + \beta - y} < 1$ whenever $\beta \geq y$. The same function also maps the open interval $(0, 1)$ to (y, ∞), and we can similarly biject $(-1, 0]$ or $(-1, 0)$ to $(-\infty, z]$ or $(-\infty, z)$, respectively, by

$$h_- : \alpha \mapsto z + \frac{\alpha}{1 + \alpha}.$$

Finally, combining a bijection from $[0, 1)$ to $[0, \infty)$ with a bijection from $(-1, 0)$ to $(-\infty, 0)$ gives a bijection from the bounded interval $(-1, 1)$ to the whole of $\mathbb{R}$, meaning that all bounded and semi-infinite intervals have bijections to $\mathbb{R}$. $\square$

The cardinality of $\mathbb{R}$ is often denoted as **c** (for "continuum"), but alternatively it may be denoted $2^{\aleph_0}$, where we recall that $\aleph_0$ is the cardinality of $\mathbb{N}$. The justification for the latter notation is as follows. In base 2 there are 2^k rational numbers with terminating representations having zero before the point and up to k digits after the point, where each of those k digits can be either 0 or 1. All real numbers have base-2 representations with up to $\aleph_0$ digits after the point; there are also finitely many digits before the point, but that leaves the total digit-count at $\aleph_0$. Hence by extension from the case of finitely many digits, there are $2^{\aleph_0}$ such base-2 representations. But, you may say, if I find representations of all real numbers in some other base b, I would obtain the cardinality of ℝ as $b^{\aleph_0}$. This does not invalidate the argument; it just shows that the cardinal numbers $b^{\aleph_0}$ are equal for all $b \in \mathbb{N}\setminus\{1\}$.

Exercise 7.8.6. *Prove by induction on k that there are b^k numbers with terminating base-b representations having zero before the point and up to k non-zero digits after the point.*

Theorem 7.8.4 shows that every interval in ℝ, however small, has the same cardinality **c**, which is greater than the cardinality of the whole of ℚ, which is $\aleph_0$. But is there any cardinality between $\aleph_0$ and **c**? That is, does there exist any uncountable set (i.e. with cardinality greater than $\aleph_0$) that has a bijection to a subset of ℝ but not to ℝ itself (so with cardinality less than **c**)? Georg Cantor thought that no such set existed, so that **c** is the "next" cardinal number after $\aleph_0$: this is the **Continuum Hypothesis**. But despite much effort, Cantor was unable to prove (or disprove) the hypothesis. This was not because he was insufficiently clever. It is because the Continuum Hypothesis is *undecidable*: starting from the axioms of Set Theory (which are now generally accepted to form the foundation for all of mathematics, although they have not been discussed in this book), Kurt Gödel found in 1940 that the Continuum Hypothesis could not be disproved, and then in 1963 Paul Cohen showed that it cannot be proved to be true. The existence of undecidable hypotheses, truths which cannot be proved by a process of logical deduction from axioms, was first demonstrated by Gödel in 1930 in his Incompleteness Theorem. In fact, this takes us back to where we started on number systems in Chapter 3, because Gödel's proof was based on Peano Arithmetic: the five axioms and the definitions of addition and multiplication for ℕ. Gödel showed that if Peano Arithmetic is consistent (i.e. cannot produce contradictions), then it must be possible to formulate undecidable propositions about it (recall that a proposition is a statement that can only be true or false, so is not self-contradictory). Peano Arithmetic is possibly the simplest mathematical system you can imagine, and Gödel's Theorem carries the implication that any more advanced mathematical system must also allow undecidable hypotheses. This has the important implication that it will never be possible to construct a machine to verify all mathematical theory: mathematicians will never be out of work!

7.9 Algebraic and Transcendental Numbers

We have tended to discuss the deficiencies of number systems in terms of the operations, or inverses of operations, that are not closed in those systems. The inverse of addition is not closed in ℕ, the inverse of multiplication is not closed in ℤ, and exponentiation is not closed in ℚ, nor in ℝ although it is closed in $\mathbb{R}_+$. However, we can also consider those deficiencies in terms of equations that cannot be solved in the number systems. Consider the following list of equations, involving only the operations of addition and multiplication (closed in ℕ) and the first two Natural Numbers, 1 and 2, and an unknown x for which we seek a solution.

- $x + 2 = 1$ has no solution in ℕ, but $x = -1 \in \mathbb{Z}$;
- $x \times 2 = 1$ has no solution in ℤ, but $x = \frac{1}{2} \in \mathbb{Q}$;

- $x \times x = 2$ has no solution in $\mathbb{Q}$, but $x = 2^{1/2} \in \mathbb{R}$;

- $x \times x + 2 = 1$ has no solution in $\mathbb{R}$.

The equations listed here are simple examples of **algebraic equations**, in which a polynomial with integer coefficients is equated to zero: specifically, an algebraic equation of **degree** n is of the form

$$a_0 + a_1 x + \cdots + a_{n-1} x^{n-1} + a_n x^n = 0 \qquad (7.17)$$

with $a_0, a_1, \ldots, a_{n-1}, a_n \in \mathbb{Z}$ and $a_n \neq 0$. So in the list above, the first two equations are of degree 1 and can be written in the form (7.17) as $1 + x = 0$ and $(-1) + 2x = 0$, while the remaining two are of degree 2 and may be written $(-2) + x^2 = 0$ and $1 + x^2 = 0$.

Definition 7.9.1. An **algebraic number** is a number x satisfying an algebraic equation.

Clearly rational numbers are algebraic: a rational number can be written as $\frac{p}{q}$ where $p, q \in \mathbb{Z}$, and satisfies $(-p) + qx = 0$, an algebraic equation of degree 1.

The procedure for finding a solution of a quadratic (degree 2) equation is well known: the equation

$$a_0 + a_1 x + a_2 x^2 = 0 \qquad (7.18)$$

can be rewritten by dividing through by a_2 and "completing the square" as

$$\left(x + \frac{a_1}{2a_2} \right)^2 + \left(\frac{a_0}{a_2} - \frac{a_1^2}{4a_2^2} \right) = 0,$$

which is satisfied when

$$x = -\frac{a_1}{2a_2} \pm \left(\frac{a_1^2}{4a_2^2} - \frac{a_0}{a_2} \right)^{1/2}$$

$$= \frac{1}{2a_2} \left(-a_1 \pm \sqrt[2]{a_1^2 - 4a_0 a_2} \right).$$

This final formula is usually remembered with the notations a, b, c in place of our a_2, a_1, a_0, respectively. So there are two real algebraic numbers (obtained using the $+$ and $-$ signs) satisfying the quadratic equation (7.18) if the **discriminant** $a_1^2 - 4a_0 a_2$ is positive, but no such numbers if this quantity is negative. This generalises our earlier observation that the equation $(-2) + x^2 = 0$ can be solved in $\mathbb{R}$ but the equation $1 + x^2 = 0$ has no solution in $\mathbb{R}$. In further generalisation, it can be shown that every algebraic equation of odd degree has at least one solution in $\mathbb{R}$,[xv] but equations of even degree may have no solution in $\mathbb{R}$;[xvi] obvious examples with no real solution are of the form $a_0 + x^{2m} = 0$ for any $m \in \mathbb{N}$ and any $a_0 > 0$.

This situation seems as unsatisfactory as being unable to exponentiate negative reals; and in Chapter 9 we shall meet the Complex Numbers, in which every algebraic equation does have a solution, and in which exponentiation is closed except for powers of zero. But for now there is one further question: although not all algebraic equations have solutions

---

[xv]This is usually proved using the Intermediate Value Theorem, an important theorem in Analysis but beyond the scope of this book; but it can also be obtained as a consequence of the Fundamental Theorem of Algebra: see Section 9.6.

[xvi]To find solutions for algebraic equations of degrees 3 and 4, there exist explicit procedures and formulae, more complicated than that for quadratic equations but still only involving addition, subtraction, multiplication, division, integer powers and roots. However, it has been shown that it is impossible to derive such a procedure for general algebraic equations of degree 5, although for simple examples (e.g. where one can factorise the polynomial by inspection), one may be able to find one or more solutions explicitly. In any case, inability to find solutions does not negate their existence!

in $\mathbb{R}$, do all Real Numbers satisfy algebraic equations? We shall answer this in a rather indirect way, by showing that real algebraic numbers are countable. To do this, we first need to establish some theory about polynomials and algebraic equations.

Theorem 7.9.2 (The Division Theorem for Polynomials). *If $P(x)$ is a polynomial of degree n with real coefficients, then, given any $\alpha \in \mathbb{R}$, there exists a unique polynomial $Q(x)$ of degree $n - 1$ and a unique real number R such that*

$$P(x) = (x - \alpha)Q(x) + R.^{\text{xvii}} \tag{7.19}$$

Proof. The existence of the polynomial $Q(x)$ and the number R that satisfy the given equation is best proved by constructing them. $P(x)$ has the form

$$P(x) = a_0 + a_1 x + \cdots + a_{n-1}x^{n-1} + a_n x^n,$$

and we need to find coefficients $b_0, b_1, \ldots, b_{n-2}, b_{n-1}$ in the polynomial

$$Q(x) = b_0 + b_1 x + \cdots + b_{n-2}x^{n-2} + b_{n-1}x^{n-1}$$

such that $P(x) - (x - \alpha)Q(x)$ is a constant, the required R in (7.19). Now,

$$P(x) - (x - \alpha)Q(x) = (a_0 + \alpha b_0) + (a_1 - b_0 + \alpha b_1)x + \cdots$$
$$+ (a_{n-1} - b_{n-2} + \alpha b_{n-1})x^{n-1} + (a_n - b_{n-1})x^n$$

and we can set the coefficients of each power of x to zero by letting $b_{n-1} = a_n$ (for the coefficient of x^n) and then $b_{k-1} = a_k + \alpha b_k$ successively for coefficients of x^k with k decreasing from $n - 1$ to 1, to finally leave $R = a_0 + \alpha b_0.^{\text{xviii}}$

To verify uniqueness, suppose that there is another polynomial $Q'(x)$ and another number R' satisfying $P(x) = (x - \alpha)Q'(x) + R'$. Subtracting this from (7.19), we obtain

$$0 = (x - \alpha)(Q(x) - Q'(x)) + (R - R').$$

Since $Q'(x)$ must be of the form $Q'(x) = b'_0 + b'_1 x + \cdots + b'_{n-2}x^{n-2} + b'_{n-1}x^{n-1}$, this becomes

$$0 = -\alpha(b_0 - b'_0) + ((b_0 - b'_0) - \alpha(b_1 - b'_1))x + \cdots$$
$$+ ((b_{n-2} - b'_{n-2}) - \alpha(b_{n-1} - b'_{n-1}))x^{n-1} + (b_{n-1} - b'_{n-1})x^n$$
$$+ (R - R').$$

It is important to note here that we are not trying to solve an equation for x; we are requiring the equation to be true *for all* real x. So, similarly to our earlier calculation of b_k, we need the coefficients of each power of x to be zero. This yields $b'_{n-1} = b_{n-1}$ (for the coefficient of x^n), and the requirement that $((b_{k-1} - b'_{k-1}) - \alpha(b_k - b'_k)) = 0$ then implies $b'_{k-1} = b_{k-1}$ successively for k decreasing from $n - 1$ to 1. Hence $Q'(x) = Q(x)$, and we are left with $R - R' = 0$. $\qquad \square$

Exercise 7.9.3. *Find the polynomial $Q(x)$ and the remainder R satisfying each of the following equations:*

(i) $2x^3 - 5x^2 + x - 3 = (x + 2)Q(x) + R$ *(ii) $x^4 + 3x^3 - 6x + 8 = (x - 3)Q(x) + R$*

---

^{xvii}A more general form of the Division Theorem for Polynomials is the subject of Investigation 2 at the end of this chapter.

^{xviii}It would perhaps be more in keeping with the spirit of this book to prove by induction that the formulae claimed for R and $Q(x)$ (in terms of the coefficients b_k) satisfy equation (7.19). Such a proof is left to the interested reader.

Corollary 7.9.4. *If $x = \alpha$ satisfies the equation $P(x) = 0$, then $(x - \alpha)$ is a factor of $P(x)$.*

Proof. Set $x = \alpha$ and $P(x) = 0$ in (7.19), and we obtain $R = 0$. So $P(x) = (x - \alpha)Q(x)$, which is what it means for $(x - \alpha)$ to be a factor of $P(x)$. $\qquad\square$

Exercise 7.9.5. *Check that $x = -2$ satisfies the equation*

$$4x^5 + 5x^4 - 3x^3 - 7x + 10 = 0.$$

Find the polynomial $Q(x)$ such that

$$4x^5 + 5x^4 - 3x^3 - 7x + 10 = (x + 2)Q(x).$$

We can now take the first step towards demonstrating the countability of the algebraic numbers.

Theorem 7.9.6. *An algebraic equation of degree n has at most n real solutions.*

Proof. We prove this by induction on the degree of the equation. For $n = 1$, we have an equation of the form $a_0 + a_1 x = 0$, with unique solution $x = -a_0/a_1$.

Now suppose as our inductive hypothesis that an equation of degree n has no more than n solutions, and let $P(x)$ be a polynomial of degree $S(n)$. If the equation $P(x) = 0$ has a solution $x = \alpha$, then from Corollary 7.9.4 we can write $P(x) = (x - \alpha)Q(x)$ where $Q(x)$ is of degree n. So $(x - \alpha)Q(x) = 0$, which requires either $x - \alpha = 0$ or $Q(x) = 0$. The equation $x - \alpha = 0$ has one solution, and the equation $Q(x) = 0$ has no more than n solutions (by the inductive hypothesis), so in total there are no more than $1 + n = S(n)$ solutions. $\qquad\square$

Recalling that algebraic numbers are simply defined to be solutions of algebraic equations, we now know enough about such solutions to prove:

Theorem 7.9.7. *The real algebraic numbers are countable.*

Proof. First we show that for any given $n \in \mathbb{N}$, there are countably many real solutions of algebraic equations of degree n. To verify this, we need to know how many equations of degree n can be constructed. Such an equation has $n + 1$ coefficients, each taken from the countable set $\mathbb{Z}$; so the set of all such equations bijects to the Cartesian product of $n + 1$ copies of $\mathbb{Z}$, which is countable [*Theorem 6.3.2*]. The set of real solutions of all these degree-n equations is thus a countable union of the sets of solutions of each equation. But each equation has finitely many solutions [*Theorem 7.9.6*], and a countable union of finite sets is countable [*Theorem 6.3.1*].

The set of all solutions of all algebraic equations is then the union of the sets of solutions for each degree n. Since $n \in \mathbb{N}$, this is a countable union of sets already shown to be countable, and hence is countable itself. $\qquad\square$

Note that all algebraic numbers are counted more than once in this process; in fact, any number that is a solution of an equation of degree n is also a solution of infinitely many equations of degree higher than n [Why?]. But this multiple counting means that the set of real algebraic numbers is smaller than we may have supposed, so does not negate the countability of the set.

The important consequence of Theorem 7.9.7 is that not all real numbers are algebraic. This is because we have already shown that $\mathbb{R}$ is uncountable; so there must actually be uncountably many non-algebraic real numbers (since if there were countably many non-algebraic real numbers, the union of the algebraic and non-algebraic numbers would still be countable). These numbers that are not the solution of any algebraic equation are called **transcendental numbers**. Although there are "far more" transcendentals than algebraics (uncountable *vs.* countable), the transcendentals are typically much more difficult to specify; and proving that a number is transcendental is typically much more difficult than proving a number to be irrational. The first instance of such a proof was in 1844 when Joseph Liouville proved that the series

$$\mathcal{L}(n) = \sum_{k=1}^{n} \frac{1}{10^{k!}}$$

(where $k!$ is the factorial of k) converges to a transcendental number. [The proof works for the sequence of partial sums of $1/b^{k!}$ for any base $b \in \mathbb{N}\backslash\{1\}$, but Liouville chose $b = 10$ simply because base ten is the default for representing numbers. In any case, the digits of the number when represented in the appropriate base are 1's at positions $k!$ for $k = 1, 2, \ldots$ after the point, and zeroes elsewhere; so $0 \cdot 110001000\ldots$, since $1! = 1, 2! = 2, 3! = 6$, with the next 1 occurring at the 24'th position since $4! = 24$.] More usefully, in 1934 Aleksandr Gelfond and Theodor Schneider independently proved that if α is algebraic (but not 0 or 1) and β is irrational and algebraic, then α^β is transcendental. So, for example with $\alpha = \beta = 2^{1/2}$ we have $\alpha^\beta = \theta$, the number defined in equation (7.10) and used in some of our examples of irrationality; the Gelfond-Schneider theorem shows θ to be transcendental. However, there are still only countably many numbers that may be shown to be transcendental according to the Gelfond-Schneider Theorem; and while many other transcendental numbers are defined as limits of sequences and series (like the Liouville number), there are again only countably many sequences that can be specified by finite formulae. This leaves uncountably many other transcendental numbers, which all have representations in any base, but with no finite way of specifying or determining the representations. So we conclude this chapter with the philosophical conundrum: how real is a Real Number that cannot be specified?

Investigations

1. A subset of $\mathbb{R}$, sometimes given the symbol $\mathbb{A}$ and known as the **Arithmetic Numbers** (although that term is also used for a different concept in Number Theory), consists of all numbers produced by combining finitely many natural numbers by finitely many arithmetic operations (additions, subtractions, multiplications, divisions, and exponentiations). For example, the number

$$\frac{7 - (9 - 4^{2/7})^{1/5}}{\left(8 + \frac{4}{29}\right)^{3/7}}.$$

is a member of $\mathbb{A}$, produced by eleven operations combining twelve natural numbers.

(a) Why was it not necessary to specify "taking roots" among the list of arithmetic operations?

(b) Obviously the definition of $\mathbb{A}$ should exclude the possibility of division by zero; for example, $1/(2 - 4^{1/2})$ is not valid as a member of $\mathbb{A}$. What other restrictions should be made, regarding exponentiation? In particular, what is wrong with the examples,

$$(2 - 4^{1/2})^{3 - 25^{1/2}} \quad \text{and} \quad \left(7^{2/3} - \frac{8}{1 - \left(\frac{2}{5}\right)^{9/7}}\right)^{5/4}$$

(c) Prove that $\mathbb{A}$ is countable. [Theorem 6.3.2 will be useful for this.]

2. Theorem 7.9.2 is a special case of the Division Theorem for Polynomials. The more general theorem is:

> If $P(x)$ is a polynomial of degree n, then, given any polynomial $D(x)$ of degree $m < n$, there exists a unique polynomial $Q(x)$ of degree $n - m$ and a unique polynomial $R(x)$ of degree $\leq m - 1$ such that
>
> $$P(x) = D(x)Q(x) + R(x)$$
>
> (with all coefficients real in all polynomials).

Prove this general Division Theorem.

3. Roots of rational numbers are algebraic: $\sqrt[n]{a}$ is a solution of $a - x^n = 0$. But in general, given an Arithmetic Number (as defined in question 1 above), it is not easy to find an algebraic equation that it satisfies, although there is a theorem that all Arithmetic Numbers are indeed algebraic. In this question you will consider some cases where it is not too difficult to find an algebraic equation satisfied by a given number.

(a) First use induction on n to prove the **Binomial Theorem**: for $a, b \in \mathbb{R}$ and $n \in \mathbb{N}$,

$$(a + b)^n = \sum_{k=0}^{n} \frac{n!}{(n-k)!\,k!} a^{n-k} b^k$$

where $k!$ is defined in Definition 3.7.9 for $k \in \mathbb{N}$, and $0! = 1$ (why is this consistent with Definition 3.7.9?).

(b) Find the algebraic equations satisfied by: (i) $x = a + \sqrt[2]{b}$, and (ii) $x = a + \sqrt[n]{b}$, where a, b are integers and $\sqrt[n]{b}$ is irrational for $n \geq 2$. [For (ii), use the Binomial Theorem to give explicit formulae for the coefficients in the algebraic equation.]

(c) Now suppose a and b are rational numbers, with $a = \frac{p}{q}$ and $b = \frac{r}{s}$ in lowest terms. Find the algebraic equations satisfied by: (i) $x = a + \sqrt[2]{b}$, and (ii) $x = a + \sqrt[n]{b}$, obtaining integer coefficients in the equations in terms of the integers p, q, r, s.

(d) Find the algebraic equation satisfied by $x = \sqrt[2]{2} + \sqrt[2]{3}$. [Start by squaring the equation $x = \sqrt[2]{2} + \sqrt[2]{3}$, then put the term involving a square root on one side of the resulting equation and everything else on the other side, and square again.]

(e) If $x = a + \sqrt[2]{b}$ and $y = c + \sqrt[2]{d}$, where a, b, c, d are integers, find the algebraic equation satisfied by $x + y$.

(f) It can be shown that if x and y are any real algebraic numbers, then so are $x + y$ and $x \times y$ (the proof requires linear algebra and abstract algebra outside the scope of this book). Why does this imply that the set of real algebraic numbers with the operations of addition and multiplication constitutes a field?

Chapter 8

Quadratic Extensions I: General Concepts and Extensions of $\mathbb{Z}$ and $\mathbb{Q}$

8.1   General Concepts of Quadratic Extensions

We have already hinted that the deficiencies of the Real Number system (the inabilities to exponentiate negative numbers and to solve all algebraic equations) will be remedied by a system called the Complex Numbers. But the Complex Numbers are a particular case of a larger class of number systems, obtained from previously defined number systems by a process called **quadratic extension**. Complex Numbers will be obtained by applying this process to $\mathbb{R}$ in the next chapter, but we shall first consider some number systems which are quadratic extensions of $\mathbb{Z}$ and $\mathbb{Q}$. While the extensions of $\mathbb{Z}$ and $\mathbb{Q}$ do not have the broad relevance throughout pure and applied mathematics that the Complex Numbers have, they are nevertheless important in Number Theory (where theorems about ordinary integers are often proved most easily by considering quadratic extensions of $\mathbb{Z}$) and in Abstract Algebra (where they provide important examples of rings and fields).

We start with some general concepts that are common to all quadratic extensions. We have previously defined two number systems, $\mathbb{Z}$ and $\mathbb{Q}$, as equivalence classes of ordered pairs from the previous number systems ($\mathbb{N}$ and $\mathbb{Z}$, respectively), with addition and multiplication operations defined to ensure isomorphisms with the previous number systems. Our procedure now starts by taking a number system $\mathbb{S}$ (which will be $\mathbb{Z}$ or $\mathbb{Q}$ in the cases examined in this chapter, and $\mathbb{R}$ in the next chapter), together with a number $k \in \mathbb{S}$ which has *no square root in* $\mathbb{S}$ (so there does not exist $x \in \mathbb{S}$ such that $x^2 = k$). Then the number system $\mathbb{S}[\sqrt{k}]$[i] is defined to consist of ordered pairs $(a, b) \in \mathbb{S} \times \mathbb{S}$, with addition and multiplication defined by supposing that (a, b) behaves as if it is $a + b\sqrt{k}$, with the usual commutative, associative, and distributive laws applying, and with $\sqrt{k}$ treated as just another number subject to these laws, with $(\sqrt{k})^2 = k$.[ii] Applying these laws to $a + b\sqrt{k}$, we have

$$(a + b\sqrt{k}) + (c + d\sqrt{k}) = (a + c) + (b + d)\sqrt{k}$$

and

$$(a + b\sqrt{k}) \times (c + d\sqrt{k}) = ac + ad\sqrt{k} + bc\sqrt{k} + bdk = (ac + bdk) + (ad + bc)\sqrt{k}.$$

Thus we define operations of addition $\oplus$ and multiplication $\otimes$ on $\mathbb{S}[\sqrt{k}]$:

Definition 8.1.1.

$$(a, b) \oplus (c, d) := (a + c, b + d).$$

---

[i]In this chapter, we use the common notation $\sqrt{k}$ rather than tediously insisting on $\sqrt[2]{k}$, since square roots are the only kind of root we shall meet here.

[ii]In Section 8.6, we shall find that in certain circumstances we need a more general definition, that elements (a, b) in a number system $\mathbb{S}[\omega]$ are required to behave like $a + b\omega$ where $\omega \notin \mathbb{S}$ but ω formally satisfies some quadratic equation involving a given number $k \in \mathbb{S}$. In the definition given for $\mathbb{S}[\sqrt{k}]$, the quadratic equation satisfied by ω is $\omega^2 - k = 0$.

Definition 8.1.2.

$$(a, b) \otimes (c, d) := (ac + bdk, ad + bc).$$

Most textbooks write elements of $\mathbb{S}[\sqrt{k}]$ as $a + b\sqrt{k}$ because that is how the arithmetic in $\mathbb{S}[\sqrt{k}]$ is defined. However, in this book we use the ordered-pair format, to emphasise that we are dealing with newly defined number systems, not merely particular combinations of old numbers; a calculation is done using both formats in Example 8.1.3. Note that, unlike when defining $\mathbb{Z}$ and $\mathbb{Q}$, we have not invoked any equivalence relation between ordered pairs: $(a_1, b_1) = (a_2, b_2)$ only if both $a_1 = a_2$ and $b_1 = b_2$ (or one could say that the equivalence classes in $\mathbb{S}[\sqrt{k}]$ consist of a single ordered pair each).

The operations $\oplus$ and $\otimes$ are clearly closed on $\mathbb{S}[\sqrt{k}]$, since if $a, b, c, d, k \in \mathbb{S}$ then the right-hand sides of the definitions are ordered pairs in $\mathbb{S} \times \mathbb{S}$ because addition and multiplication are closed in $\mathbb{S}$ (for any number system $\mathbb{S}$ that we have considered). Observe that addition looks the same in all $\mathbb{S}[\sqrt{k}]$, but multiplication depends on the particular k.

Example 8.1.3. Find $(1, -3) \otimes (3, 4)$ in $\mathbb{Z}[\sqrt{-2}]$.

Solution. Here we have $k = -2$, and we want to multiply $(a, b) = (1, -3)$ by $(c, d) = (3, 4)$. Simply substituting $a = 1, b = -3, c = 3, d = 4, k = -2$ into Definition 8.1.2, we obtain

$$(1, -3) \otimes (3, 4) = (1 \times 3 + (-3) \times 4 \times (-2), 1 \times 4 + (-3) \times 3) = (27, -5).$$

Alternatively, we could write $(1, -3)$ as $1 - 3\sqrt{-2}$, and $(3, 4)$ as $3 + 4\sqrt{-2}$ and set out the calculation as

$$(1 - 3\sqrt{-2}) \times (3 + 4\sqrt{-2}) = 3 + 4\sqrt{-2} - 9\sqrt{-2} - 12(\sqrt{-2})^2$$
$$= 27 - 5\sqrt{-2}$$

where we have used the intuitive idea that $(\sqrt{-2})^2 = -2$, even though $\sqrt{-2}$ has no meaning in any number system that we have encountered so far. Doing the calculation using the "$a + b\sqrt{k}$" format has the advantage that the value of k appears explicitly in numbers, unlike in ordered-pair form where it only appears in the multiplication formula (Definition 8.1.2) for the number system. $\square$

Exercise 8.1.4. *With $\mathbb{S} = \mathbb{Z}$, do the following additions and multiplications in the given quadratic extensions of $\mathbb{Z}$:*
(a) In $\mathbb{Z}[\sqrt{5}]$:   *(i) $(2, -1) \oplus (-3, 4)$*   *(ii) $(2, -1) \otimes (-3, 4)$*
(b) In $\mathbb{Z}[\sqrt{-3}]$:   *(i) $(2, -1) \oplus (-3, 4)$*   *(ii) $(2, -1) \otimes (-3, 4)$*

Because (a, b) "behaves like" $a + b\sqrt{k}$, putting $b = 0$ gives us a number $(a, 0) \in \mathbb{S}[\sqrt{k}]$ which we expect to behave like the number $a \in \mathbb{S}$. Formally, we expect to find isomorphisms between the set

$$\mathbb{S}[\sqrt{k}]_0 := \{(a, b) \in \mathbb{S}[\sqrt{k}] : b = 0\}$$

and $\mathbb{S}$. There is indeed the obvious bijection $f : \mathbb{S}[\sqrt{k}]_0 \to \mathbb{S}$ defined by

$$f : (a, 0) \mapsto a;$$

then from Definitions 8.1.1 and 8.1.2, for any $(a, 0), (c, 0) \in \mathbb{S}[\sqrt{k}]_0$ we find

$$(a, 0) \oplus (c, 0) = (a + c, 0)$$

and

$$(a, 0) \otimes (c, 0) = (a \times c, 0),$$

which establish isomorphisms between the operations $\oplus$ and $\otimes$ on $\mathbb{S}[\sqrt{k}]_0$ and the respective operations $+$ and $\times$ on $\mathbb{S}$. Thus we shall follow our usual practice of using the standard symbols $+$ and $\times$ for the operations on the new number system, except where we want to emphasise the definitions in $\mathbb{S}[\sqrt{k}]$.

Now observe that

$$(0,1) \times (0,1) = (k,0) \tag{8.1}$$

and

$$(a,b) = (a,0) + (b,0) \times (0,1), \tag{8.2}$$

where we have used Definitions 8.1.1 and 8.1.2 but with the standard symbols for addition and multiplication. The numbers $(k,0)$, $(a,0)$, and $(b,0)$ in $\mathbb{S}[\sqrt{k}]_0$ correspond to k, a and b in $\mathbb{S}$, so equation (8.1) suggests representing $(0,1)$ with the symbol $\sqrt{k}$, and we can then write equation (8.2) as

$$(a,b) = a + b\sqrt{k}.$$

This formally identifies our ordered pairs with the common notation "$a + b\sqrt{k}$". Nevertheless, since $\sqrt{k}$ is not a number in $\mathbb{S}$, we cannot assume that our arithmetic operations in $\mathbb{S}[\sqrt{k}]$ will inherit all the same properties as the corresponding operations in $\mathbb{S}$ simply because we operate with (a,b) as if it was $a + b\sqrt{k}$. In $\mathbb{Z}$, $\mathbb{Q}$, and $\mathbb{R}$, the arithmetic operations have properties which constitute the commutative ring structure, so we need to verify:

Theorem 8.1.5. *If* $\mathbb{S}$ *is a commutative ring, then so is* $\mathbb{S}[\sqrt{k}]$. *The identity elements for addition and multiplication in* $\mathbb{S}[\sqrt{k}]$ *are those which correspond to the identity elements in* $\mathbb{S}$ *under the isomorphisms, i.e.* $(0,0)$ *for addition and* $(1,0)$ *for multiplication. The additive inverse of* (a,b) *is* $(-a,-b)$.

The properties that constitute a commutative ring are listed in full in Definition 4.3.5 (with reference to Definition 4.3.1); we shall prove just one property for $\mathbb{S}[\sqrt{k}]$, leaving the rest for readers to prove.

Proof that multiplication is associative in $\mathbb{S}[\sqrt{k}]$.

$$\begin{aligned}
((a,b) \times (c,d)) \times (e,f) &= (ac+bdk, ad+bc) \times (e,f) & [\textit{Def. 8.1.2}] \\
&= ((ac+bdk)e + (ad+bc)fk, (ac+bdk)f + (ad+bc)e) & [\textit{Def. 8.1.2}] \\
&= (ace+bdke+adfk+bcfk, acf+bdkf+ade+bce). & (*)
\end{aligned}$$

Also,

$$\begin{aligned}
(a,b) \times ((c,d) \times (e,f)) &= (a,b) \times (ce+dfk, cf+de) & [\textit{Def. 8.1.2}] \\
&= (a(ce+dfk) + b(cf+de)k, a(cf+de) + b(ce+dfk)) & [\textit{Def. 8.1.2}] \\
&= (ace+adfk+bcfk+bdek, acf+ade+bce+bdfk)
\end{aligned}$$

where the last expression is equal to that marked $(*)$, so that

$$((a,b) \times (c,d)) \times (e,f) = (a,b) \times ((c,d) \times (e,f)). \quad \square$$

Exercise 8.1.6. *Verify the remaining properties of a commutative ring for* $\mathbb{S}[\sqrt{k}]$: *that addition is commutative and associative, multiplication is commutative, multiplication distributes over addition,* $(0,0)$ *and* $(1,0)$ *are the respective identity elements for addition and multiplication, and* $(-a,-b)$ *is the additive inverse of* (a,b).

As usual we can subtract a number in $\mathbb{S}[\sqrt{k}]$ by adding its additive inverse, so that $(a,b) - (c,d) = (a-c, b-d)$. We also have all the other properties and formulae that follow from the commutative ring structure, in particular, that multiplication by the additive identity ("zero") always yields the additive identity:

$$(a,b) \times (0,0) = (0,0) \times (a,b) = (0,0) \quad \text{for all } (a,b) \in \mathbb{S}[\sqrt{k}].$$

But a further important property of multiplication that is specific to $\mathbb{S}[\sqrt{k}]$ and easily verified from Definition 8.1.2 is the following:

$$(c,0) \times (a,b) = (ca, cb). \tag{8.3}$$

Here $(c,0)$ is the member of $\mathbb{S}[\sqrt{k}]_0$ which corresponds to a number $c \in \mathbb{S}$, and when it multiplies any $(a,b) \in \mathbb{S}[\sqrt{k}]$ the result is as if we had multiplied each member of the ordered pair (a,b) by c. So we define:

Definition 8.1.7. The **scalar multiplication** of $(a,b) \in \mathbb{S}[\sqrt{k}]$ by $c \in \mathbb{S}$ is

$$c(a,b) := (ca, cb).$$

Readers who are familiar with vectors will understand where the inspiration for this definition comes from, but bear in mind that here it is really just a convenient shorthand for (8.3), the multiplication of two members of $\mathbb{S}[\sqrt{k}]$ where one of those numbers corresponds to a member of $\mathbb{S}$.[iii] If $\mathbb{S}$ is a field, we can also do **scalar division**, noting that in $\mathbb{S}$ dividing by c is the same as multiplying by $\frac{1}{c}$:

$$\frac{(a,b)}{c} := \left(\frac{a}{c}, \frac{b}{c} \right), \tag{8.4}$$

in which the left-hand side is really shorthand for $\left(\frac{1}{c}, 0 \right) \times (a,b)$. The associative and commutative properties of multiplication then allow us to write

$$(c(a,b)) \times (d,e) = c((a,b) \times (d,e)) = (a,b) \times (c(d,e))$$

and

$$\frac{(a,b)}{c} \times (d,e) = \frac{(a,b) \times (d,e)}{c} = (a,b) \times \frac{(d,e)}{c}. \tag{8.5}$$

Now, among the number systems to which we are applying quadratic extension, $\mathbb{Q}$ and $\mathbb{R}$ are fields whereas $\mathbb{Z}$ is only a commutative ring. The difference is that in a field, all elements except zero have a multiplicative inverse; and this property does extend to their quadratic extensions:

Theorem 8.1.8. *If $\mathbb{S}$ is a field, then so is $\mathbb{S}[\sqrt{k}]$.*

To prove this, by finding the multiplicative inverse of a number in $\mathbb{S}[\sqrt{k}]$, it will be useful to have definitions of some further concepts specific to $\mathbb{S}[\sqrt{k}]$.

Definition 8.1.9. The **conjugate** of (a,b) is the number $(a,-b)$ in $\mathbb{S}[\sqrt{k}]$. We denote conjugates in $\mathbb{S}[\sqrt{k}]$ by an overbar: $\overline{(a,b)} := (a,-b)$.[iv]

[iii] The terms *scalar multiplication* and *scalar division* are not generally applied to quadratic extensions by other authors, who tend to ignore the distinction between a number in $\mathbb{S}$ and the corresponding number in $\mathbb{S}[\sqrt{k}]$; but we use these terms in this book to remind you of what it means when we show a number in an extension of $\mathbb{S}$ being multiplied or divided by a number in $\mathbb{S}$ (here and in Chapters 9 and 10).

[iv] We have previously used overbars to denote inverse elements under a binary operation. That will not be done in this chapter, so there should be no confusion between the two uses of the notation.

Conjugates behave in a simple way when added or multiplied:

Theorem 8.1.10. *(i)* $\overline{(a,b)} + \overline{(c,d)} = \overline{(a,b) + (c,d)}$
(ii) $\overline{(a,b)} \times \overline{(c,d)} = \overline{(a,b) \times (c,d)}$

Proof of (i).

$$
\begin{aligned}
\overline{(a,b)} + \overline{(c,d)} &= (a,-b) + (c,-d) && [Def.\ 8.1.9]\\
&= (a+c, (-b)+(-d)) && [Def.\ 8.1.1]\\
&= (a+c, -(b+d))\\
&= \overline{(a+c, b+d)} && [Def.\ 8.1.9]\\
&= \overline{(a,b)+(c,d)}. && [Def.\ 8.1.1] \qquad \square
\end{aligned}
$$

Exercise 8.1.11. *Prove the multiplication rule, part (ii) of Theorem 8.1.10.*

So taking conjugates and then adding or multiplying gives the same result as when you add or multiply the original numbers and then take the conjugate of the result. This sort of behaviour should be familiar: looking at Definition 4.2.4, you will see that conjugation yields an *isomorphism* from $\mathbb{S}[\sqrt{k}]$ to itself with both the addition and multiplication operations. Such an isomorphism from a set with a binary operation to the same set with the same operation is called an **automorphism.**

However, the most important property of conjugates for the development of the theory is found by substituting $c = a$ and $d = -b$ in Definition 8.1.2, to multiply a number by its own conjugate:

$$(a,b) \times (a,-b) = (a^2 - b^2 k, 0). \tag{8.6}$$

Thus multiplying a number $(a,b) \in \mathbb{S}[\sqrt{k}]$ by its conjugate yields a number corresponding to $a^2 - b^2 k \in \mathbb{S}$. This motivates a further definition:

Definition 8.1.12. The **norm** in $\mathbb{S}[\sqrt{k}]$ is the function $\mathcal{N} : \mathbb{S}[\sqrt{k}] \to \mathbb{S}$ given by

$$\mathcal{N} : (a,b) \mapsto a^2 - b^2 k.$$

We shall refer to $a^2 - b^2 k$ as "the norm of (a,b)", and will denote it as $\mathcal{N}(a,b)$ (using only one parenthesis, although strictly we should have one parenthesis for "function of" and one for "ordered pair").[v] So equation (8.6) can be written

$$(a,b) \times \overline{(a,b)} = (\mathcal{N}(a,b), 0). \tag{8.7}$$

An important property of norms, which will be required for the verification of Theorem 8.1.8, is the following:

Theorem 8.1.13. $\mathcal{N}(a,b) = 0$ *if and only if* $(a,b) = (0,0)$.

Proof. It is obvious that if $a = 0$ and $b = 0$, then $a^2 - b^2 k = 0$. Conversely, if $b = 0$, we clearly require $a = 0$ to satisfy $a^2 - b^2 k = 0$. So suppose $b \neq 0$ and consider separately the cases where $\mathbb{S}$ is a field or a ring.

If $a^2 - b^2 k = 0$ where a, b are members of a field (in which exact division is allowed, except by zero), then

$$k = \frac{a^2}{b^2} = \left(\frac{a}{b}\right)^2,$$

---

[v]The term "norm" is also used in other ways in Analysis, some of which resemble the square root of the quantity we have defined as the norm; but our definition is the most suitable for the study of quadratic extensions.

which contradicts the definition of $\mathbb{S}[\sqrt{k}]$ in which k must have no square root in $\mathbb{S}$.

If $a^2 - b^2 k = 0$ where a, b are members of the ring $\mathbb{Z}$, we can divide out any common factor d, writing $a = a'd$ and $b = b'd$ where a', b' are coprime; then $a'^2 = b'^2 k$. [If a, b are already coprime, then $d = 1$.] Then as in the proof of Theorem 6.4.6 we obtain $k = a'^2$, again contradicting the definition of $\mathbb{S}[\sqrt{k}]$. [This proof relies on uniqueness of prime factorisations, which applies in $\mathbb{Z}$ but not necessarily to rings in general; but $\mathbb{Z}$ is the only ring to which we shall apply quadratic extension and which is not a field.]   □

We can now verify Theorem 8.1.8 constructively, by identifying the multiplicative inverse of a number in $\mathbb{S}[\sqrt{k}]$.

Theorem 8.1.14. *For any $(a, b) \neq (0, 0)$ in $\mathbb{S}[\sqrt{k}]$ where $\mathbb{S}$ is a field, the multiplicative inverse is the number*

$$\frac{\overline{(a, b)}}{\mathcal{N}(a, b)}.$$

Thus we can say that "multiplicative inverse = conjugate divided by norm", where we have used the scalar division notation of equation (8.4). The advantage of this notation is now obvious: the multiplicative inverse written out in longhand would be

$$\left(\frac{a}{\mathcal{N}(a, b)}, \frac{-b}{\mathcal{N}(a, b)} \right).$$

Proof of Theorem 8.1.14.

$$(a, b) \times \frac{\overline{(a, b)}}{\mathcal{N}(a, b)} = \frac{(a, b) \times \overline{(a, b)}}{\mathcal{N}(a, b)} \qquad \qquad [\textit{Equation (8.5)}]$$

$$= \frac{(\mathcal{N}(a, b), 0)}{\mathcal{N}(a, b)} \qquad \qquad [\textit{Equation (8.7)}]$$

$$= (1, 0), \qquad \qquad [\textit{Equation (8.4)}]$$

where $(1, 0)$ is the multiplicative identity element. The commutative property of multiplication then ensures that we also have

$$\frac{\overline{(a, b)}}{\mathcal{N}(a, b)} \times (a, b) = (1, 0),$$

so that the definition of "multiplicative inverse" is fully satisfied.   □

Now that we have obtained the multiplicative inverse and so verified that $\mathbb{S}[\sqrt{k}]$ is a field whenever $\mathbb{S}$ is a field, we can do exact division in such a field in the usual way, by multiplying by the multiplicative inverse:

$$(a, b) \div (c, d) = \frac{(a, b) \times \overline{(c, d)}}{\mathcal{N}(c, d)}. \qquad \qquad (8.8)$$

Exercise 8.1.15. *(i) Find the multiplicative inverse of $\left(\frac{1}{2}, -1 \right)$ in $\mathbb{Q}[\sqrt{3}]$. Hence evaluate $\left(-3, \frac{3}{4} \right) \div \left(\frac{1}{2}, -1 \right)$ in $\mathbb{Q}[\sqrt{3}]$.*
(ii) Find the multiplicative inverse of $\left(\frac{1}{2}, -1 \right)$ in $\mathbb{Q}[\sqrt{-1}]$. Hence evaluate $\left(-3, \frac{3}{4} \right) \div \left(\frac{1}{2}, -1 \right)$ in $\mathbb{Q}[\sqrt{-1}]$.

A further important property of norms is that they **preserve multiplication**:

Theorem 8.1.16. *For any* $(a,b), (c,d) \in \mathbb{S}[\sqrt{k}]$,

$$\mathcal{N}((a,b) \otimes (c,d)) = \mathcal{N}(a,b) \times \mathcal{N}(c,d).^{\text{vi}}$$

We have reverted to the $\otimes$ notation to emphasise that on the left-hand side we are multiplying in $\mathbb{S}[\sqrt{k}]$ and then taking the norm, whereas on the right we are taking norms and then doing a multiplication in $\mathbb{S}$.

Proof.

$$
\begin{aligned}
\mathcal{N}((a,b) \otimes (c,d)) &= \mathcal{N}(ac + bdk, ad + bc) && \text{[\textit{Def. (8.1.2)}]} \\
&= (ac + bdk)^2 - (ad + bc)^2 k && \text{[\textit{Def. (8.1.12)}]} \\
&= a^2 c^2 + b^2 d^2 k^2 + 2acbdk - a^2 d^2 k - b^2 c^2 k - 2adbck \\
&= a^2 c^2 + b^2 d^2 k^2 - a^2 d^2 k - b^2 c^2 k. && (*)
\end{aligned}
$$

Also,

$$
\begin{aligned}
\mathcal{N}(a,b) \times \mathcal{N}(c,d) &= (a^2 - b^2 k) \times (c^2 - d^2 k) && \text{[\textit{Def. (8.1.12)}]} \\
&= a^2 c^2 - a^2 d^2 k - b^2 c^2 k + b^2 d^2 k^2,
\end{aligned}
$$

which is equal to the expression labelled $(*)$. $\qquad\square$

This is a "brute force" proof based only on Definition 8.1.12. A much more elegant proof is possible using the relation (8.7) between norms and conjugates and the multiplication property of conjugates, Theorem 8.1.10(ii).

Exercise 8.1.17. *Prove that norms preserve multiplication, using equation (8.7) and Theorem 8.1.10(ii).*

Having dealt with addition and multiplication (and their inverses) in $\mathbb{S}[\sqrt{k}]$, we may consider exponentiation. In $\mathbb{Z}$ we were only able to take positive integer powers of integers, while in $\mathbb{Q}$ we could take any integer power of a rational, but in general non-integer powers of rationals did not exist within $\mathbb{Q}$. In both number systems, positive integer powers were defined by induction (treating $\mathbb{Z}_+$ as identical to $\mathbb{N}$), and then for non-positive powers of $x \in \mathbb{Q}$ we had $x^0 = 1$ and $x^{-n} = 1/x^n$. The same ideas apply to quadratic extensions of $\mathbb{Z}$ and $\mathbb{Q}$: positive integer powers of numbers in $\mathbb{Z}[\sqrt{k}]$ and $\mathbb{Q}[\sqrt{k}]$ will be defined by induction, and non-positive powers of numbers in $\mathbb{Q}[\sqrt{k}]$ by similar formulae to those in $\mathbb{Q}$:

$$(a,b)^0 := (1,0), \quad (a,b)^{-n} := (1,0) \div (a,b)^n.$$

We do not define non-integer powers in these quadratic extensions. However, when we come to the Complex Numbers, a quadratic extension of $\mathbb{R}$, we shall see that it is possible to find all complex powers of all non-zero complex numbers.

Exercise 8.1.18. *(a) (i) Evaluate* $(2,-1)^3$ *in* $\mathbb{Z}[\sqrt{5}]$.
 (ii) Evaluate $(2,-1)^3$ *in* $\mathbb{Z}[\sqrt{-3}]$.
 (iii) Confirm that $\mathcal{N}((2,-1)^3) = (\mathcal{N}(2,-1))^3$ *in* $\mathbb{Z}[\sqrt{5}]$. *[On the left, you are taking the cube of* $(2,-1)$ *in* $\mathbb{Z}[\sqrt{5}]$ *and then finding its norm; on the right you are taking the norm of* $(2,-1)$ *in* $\mathbb{Z}[\sqrt{5}]$ *and then finding the cube in* $\mathbb{Z}$.*]*
 (b) (i) Evaluate $\left(\frac{1}{2}, -1\right)^{-2}$ *in* $\mathbb{Q}[\sqrt{3}]$
 (ii) Evaluate $\left(\frac{1}{2}, -1\right)^{-2}$ *in* $\mathbb{Q}[\sqrt{-1}]$
 (iii) Confirm that $\mathcal{N}\left(\left(\frac{1}{2}, -1\right)^{-2}\right) = \left(\mathcal{N}\left(\frac{1}{2}, -1\right)\right)^{-2}$ *in* $\mathbb{Q}[\sqrt{-1}]$.

[vi]Discovered by Brahmagupta in the 7'th Century, although he did not use the terminology of norms.

After arithmetic, you may expect that we will now discuss order in $\mathbb{S}[\sqrt{k}]$. It is possible to define order in $\mathbb{S}[\sqrt{k}]$ when $\mathbb{S}$ is an ordered ring and $k > 0$, although this has limited usefulness and will only be introduced when it is needed. On the other hand, with $k < 0$ it is easy to demonstrate that it is not possible to define order relations such that $\mathbb{S}[\sqrt{k}]$ is an ordered ring, with order isomorphism between $\mathbb{S}[\sqrt{k}]_0$ and $\mathbb{S}$. Theorem 4.4.13 states that $x^2 > 0$ (where "0" stands for the additive identity element) for any non-zero element x of an ordered ring; but equation (8.1) gives

$$(0,1)^2 = (k,0) \quad \text{in } \mathbb{S}[\sqrt{k}]$$

and $(k,0) \in \mathbb{S}[\sqrt{k}]_0$. If $k < 0$ in $\mathbb{S}$, order isomorphism requires $(k,0) < (0,0)$ in $\mathbb{S}[\sqrt{k}]$, while the "$x^2 > 0$" theorem for ordered rings requires $(k,0) > (0,0)$.

With regard to cardinality, Theorem 6.3.2 immediately shows that any quadratic extension of $\mathbb{Z}$ or $\mathbb{Q}$ is countable, since a quadratic extension of $\mathbb{S}$ is the set $\mathbb{S} \times \mathbb{S}$ with certain definitions of arithmetic operations. But Theorem 6.3.2 does not apply to Cartesian products of uncountable sets, and we shall deal with the cardinality of the set $\mathbb{R} \times \mathbb{R}$ in Section 9.7.

8.2   Introduction to Quadratic Rings: Extensions of $\mathbb{Z}$

Primes and factorisation are a major theme in the study of integers. In Section 5.3, we initially defined primes in $\mathbb{N}$; we then saw that if we wanted to generalise the concept of primes to $\mathbb{Z}$ and other number systems, it would only make sense if we first identified **units** (numbers with a multiplicative inverse) and then considered equivalence classes in which every number is a unit multiple of the other numbers. Units, primes, and factorisation are the main topics of interest in $\mathbb{Z}[\sqrt{k}]$, the quadratic extensions of $\mathbb{Z}$, and we shall examine their properties, concluding with the finding that the Fundamental Theorem of Arithmetic (uniqueness of prime factorisations) does not apply in $\mathbb{Z}[\sqrt{k}]$ for some values of k. To gain a full theoretical understanding of $\mathbb{Z}[\sqrt{k}]$ requires us to examine $\mathbb{Q}[\sqrt{k}]$, the quadratic extensions of $\mathbb{Q}$; and it is our study of $\mathbb{Q}[\sqrt{k}]$ that will lead ultimately to an explanation of when the Fundamental Theorem of Arithmetic must or may not apply in $\mathbb{Z}[\sqrt{k}]$.

From our definition of quadratic extensions, the number k in $\mathbb{Z}[\sqrt{k}]$ must be an integer without a square root in $\mathbb{Z}$. This allows k to be any positive integer which is not the square of an integer, or any negative integer at all. With $k = -1$, the number system $\mathbb{Z}[\sqrt{-1}]$ is called the **Gaussian integers**, and is particularly important in Number Theory because of the norm being equal to the sum of squares: $\mathcal{N}(a,b) = a^2 + b^2$ in $\mathbb{Z}[\sqrt{-1}]$. Number systems $\mathbb{Z}[\sqrt{k}]$ with positive k are called **real quadratic rings** since $\sqrt{k}$ does exist in $\mathbb{R}$, while with negative k they are **imaginary quadratic rings**.

It is usually insisted that k should be **squarefree**, meaning that its prime factorisation should not include squares (or higher powers) of any prime. This is because if $k = lq^2$ for some $l, q \in \mathbb{Z}$, then $\mathbb{Z}[\sqrt{k}]$ is just a **subring** of $\mathbb{Z}[\sqrt{l}]$, i.e. a subset with the same ring properties. More precisely:

Theorem 8.2.1. *The quadratic ring $\mathbb{Z}[\sqrt{lq^2}]$ is isomorphic under both addition and multiplication to the subset of $\mathbb{Z}[\sqrt{l}]$,*

$$\mathbb{Z}[\sqrt{l}]_q := \{(a, b') \in \mathbb{Z}[\sqrt{l}] : q|b'\}.$$

Proof. The subset $\mathbb{Z}[\sqrt{l}]_q$ is defined to consist of ordered pairs in which the second member is divisible by q, so $b' = bq$ for some $b \in \mathbb{Z}$. Thus there is an obvious bijection,

$$f : \mathbb{Z}[\sqrt{lq^2}] \to \mathbb{Z}[\sqrt{l}]_q \quad \text{defined by} \quad f : (a,b) \mapsto (a,bq).$$

Definition 8.1.1 for addition yields

$$(a,b) \oplus (c,d) = (a+c, b+d) \quad \text{in } \mathbb{Z}[\sqrt{lq^2}],$$

and

$$(a,bq) \oplus (c,dq) = (a+c, (b+d)q) \quad \text{in } \mathbb{Z}[\sqrt{l}]_q,$$

which gives the isomorphism under addition with the bijection f. Definition 8.1.2 for multiplication gives

$$(a,b) \otimes (c,d) := (ac + bdlq^2, ad + bc) \quad \text{in } \mathbb{Z}[\sqrt{lq^2}],$$

whereas multiplying numbers (a,bq) and (c,dq) in $\mathbb{Z}[\sqrt{l}]_q$ yields

$$(a,bq) \otimes (c,dq) := (ac + bdlq^2, (ad + bc)q),$$

which demonstrates the isomorphism under multiplication with the bijection f. $\square$

The isomorphisms allow us to consider $\mathbb{Z}[\sqrt{lq^2}]$ to be a subring of $\mathbb{Z}[\sqrt{l}]$: any theorem or calculation in $\mathbb{Z}[\sqrt{lq^2}]$ would be equivalent to a theorem or calculation in $\mathbb{Z}[\sqrt{l}]$, so it is generally considered pointless to study $\mathbb{Z}[\sqrt{lq^2}]$ (although calculations in such a ring are still valid as long as l is not the square of an integer).

8.3   Units in $\mathbb{Z}[\sqrt{k}]$

The units in $\mathbb{Z}$ are 1 and -1; Theorems 8.1.14 and 8.1.16 then give us criteria for identifying units in $\mathbb{Z}[\sqrt{k}]$:

Theorem 8.3.1. *In* $\mathbb{Z}[\sqrt{k}]$, *a number is a unit if and only if its norm is* ± 1.

Proof. The multiplicative identity element in $\mathbb{Z}[\sqrt{k}]$ is $(1,0)$, with norm $\mathcal{N}(1,0) = 1$. So if (u,v) is a unit, with multiplicative inverse (s,t), Theorem 8.1.16 implies that $\mathcal{N}(u,v) \times \mathcal{N}(s,t) = 1$. Hence $\mathcal{N}(u,v)$ is a unit in $\mathbb{Z}$, i.e. 1 or -1.
Conversely, if $\mathcal{N}(u,v) = \pm 1$, then

$$\frac{\overline{(u,v)}}{\mathcal{N}(u,v)} = (\pm u, \mp v) \in \mathbb{Z}[\sqrt{k}]$$

(with the upper or lower sign in $\pm$ and $\mp$ chosen according to whether $\mathcal{N}(u,v)$ is 1 or -1), and this is the multiplicative inverse of (u,v) according to Theorem 8.1.14.[vii] $\square$

Theorem 8.3.1 says that (u,v) is a unit in $\mathbb{Z}[\sqrt{k}]$ if

$$u^2 - v^2 k = \pm 1.^{\text{viii}} \tag{8.9}$$

[vii] This theorem was stated as applying to fields, but is clearly valid whenever a multiplicative inverse exists.

[viii] If k is positive, the equation $u^2 - v^2 k = 1$ is known as Pell's equation; the equation $u^2 - v^2 k = -1$ is often called the negative Pell's equation. But the equation was known in ancient Greece (in the case $k = 2$), and was studied more extensively in India, long before the 17th-century English mathematician John Pell – whose contribution to its theory may be no more than as the author of a textbook in which it is discussed.

So the numbers $(1,0)$ and $(-1,0)$, which are just the numbers in $\mathbb{Z}[\sqrt{k}]_0$ corresponding to the units in $\mathbb{Z}$, are certainly units in $\mathbb{Z}[\sqrt{k}]$ for any k. Determining whether there are any other units in $\mathbb{Z}[\sqrt{k}]$ is easy for negative k, but more interesting for positive k.

Consider first the Gaussian integers, with $k = -1$, so that $u^2 - v^2 k = u^2 + v^2$. Since $u^2 + v^2 > 1$ if either $u > 1$, $v > 1$, $u < -1$ or $v < -1$, the only pairs of integers satisfying equation (8.9) are $u = \pm 1$ with $v = 0$, and $u = 0$ with $v = \pm 1$. So the units in $\mathbb{Z}[\sqrt{-1}]$ are $(1,0)$, $(-1,0)$, $(0,1)$ and $(0,-1)$.

If $k < -1$, then either $v = 0$ in which case (8.9) is satisfied by $u = \pm 1$, yielding the units $(1,0)$, $(-1,0)$; or else $v \geq 1$ or $v \leq -1$ in which case $u^2 - v^2 k \geq -k > 1$ so that (8.9) cannot be satisfied. So there are no further units in $\mathbb{Z}[\sqrt{k}]$ with $k < -1$.

For positive k we shall denote $(1,0)$ and $(-1,0)$ as the "trivial units", and then seek solutions of equation (8.9) with $v \neq 0$ to find non-trivial units. We first note that if (u,v) is a unit, then so are $(u,-v)$, $(-u,-v)$ and $(-u,v)$, since if u and v satisfy (8.9), then so will $-u$ and/or $-v$: so if we have found a unit, then its conjugate, its additive inverse and the additive inverse of its conjugate are also units; and from the general theory of units in Section 5.3, these four are all associates of each other.

Can we be certain that any non-trivial units exist? Joseph-Louis Lagrange proved a theorem that a solution of the equation $u^2 - v^2 k = 1$ exists with $v > 0$ for every $k \in \mathbb{Z}_+$ that does not have a square root in $\mathbb{Z}$. For the equation $u^2 - v^2 k = -1$, solutions exist for some but not all values of k:

Exercise 8.3.2. *Using modular arithmetic in* $\mathbb{Z}_4$, *prove that there are no integers* u, v *satisfying* $u^2 - v^2 k = -1$ *if* $k \equiv 0 \,(\mathrm{mod}\ 4)$ *or* $k \equiv 3 \,(\mathrm{mod}\ 4)$.[ix]

So there may or may not be a unit with norm equal to -1, but Lagrange's theorem assures us that there definitely will exist a non-trivial unit with norm equal to 1.

Searching for a unit is simple in principle: multiply the given k by v^2 successively for $v = 1, 2, 3, \ldots$ until you obtain a number that is within ± 1 of another square; this square is then u^2. The first pair (u, v) obtained by this process is called the **fundamental unit** for the given k, and obviously has the smallest positive value of v for any unit in $\mathbb{Z}[\sqrt{k}]$, and the smallest positive u apart from the unit $(1,0)$. For example, with $k = 2$, the fundamental unit is $(1,1)$, with norm -1, since $1^2 - 1^2 \times 2 = -1$. With $k = 3$, there is no unit with norm -1, but the fundamental unit with norm 1 is $(2,1)$, since $2^2 - 1^2 \times 3 = 1$. Which all seems fine, until you get to the notorious case of $k = 61$, for which the fundamental unit is $(29718, 3805)$ (with norm -1); and further cases where the values of u and v in the fundamental unit are surprisingly large do continue to appear sporadically for larger k, so computing units in $\mathbb{Z}[\sqrt{k}]$ may require either substantial computing time or a more sophisticated number-theoretic approach.

Once a unit (u, v) has been found, all integer powers of that unit are also units, because $\mathcal{N}(u, v) = \pm 1$ so that

$$\mathcal{N}((u,v)^n) = (\mathcal{N}(u,v))^n = (\pm 1)^n,$$

and hence $\mathcal{N}((u,v)^n) = -1$ if $\mathcal{N}(u,v) = -1$ and n is odd, while $\mathcal{N}((u,v)^n) = 1$ if $\mathcal{N}(u,v) = 1$ or n is even (or both). The negative integer powers of the unit are plus or minus the conjugates of the positive integer powers, since

$$(u,v)^{-n} = (1,0) \div (u,v)^n = \frac{(1,0) \times \overline{(u,v)^n}}{\mathcal{N}((u,v)^n)} = \pm\overline{(u,v)^n}, \qquad (8.10)$$

with the plus or minus sign chosen according to whether the norm of the positive integer power is 1 or -1; note that exact division is valid in $\mathbb{Z}[\sqrt{k}]$ precisely when we are dividing

---

[ix]In fact, it can be shown that there are no integer solutions to $u^2 - v^2 k = -1$ if k has any prime factors which are congruent to 3 (mod 4), but this requires more advanced number theory.

by a unit. It is useful to denote the fundamental unit in a real quadratic ring as (u_1, v_1) and to introduce the notation $(u_n, v_n) := (u_1, v_1)^n$ and $(u_{-n}, v_{-n}) := (u_1, v_1)^{-n}$ (so u_n and v_n are the members of the ordered pair obtained by taking the n'th power of (u_1, v_1)). So we have $(u_{-n}, v_{-n}) = \overline{(u_n, v_n)} = (u_n, -v_n)$ if $\mathcal{N}(u, v) = 1$ or n is even (or both), and $(u_{-n}, v_{-n}) = -\overline{(u_n, v_n)} = (-u_n, v_n)$ if $\mathcal{N}(u, v) = -1$ and n is odd.

We can prove by induction that the positive integer powers of the fundamental unit are all distinct, so that the set of units in a real quadratic ring $\mathbb{Z}[\sqrt{k}]$ is countably infinite:

Exercise 8.3.3. *Noting from our definition of fundamental unit that* $u_1 > 0$ *and* $v_1 > 0$, *and using the notation introduced above:*

(i) *By squaring* (u_1, v_1), *show that* $u_2 > u_1$ *and* $v_2 > v_1$.

(ii) *Show that if* $u_n > 0$ *and* $v_n > 0$, *then* $u_{S(n)} > u_n$ *and* $v_{S(n)} > v_n$.

(iii) *Deduce that there are infinitely many distinct units* (u_n, v_n) *for* $n \in \mathbb{Z}_+$, *with* $u_n > 0$ *and* $v_n > 0$ *for each* n.

Exercise 8.3.4. (i) *Find four units in* $\mathbb{Z}[\sqrt{2}]$ *that are positive powers of the fundamental unit* $(1, 1)$.

(ii) *Given the fundamental unit* $(29718, 3805)$ *in* $\mathbb{Z}[\sqrt{61}]$, *do not find four more units in* $\mathbb{Z}[\sqrt{61}]$ *by this method, unless you have computer software capable of handling very large integers.*

This procedure of finding powers of the fundamental unit captures *all* the units in $\mathbb{Z}[\sqrt{k}]$, if we also include the additive inverses of the units found this way. To prove this, we need an ordering of $\mathbb{Z}[\sqrt{k}]$ for positive k: this is provided by the correspondence between ordered pairs $(a, b) \in \mathbb{Z}[\sqrt{k}]$ and real numbers $a + b\sqrt{k}$. Formally:

Definition 8.3.5. For positive k, we define the **real magnitude** of (a, b) as the function $\mathcal{M} : \mathbb{Z}[\sqrt{k}] \to \mathbb{R}$,

$$\mathcal{M} : (a, b) \mapsto a + b\sqrt{k},$$

and we also use the notation $\mathcal{M}(a, b) := a + b\sqrt{k}$ (similarly to the notation for norms).

Thus, even though there is no order relation between elements of $\mathbb{Z}[\sqrt{k}]$, we can use the ordering of their real magnitudes in $\mathbb{R}$.

Since the arithmetic of (a, b) in $\mathbb{Z}[\sqrt{k}]$ is defined to behave like $a + b\sqrt{k}$, the real magnitude preserves multiplication (in the same sense that the norm does so), and hence preserves exponentiation by natural number powers. A further important property of real magnitudes is:

Theorem 8.3.6. *The function* $\mathcal{M}$ *is injective:* $a + b\sqrt{k} = c + d\sqrt{k}$ *only if* $a = c$ *and* $b = d$.

Proof. If $a + b\sqrt{k} = c + d\sqrt{k}$, then $a - c = (b - d)\sqrt{k}$, which equates a rational to an irrational number unless $a - c = 0 = b - d$. $\qquad\square$

We also need the following lemma on the real magnitude of units:

Lemma 8.3.7. *If* (u, v) *is a unit with* $\mathcal{M}(u, v) > 1$, *then both* $u \in \mathbb{Z}_+$ *and* $v \in \mathbb{Z}_+$.

Proof. Working in $\mathbb{R}$,

$$(u + v\sqrt{k})(u - v\sqrt{k}) = u^2 - v^2 k = \mathcal{N}(u, v) = \pm 1 \tag{8.11}$$

when (u, v) is a unit.

Consider the case where $\mathcal{N}(u, v) = +1$. If $\mathcal{M}(u, v) > 1$, i.e. $u + v\sqrt{k} > 1$, equation (8.11) then requires $0 < u - v\sqrt{k} < 1$. So $u + v\sqrt{k} > u - v\sqrt{k}$, from which $2v\sqrt{k} > 0$ and hence $v > 0$ (since $\sqrt{k} > 0$). The condition $0 < u - v\sqrt{k}$ with $v > 0$ then yields $u > 0$. Since u and v are integers, being greater than zero means that they are members of $\mathbb{Z}_+$.

Exercise 8.3.8. *Complete the proof by considering the case where $\mathcal{N}(u, v) = -1$.* $\qquad\square$

We can now prove:

Theorem 8.3.9. *For positive k, every unit $(u, v) \in \mathbb{Z}[\sqrt{k}]$ with $u, v \in \mathbb{Z}_+$ is a positive integer power of the fundamental unit.*

Proof. The fundamental unit (u_1, v_1) has the least positive real magnitude of all non-trivial units (u, v) with $u, v \in \mathbb{Z}_+$ (since it has the least u and the least v). Also, since (u_1, v_1) is non-trivial, $u_1 \geq 1$ and $v_1 \geq 1$ so that $u_1 + v_1\sqrt{k} > 1$. Hence $(u_1 + v_1\sqrt{k})^n < (u_1 + v_1\sqrt{k})^{n+1}$ for all $n \in \mathbb{N}$.

Suppose there is a unit (r, s) which is *not* a power of (u_1, v_1), with $r, s \in \mathbb{Z}_+$. Then for some $m \in \mathbb{N}$,

$$(u_1 + v_1\sqrt{k})^m < r + s\sqrt{k} < (u_1 + v_1\sqrt{k})^{m+1}. \tag{8.12}$$

This is because: (a) Theorem 7.6.10 verifies the existence of $m \in \mathbb{N}$ such that $r + s\sqrt{k} < (u_1 + v_1\sqrt{k})^{m+1}$, and the well-ordering of $\mathbb{N}$ allows us to specify the least such m; (b) if $r + s\sqrt{k} < (u_1 + v_1\sqrt{k})^n$ for all $n \in \mathbb{N}$ it would contradict (u_1, v_1) having the least real magnitude among non-trivial units with $u, v \in \mathbb{Z}_+$; (c) injectivity of $\mathcal{M}$ ensures that $r + s\sqrt{k} \neq (u_1 + v_1\sqrt{k})^n$ for all $n \in \mathbb{N}$.

Now observe that units can be divided (because they have multiplicative inverses), and the result of such a division will be a unit (it will have norm ± 1). So $(r, s) \div (u_1, v_1)^m$ will be a unit, and since the real magnitude preserves division, inequalities (8.12) yield

$$1 < \mathcal{M}((r, s) \div (u_1, v_1)^m) < \mathcal{M}(u_1, v_1).$$

Lemma 8.3.7 with the left-hand inequality here verifies that $u, v \in \mathbb{Z}_+$ for the supposed unit $(u, v) = (r, s) \div (u_1, v_1)^m$, and the right-hand inequality then contradicts (u_1, v_1) having the least real magnitude among non-trivial units with $u, v \in \mathbb{Z}_+$. The contradiction implies that our unit (r, s) cannot exist. $\qquad\square$

This theorem accounts for all units (u, v) with u and v both positive: they are all of the form $(u_1, v_1)^n$ for some $n \in \mathbb{Z}_+$. The remaining units are: $(u_1, v_1)^n$ with $n \in \mathbb{Z}_-$, where each such negative power of (u_1, v_1) is either the conjugate or the additive inverse of the conjugate of a positive power; the trivial unit $(1, 0)$, which is the zero'th power of (u_1, v_1); and the additive inverses of all the previously mentioned units.

Corollary 8.3.10. *If the equation $u^2 - v^2k = -1$ does have a solution for a particular k, the fundamental unit in $\mathbb{Z}[\sqrt{k}]$ will have norm equal to -1.*

Proof. If the fundamental unit had norm equal to 1, then other units would have norm $1^n = 1$, so there would be no unit (u, v) satisfying $u^2 - v^2k = -1$. $\qquad\square$

We conclude this section with an important application of units in real quadratic rings: to find rational approximations to irrational square roots. Finding a rational a/b equal to $\sqrt{k}$ (for some $k \in \mathbb{Z}_+$) would be equivalent to finding $a, b \in \mathbb{Z}_+$ such that $a^2 - b^2k = 0$; so if this is not possible, the best rational approximation to $\sqrt{k}$ would intuitively be obtained by finding a solution to $a^2 - b^2k = m$ with m as near as possible to zero. Working in $\mathbb{R}$,

$$a^2 - b^2k = m \Leftrightarrow a - b\sqrt{k} = \frac{m}{a + b\sqrt{k}}$$

$$\Leftrightarrow \left| \frac{a}{b} - \sqrt{k} \right| = \left| \frac{m}{b(a + b\sqrt{k})} \right| < \frac{|m|}{b^2},$$

so the error in approximating $\sqrt{k}$ by a/b is minimised by making $|m|$ as small as possible and b as large as possible. We know that there always exists a non-trivial unit in $\mathbb{Z}[\sqrt{k}]$, which is a solution of $a^2 - b^2 k = m$ with $m = \pm 1$, so that $|m|$ is minimised; and the existence of infinitely many units (powers of the fundamental unit) allows b to be made as large as desired by choosing a sufficiently high power of the fundamental unit. If we want the error in the approximation, $\left|(a/b) - \sqrt{k}\right|$, to be smaller than some specified positive ϵ, we can achieve this by letting $(a, b) = (u_n, v_n)$, a unit in $\mathbb{Z}[\sqrt{k}]$ with $v_n^2 > 1/\epsilon$.

We may sometimes want to find an **upper approximation** with $a/b > \sqrt{k}$ and/or a **lower approximation** with $a/b < \sqrt{k}$. The former can always be found as above, since there always exists a unit with norm equal to $+1$. But if there is no unit with norm equal to -1, the best lower approximation requires a solution to $a^2 - b^2 k = -n$ with the least possible n. Given one such solution, (a_1, b_1), we can find a solution (a, b) with b as large as desired by letting

$$(a, b) = (a_1, b_1) \times (u_n, v_n),$$

where (u_n, v_n) is a unit in $\mathbb{Z}[\sqrt{k}]$ with sufficiently large v_n.

Exercise 8.3.11. *Noting that the fundamental unit in $\mathbb{Z}[\sqrt{3}]$ is $(2, 1)$, find a rational upper approximation to $\sqrt{3}$ with $(a/b) - \sqrt{3} < 1/500^2$.*

There is no unit in $\mathbb{Z}[\sqrt{3}]$ with norm equal to -1, but $\mathcal{N}(1, 1) = -2$. Find a rational lower approximation to $\sqrt{3}$ with $\left|(a/b) - \sqrt{3}\right| < 2/100^2$.

8.4 Primes in $\mathbb{Z}[\sqrt{k}]$

8.4.1 Basic Theorems about Primes

As in $\mathbb{N}$, primes are irregularly distributed in $\mathbb{Z}[\sqrt{k}]$; furthermore, in $\mathbb{Z}[\sqrt{k}]$ there is no simple algorithm such as the sieve of Eratosthenes to generate a list of them. However, there are several theorems that will help in the search for primes.

Theorem 8.4.1. *If $\mathcal{N}(a, b)$ is prime in $\mathbb{Z}$, then (a, b) is prime in $\mathbb{Z}[\sqrt{k}]$.*

Proof. If (a, b) is factorised in $\mathbb{Z}[\sqrt{k}]$ as $(a, b) = (c, d) \times (e, f)$, Theorem 8.1.16 then requires that

$$\mathcal{N}(a, b) = \mathcal{N}(c, d) \times \mathcal{N}(e, f).$$

Recalling Definition 5.3.20, if $\mathcal{N}(a, b)$ is prime in $\mathbb{Z}$, then either $\mathcal{N}(c, d)$ or $\mathcal{N}(e, f)$ is a unit in $\mathbb{Z}$. Hence by Theorem 8.3.1, either (c, d) or (e, f) is a unit in $\mathbb{Z}[\sqrt{k}]$ and so, according to Definition 5.3.20, (a, b) is prime in $\mathbb{Z}[\sqrt{k}]$. $\square$

Note that, unlike Theorem 8.3.1, this is a one-way implication: there can exist primes in $\mathbb{Z}[\sqrt{k}]$ whose norms are not prime in $\mathbb{Z}$. Nevertheless, it is still useful; in applying it, one should recall that the primes in $\mathbb{Z}$ are $\pm p$ where p is any prime in $\mathbb{N}$.

Example 8.4.2. (i) In the Gaussian integers $\mathbb{Z}[\sqrt{-1}]$, the number $(2, 1)$ is prime, since $\mathcal{N}(2, 1) = 2^2 + 1^2 = 5$, which is prime in $\mathbb{Z}$.

(ii) In $\mathbb{Z}[\sqrt{2}]$, the number $(1, 2)$ is prime since $\mathcal{N}(1, 2) = 1^2 - 2^2 \times 2 = -7$, which is prime in $\mathbb{Z}$.

An important class of numbers that are prime in $\mathbb{Z}[\sqrt{k}]$ but have a composite norm is covered by the next theorem:

Theorem 8.4.3. *If $\mathcal{N}(a,b) = p_1 p_2$ where p_1 and p_2 are prime in $\mathbb{Z}$, and if either $p = p_1$ or $p = p_2$ does not admit solutions of $c^2 - d^2 k = \pm p$ with $c, d \in \mathbb{Z}$, then (a,b) is prime in $\mathbb{Z}[\sqrt{k}]$.*

Proof. If (a,b) is composite in $\mathbb{Z}[\sqrt{k}]$, so that $(a,b) = (c_1, d_1) \times (c_2, d_2)$ where neither (c_1, d_1) nor (c_2, d_2) is a unit in $\mathbb{Z}[\sqrt{k}]$, then

$$\mathcal{N}(a,b) = \mathcal{N}(c_1, d_1) \times \mathcal{N}(c_2, d_2), \tag{8.13}$$

where neither $\mathcal{N}(c_1, d_1)$ nor $\mathcal{N}(c_2, d_2)$ is a unit in $\mathbb{Z}$. But uniqueness of prime factorisations in $\mathbb{Z}$ means that if $\mathcal{N}(a,b) = p_1 p_2$, then the only other factorisation that does not involve a unit in $\mathbb{Z}$ is $\mathcal{N}(a,b) = (-p_1)(-p_2)$. So the norms in the factorisation (8.13) must be identified with $\pm p_1$ and $\pm p_2$. Thus since $\mathcal{N}(c,d) = c^2 - d^2 k$, there must exist integers c_1, d_1, c_2, d_2 such that $c_1^2 - d_1^2 k = \pm p_1$ and $c_2^2 - d_2^2 k = \pm p_2$. So if either p_1 or p_2 does not admit integer solutions of these equations, then (a,b) cannot be composite. $\qquad\square$

An obvious question arising from this theorem is how to determine whether integer solutions of $c^2 - d^2 k = \pm p$ are admitted for any given p. *Non-existence* of solutions can often be determined using modular arithmetic, although this will not resolve all cases: see Subsection 8.4.3 and the Investigations at the end of this chapter for examples.

A particularly important case of the theorem is where $p_1 = p_2$:

Corollary 8.4.4. *If $\mathcal{N}(a,b) = p^2$ where p is prime in $\mathbb{N}$ and there do not exist solutions of $c^2 - d^2 k = \pm p$ with $c, d \in \mathbb{Z}$, then (a,b) is prime in $\mathbb{Z}[\sqrt{k}]$. In particular, since $\mathcal{N}(\pm p, 0) = p^2$, the numbers $(\pm p, 0)$ are prime in $\mathbb{Z}[\sqrt{k}]$ when the conditions of this corollary are satisfied.*

Although the numbers $(p, 0)$ and $(-p, 0)$ in $\mathbb{Z}[\sqrt{k}]$ correspond to p and $-p$ in $\mathbb{Z}$, it does not follow that $(p, 0)$ and $(-p, 0)$ are prime in $\mathbb{Z}[\sqrt{k}]$ when p and $-p$ are prime in $\mathbb{Z}$. But Corollary 8.4.4 gives a criterion for when $(p, 0)$ and $(-p, 0)$ are indeed prime in $\mathbb{Z}[\sqrt{k}]$. It also leads to a simple proof that one of the most important theorems about primes in $\mathbb{N}$ may be extended to any quadratic ring:

Theorem 8.4.5. *There are infinitely many primes in any quadratic ring $\mathbb{Z}[\sqrt{k}]$.*

Proof. Corresponding to the infinitely many primes in $\mathbb{N}$ which exist according to Theorem 5.3.3, there are infinitely many numbers $(p, 0)$ in $\mathbb{Z}[\sqrt{k}]$. For each p, either there do not exist solutions of $c^2 - d^2 k = \pm p$ with $c, d \in \mathbb{Z}$, in which case $(p, 0)$ is prime in $\mathbb{Z}[\sqrt{k}]$; or there do exist solutions of $c^2 - d^2 k = \pm p$, in which case $(p, 0) = (c, d) \times (c, -d)$ and (c, d) is prime because its norm is $\pm p$. So, corresponding to each prime p in $\mathbb{N}$, there exists in $\mathbb{Z}[\sqrt{k}]$ either a prime $(p, 0)$ or a prime (c, d) whose norm is $\pm p$. $\qquad\square$

Theorem 8.4.3 covers numbers in $\mathbb{Z}[\sqrt{k}]$ whose norm is the product of just two primes in $\mathbb{Z}$, and cannot be extended to more general composite numbers. For example, if $\mathcal{N}(a,b) = p_1 p_2 p_3$ and there are no solutions of $c^2 - d^2 k = \pm p_3$, there may still be a solution of $c^2 - d^2 k = \pm p_2 p_3$ which would allow a factorisation like (8.13) with $\mathcal{N}(c_2, d_2) = \pm p_2 p_3$.

8.4.2 Associates Classes and Conjugates of Primes

Before considering further the practicalities of searching for primes, we should recall that according to Theorem 5.3.21 and the discussion following that theorem, we are only concerned with finding *associate classes* of primes. Since an associate of a number is defined to be a unit multiple of that number, we have:

Theorem 8.4.6. *For $(a,b) \in \mathbb{Z}[\sqrt{k}]$: if $\mathcal{N}(a,b) = q$, then the norm of every associate of (a,b) is $\pm q$.*

Proof. Units (u, v) have norm equal to ± 1. So by Theorem 8.1.16, if $\mathcal{N}(a, b) = q$, then $\mathcal{N}((a, b) \times (u, v)) = q \times (\pm 1) = \pm q$.   □

In the proof of Theorem 8.4.5, the primes in $\mathbb{Z}[\sqrt{k}]$ corresponding to different $p \in \mathbb{N}$ have different norms, and so by Theorem 8.4.6 cannot be associates. So in fact we have shown that there are *infinitely many associate classes of primes* in any quadratic ring $\mathbb{Z}[\sqrt{k}]$.

Note that the converse of Theorem 8.4.6 is not in general true. In particular, conjugates in a quadratic ring have equal norms, but are not in general associates of each other. Nevertheless, we do have:

Theorem 8.4.7. *If (a, b) is prime in $\mathbb{Z}[\sqrt{k}]$, then its conjugate $(a, -b)$ is also prime in $\mathbb{Z}[\sqrt{k}]$.*

Proof. According to Theorem 8.1.10, any factorisation $(a, b) = (c, d) \times (e, f)$ is "mirrored" by a factorisation of the conjugate, $\overline{(a, b)} = \overline{(c, d)} \times \overline{(e, f)}$. In particular, if (a, b) can only be factorised with one of the factors being a unit (i.e. if (a, b) is prime), then since the conjugate of a unit is itself a unit, the same will be true for $\overline{(a, b)}$.   □

Now observe that since $(-1, 0)$ is a unit in any quadratic ring and $(-1, 0) \times (a, b) = (-a, -b)$, the additive inverse of a number in $\mathbb{Z}[\sqrt{k}]$ is an associate of that number (as in $\mathbb{Z}$). Combining this observation with Theorem 8.4.7, we conclude that if (a, b) is prime, then so are $(a, -b)$, $(-a, -b)$ and $(-a, b)$ (respectively its conjugate, its additive inverse, and the additive inverse of its conjugate). Hence we can limit the search for primes to numbers (a, b) with $a \geq 0$ and $b \geq 0$ in any quadratic ring. That is as far as we can take this line of argument in $\mathbb{Z}[\sqrt{k}]$ with $k \leq -2$, where the only units are $(1, 0)$ and $(-1, 0)$. But in the Gaussian integers $\mathbb{Z}[\sqrt{-1}]$ we also have the units $(0, 1)$ and $(0, -1)$, which multiply (a, b) to give associates $(-b, a)$ and $(b, -a)$, and multiply its conjugate $(a, -b)$ to give associates (b, a) and $(-b, -a)$. So corresponding to any prime $(a, b) \in \mathbb{Z}[\sqrt{-1}]$ with $a > 0$ and $b > 0$ there are three further associates, and also the conjugate prime $(a, -b)$ and its three associates including (b, a).

Things become much more interesting in real quadratic rings, where there are infinitely many units so that any prime has infinitely many associates. Since the units are the integer powers of the fundamental unit (u_1, v_1) and the additive inverses of those integer powers, a prime (a_0, b_0) in a real quadratic ring has an associate class consisting of the numbers

$$(a_0, b_0) \times (u_1, v_1)^n \quad \text{and} \quad (-a_0, -b_0) \times (u_1, v_1)^n \quad \text{for all} \quad n \in \mathbb{Z}.$$

As an example of such an associate class, we show in Table 8.1 some of the associates of the prime $(a_0, b_0) = (1, 2)$ in $\mathbb{Z}[\sqrt{2}]$, where the fundamental unit is $(u_1, v_1) = (1, 1)$. We use the notation,

$$(a_n, b_n) := (a_0, b_0) \times (u_1, v_1)^n \quad \text{for all} \quad n \in \mathbb{Z} \tag{8.14}$$

(so that $(a_{-n}, b_{-n}) = (a_0, b_0) \times (u_1, v_1)^{-n}$, where negative powers of the fundamental unit are found from equation (8.10)).

Note the structure of the list of associates in Table 8.1: for $n \geq 0$, both a_n and b_n are positive, with increasing values of a_n and b_n as n increases, so that a_0 and b_0 are the smallest both-positive values. For each negative n, one of a_n or b_n is positive while the other is negative. This structure can be found for the associate class of any prime in a real quadratic ring:

Theorem 8.4.8. *Each prime class in a real quadratic ring contains a **fundamental prime** (a_0, b_0) with $a_0 \geq 0$ and $b_0 \geq 0$, such that the associates (a_n, b_n) defined in (8.14) have the properties:*

    (i) For $n \in \mathbb{Z}_+$, a_n and b_n are both positive;

    (ii) For $n \in \mathbb{Z}_-$, one of a_n and b_n is positive and the other is negative.

n	(a_n, b_n)	(a_{-n}, b_{-n})
0	$(1,2)$	
1	$(5,3)$	$(3,-1)$
2	$(11,8)$	$(-5,4)$
3	$(27,19)$	$(13,-9)$
4	$(65,46)$	$(-31,22)$
$\vdots$	$\vdots$	$\vdots$

TABLE 8.1: Some associates of the prime $(1,2)$ in $\mathbb{Z}[\sqrt{2}]$.

Proof. First observe that a prime class in a real quadratic ring certainly includes a prime with positive real magnitude (see Definition 8.3.5); for if (a,b) has negative real magnitude, its associate $(-a,-b)$ will have positive real magnitude. Furthermore, the fundamental unit in any quadratic ring has positive real magnitude. Thus, since real magnitudes preserve multiplication, if we take a prime (a,b) with positive real magnitude, then $(a,b) \times (u_1,v_1)^n$ has positive real magnitude for all $n \in \mathbb{Z}$. So we shall assume positive real magnitude for all primes in the analysis below, noting that the additive inverse of each such prime is also a prime, with negative real magnitude. Note that for (a,b) to have positive real magnitude, one of a and b may be negative, but not both.

We have

$$(a_{n+1}, b_{n+1}) = (a_n, b_n) \times (u_1, v_1)$$
$$= (a_n u_1 + b_n v_1 k, a_n v_1 + b_n u_1), \qquad (8.15)$$

and

$$(a_{m-1}, b_{m-1}) = (a_m, b_m) \times (u_{-1}, v_{-1})$$
$$= (a_m u_{-1} + b_m v_{-1} k, a_m v_{-1} + b_m u_{-1}). \qquad (8.16)$$

Furthermore, the fundamental unit has $u_1 \geq 1$ and $v_1 \geq 1$ in any real quadratic ring, while its multiplicative inverse (u_{-1}, v_{-1}) has either $u_{-1} \geq 1$ and $v_{-1} \leq -1$, or $u_{-1} \leq -1$ and $v_{-1} \geq 1$ (from equation (8.10)); also, $k > 1$ in a real quadratic ring. Then if $a_n \geq 0$ and $b_n \geq 0$, equation (8.15) yields the orderings

$$a_{n+1} \geq a_n, \quad a_{n+1} > b_n, \quad b_{n+1} \geq a_n, \quad b_{n+1} \geq b_n \qquad (8.17)$$

as seen in the second column of Table 8.1; the equalities in (8.17) are only possible if either $a_n = 0$ or $b_n = 0$. If $a_m > 0$ and $b_m < 0$, with $u_{-1} \leq -1$ and $v_{-1} \geq 1$, equation (8.16) yields the orderings

$$a_{m-1} < -a_m < 0, \quad a_{m-1} < b_m, \quad b_{m-1} > a_m, \quad b_{m-1} > -b_m > 0 \qquad (8.18)$$

as seen in the third column of Table 8.1. Similar orderings to (8.18) arise in the cases where $a_m < 0$ and $b_m > 0$, and/or $u_{-1} \geq 1$ and $v_{-1} \leq -1$; in all these cases, the pair (a_{m-1}, b_{m-1}) consists of one positive and one negative member whenever this is true of the pair (a_m, b_m), as you can show:

Exercise 8.4.9. *Write down orderings for a_{m-1} and b_{m-1} relative to a_m and b_m for the cases:*

(i) $a_m < 0$ and $b_m > 0$, with $u_{-1} \leq -1$ and $v_{-1} \geq 1$;
(ii) $a_m > 0$ and $b_m < 0$, with $u_{-1} \geq -1$ and $v_{-1} \leq 1$;
(iii) $a_m < 0$ and $b_m > 0$, with $u_{-1} \geq -1$ and $v_{-1} \leq 1$.

Also, if either $a_m = 0$ or $b_m = 0$, (8.16) can be used to show that one of a_{m-1} and b_{m-1} will be positive, and the other negative.

To complete the verification of the theorem, it only remains to show that there must exist within the prime class a particular prime, which we can denote (a_0, b_0), with the property that $a_0 \geq 0$ and $b_0 \geq 0$ while one of a_{-1} and b_{-1} is positive and the other negative. If we note from (8.17) that $a_{n-1} < a_n$ and $b_{n-1} < b_n$ (if a_{n-1} and b_{n-1} are still both positive), then an *infinite descent* argument shows that, given a pair with $a_n > 0$ and $b_n > 0$, there must exist a pair (a_{n-j}, b_{n-j}) with either $a_{n-j} \leq 0$ or $b_{n-j} \leq 0$, for some $j \in \mathbb{N}$. This is because the index $n - j$ can decrease through infinitely many negative values as j increases, while there are only finitely many positive integers less than both a_n and b_n; and consideration of real magnitudes (see above) shows that it is not possible for *both* $a_{n-j} \leq 0$ and $b_{n-j} \leq 0$. A similar argument using orderings (8.18) or those found in Exercise 8.4.9 shows that, given a pair with one of a_m and b_m positive and the other negative, then for some $i \in \mathbb{N}$ we will have a_{m+i} and b_{m+i} both positive. So, as we repeatedly multiply or divide a prime by the fundamental unit, there must be a transition between pairs (a, b) with $a \geq 0$ and $b \geq 0$ and pairs with one of a or b negative; it is the prime at this transition that we can label with index 0 and describe as the fundamental prime. $\qquad\square$

Given a fundamental prime (a_0, b_0) with $b_0 > 0$, its conjugate $(a_0, -b_0)$ will also be a prime, and in most cases the conjugates will not be associates. So $(a_0, -b_0)$ will be a prime in a separate class, but not the fundamental prime in that class, since $-b_0 < 0$ if $b_0 > 0$. However, consider the numbers $\pm(a_{-1}, -b_{-1})$, which are associates of $(a_0, -b_0)$; according to Theorem 8.4.8, one of a_{-1} and b_{-1} will be negative and the other positive, so either $(a_{-1}, -b_{-1})$ or $(-a_{-1}, b_{-1})$ will have both members of the pair positive: this will then be a fundamental prime. For example, in Table 8.1, $(a_{-1}, b_{-1}) = (3, -1)$ and its conjugate, $(a_{-1}, -b_{-1}) = (3, 1)$, is the fundamental prime in a separate prime class.

Exercise 8.4.10. *Show that if* (a_0, b_0) *is a fundamental prime with its conjugate not being an associate, then* $(a_0, -b_0) \times (u_1, v_1) = \pm(a_{-1}, -b_{-1})$. *Explain why this means that either* $(a_{-1}, -b_{-1})$ *or* $(-a_{-1}, b_{-1})$ *is a fundamental prime.*

However, in some instances the conjugate of a fundamental prime may be an associate, so does not yield a separate prime class. Some cases where this happens are fairly obvious:

Exercise 8.4.11. *Explain why if either* $a_0 = 0$ *or* $b_0 = 0$, *then the conjugate of the fundamental prime will not yield a separate class of primes in any real quadratic ring. For the case* $a_0 = 0$, *show that* $(0, b_0)$ *cannot be prime unless* $b_0 = \pm 1$.

More generally, to test whether the conjugate of a number is an associate, divide the number by its conjugate according to formula (8.8). This will yield an ordered pair of rationals; but if the rationals are both integers, you will have obtained a unit in $\mathbb{Z}[\sqrt{k}]$ (since the number and its conjugate have equal norms), so the conjugate must then be an associate of the original number.

Exercise 8.4.12. *(i) In* $\mathbb{Z}[\sqrt{3}]$, *show that* $(1, 1)$ *is prime and that its conjugate* $(1, -1)$ *is in the same associate class.*

(ii) In $\mathbb{Z}[\sqrt{6}]$, *show that* $(3, 1)$ *is prime and that its conjugate* $(3, -1)$ *is in the same associate class.*

8.4.3   How to Search for Primes

In this subsection, we consider the practicalities of using the theorems in Subsection 8.4.1 to obtain lists of primes in various quadratic rings. Just as with Eratosthenes' sieve in $\mathbb{N}$, where it is necessary to set an upper limit N for the primes to be sought because we cannot find infinitely many, we need to impose finite limits for any search for primes in $\mathbb{Z}[\sqrt{k}]$. Since determining primality involves working with norms, it is sensible to set this limit on the norm: so we seek all primes in $\mathbb{Z}[\sqrt{k}]$ with norm between $-N$ and N, for some arbitrarily chosen upper limit N. Thus, for each successive Natural Number, $n = 2, 3, 4, \ldots, N$, we consider numbers $(a, b) \in \mathbb{Z}[\sqrt{k}]$ with norm equal to $\pm n$, i.e. with a and b being integers satisfying

$$a^2 - b^2 k = \pm n.^{\times} \tag{8.19}$$

The details depend on the value of k.

(a) The Gaussian integers, $k = -1$.

In this case, the norm cannot be negative, and equation (8.19) becomes $a^2 + b^2 = n$. According to Theorem 8.4.1 any (a, b) satisfying this will be prime in $\mathbb{Z}[\sqrt{-1}]$ if n is prime in $\mathbb{N}$. Consider three types of prime in $\mathbb{N}$:

(i) The only even prime, $n = 2$, yields the prime $(1, 1)$ in $\mathbb{Z}[\sqrt{-1}]$. Its conjugate $(1, -1)$ is an associate of $(1, 1)$, so does not yield a separate prime class.

(ii) For primes $n \equiv 1 \pmod 4$, a theorem mentioned but not proved at the end of Section 5.5 states that primes of this type can be expressed uniquely as a sum of squares. Thus, if n is prime with $n \equiv 1 \pmod 4$ we will be able to find two primes, (a, b) and (b, a) in $\mathbb{Z}[\sqrt{-1}]$, where $a^2 + b^2 = n$. These two primes will be in separate classes: (b, a) is an associate of the conjugate of (a, b), but not of (a, b) itself.

Exercise 8.4.13. *Prove that (b, a) is not an associate of (a, b) in $\mathbb{Z}[\sqrt{-1}]$ if $a > 0, b > 0$ and $b \neq a$.*

(iii) For primes $n \equiv 3 \pmod 4$, it was shown in Example 5.5.9 on modular arithmetic that there do not exist $a, b \in \mathbb{Z}$ such that $a^2 + b^2 = n$. But then Corollary 8.4.4 shows that $(n, 0)$ is prime in $\mathbb{Z}[\sqrt{-1}]$.

It can be shown (using more advanced results in Number Theory) that the *only* primes in $\mathbb{Z}[\sqrt{-1}]$ are those described above (and their associates); there are none whose norm is the product of two (or more) distinct primes, or whose norm is n^2 but the prime is not of form $(n, 0)$.

Exercise 8.4.14. *Find all the primes $(a, b) \in \mathbb{Z}[\sqrt{-1}]$ with norms less than 120, and with $a > 0$ and $b \geq 0$; note that this will capture every associate class of primes with norm less than 120.*

(b) Other imaginary quadratic rings, $k \leq -2$.

Again, $a^2 - b^2 k$ must be positive. Modular arithmetic may identify values of n for which no integer solutions to (8.19) exist: see Investigation 2 at the end of this chapter for an example. If solutions are thought to exist, a systematic way to search for them is to evaluate $n + b^2 k$ successively for $b = 0, 1, 2, 3, \ldots$: whenever the result is the square of an integer, that integer is a in a solution (a, b). With $k < 0$, the process terminates when b has become large enough that $n + b^2 k < 0$, so that this quantity can no longer be a square. When n is prime, (a, b) is prime in $\mathbb{Z}[\sqrt{k}]$, but for composite n further primes in $\mathbb{Z}[\sqrt{k}]$ may be found satisfying the conditions of Theorem 8.4.3 or Corollary 8.4.4; to be certain of finding all primes, one should go through the procedure to obtain the prime factorisation for each (a, b) (see Section 8.5). The conjugate of a prime will also be prime, and not an associate unless $a = 0$ or $b = 0$.

$^{\times}$Known as the Generalised Pell Equation.

(c) Real quadratic rings, $k \geq 2$.

If $k > 1$, there may be solutions of either or both $a^2 - b^2k = n$ and $a^2 - b^2k = -n$. To find solutions, methods similar to those for $k < -1$ may be employed: modular arithmetic may be used to reveal values of n for which no solution exists, while for other values of n successive evaluation of both $n + b^2k$ and $-n + b^2k$ for $b = 0, 1, 2, 3, \ldots$ will yield a solution (a, b) whenever the result is an exact square, a^2. The difficulty is that there is no obvious upper limit to b; however, it has been proved that for any given k and n, there are finitely many associate classes of solutions (a, b) to equation (8.19), and that if

$$b > \frac{\sqrt{n}\left(1 + \sqrt{u_1 + v_1\sqrt{k}}\right)}{2\sqrt{k}}, \,^{\text{xi}} \tag{8.20}$$

(where $u_1 + v_1\sqrt{k}$ is the real magnitude of the fundamental unit (u_1, v_1) in $\mathbb{Z}[\sqrt{k}]$), then (a, b) is an associate of a solution (a_1, b_1) with $0 \leq b_1 < b$. So in seeking separate associate classes of solutions, the search can stop when b exceeds the real number on the right of (8.20). If more than one solution (a, b) is found within this bound for a given n, exact division will reveal whether they are associates; if two or more prime associates are found, we are only interested in the fundamental prime (with the least non-negative b).

The use of modular arithmetic and the theorems of Subsection 8.4.1 in seeking primes in real quadratic rings is exemplified in the following exercise and in Investigation 3 at the end of this chapter.

Exercise 8.4.15. *(i) Show that there do not exist* $c, d \in \mathbb{Z}$ *such that*

$$c^2 - 2d^2 \equiv 3 \text{ or } 5 \,(\text{mod } 8).$$

Deduce that there also do not exist $c, d \in \mathbb{Z}$ *such that*

$$c^2 - 2d^2 \equiv -3 \text{ or } -5 \,(\text{mod } 8).$$

(ii) Why does it follow immediately that there do not exist $c, d \in \mathbb{Z}$ *such that*

$$c^2 - 10d^2 \equiv \pm 3 \text{ or } \pm 5 \,(\text{mod } 8)?$$

(iii) Explain why any number with norm equal to ± 15 *is prime in both* $\mathbb{Z}[\sqrt{2}]$ *and* $\mathbb{Z}[\sqrt{10}]$.
(iv) Find two fundamental primes in $\mathbb{Z}[\sqrt{10}]$, *one with norm equal to 15 and the other with norm equal to* -15.
(v) In $\mathbb{Z}[\sqrt{2}]$, *modular arithmetic in* $\mathbb{Z}_8$ *does not give any information on the existence of numbers with norm equal to* ± 15. *Use modular arithmetic in* $\mathbb{Z}_9$ *to show that there do not exist* $a, b \in \mathbb{Z}$ *such that* $a^2 - 2b^2 = \pm 15$. *Hence there are no primes in* $\mathbb{Z}[\sqrt{2}]$ *with norm equal to* ± 15.
(vi) Using the result from (ii) above, find four primes of type $(p, 0)$ *in* $\mathbb{Z}[\sqrt{10}]$.

In fact there are theorems relating to $\mathbb{Z}[\sqrt{2}]$ that simplify the situation similarly to that in $\mathbb{Z}[\sqrt{-1}]$. The result in Exercise 8.4.15(i) implies that primes of form $(n, 0)$ exist corresponding to primes $n \in \mathbb{N}$ when $n \equiv 3 \,(\text{mod } 8)$ or $n \equiv 5 \,(\text{mod } 8)$, according to Corollary 8.4.4. Corresponding to $n = 2$ there is the fundamental prime $(0, 1)$ with norm equal to -2; and corresponding to any prime $n \in \mathbb{N}$ with $n \equiv 1 \,(\text{mod } 8)$ or $n \equiv 7 \,(\text{mod } 8)$, it can be shown that there are just two associate classes of primes with norms equal to n or $-n$, one of these classes consisting of the conjugates of the primes in the other class; and no primes exist in $\mathbb{Z}[\sqrt{2}]$ other than those described above.

Exercise 8.4.16. *Find all fundamental primes in* $\mathbb{Z}[\sqrt{2}]$ *with norm between* -50 *and* 50.

---

^{xi}This formula obviously requires some sophisticated Number Theory to derive.

8.5   Prime Factorisation in $\mathbb{Z}[\sqrt{k}]$

The Fundamental Theorem of Arithmetic (Theorem 5.3.2) states that every Natural Number greater than 1 can be expressed uniquely as the product of one or more prime factors. The theorem asserts two properties: the *existence* and the *uniqueness* of prime factorisation. Because factorisations of numbers in $\mathbb{Z}[\sqrt{k}]$ correspond to factorisations of their norms in $\mathbb{Z}$ according to Theorem 8.1.16, it is easy to verify that the existence of prime factorisations extends to $\mathbb{Z}[\sqrt{k}]$, although to allow for the fact that norms can be negative in $\mathbb{Z}$, we need to consider the *modulus* of norms in the infinite descent argument.

Exercise 8.5.1. *Prove that every number in $\mathbb{Z}[\sqrt{k}]$ can be expressed as the product of one or more primes.*

Unlike the existence proof, if we wanted to use norms in a proof of the *uniqueness* of prime factorisations along the lines of the proof for Theorem 5.3.2, we would need a one-to-one correspondence between norms and numbers in $\mathbb{Z}[\sqrt{k}]$, or at least between norms and associate classes of numbers; whereas in $\mathbb{Z}[\sqrt{k}]$ it is possible for two numbers which are not associates to have the same norm. If uniqueness of prime factorisation does not apply in certain rings $\mathbb{Z}[\sqrt{k}]$, further theorems which follow from it may then fail: most importantly, Theorem 5.3.10, which states that a prime which is a divisor of a product yz must be a divisor of either y or z, need not be true in $\mathbb{Z}[\sqrt{k}]$. Indeed, more advanced texts on the algebra of rings reserve the term "prime" to denote a member of a ring that does satisfy Theorem 5.3.10, while using the word **irreducible** to mean that it can only be factorised with one of the factors being a unit (the property which more elementary texts like ours refer to as being "prime").[xii] The uniqueness of prime factorisations is a major topic of ring theory; some quadratic rings do have the uniqueness property, while others do not. In this section we will take an experimental approach, finding examples of prime factorisation that are or are not unique; to discover the theory underlying the experimental findings, you will need to read on until the end of the chapter.

We now consider the practicalities of finding prime factorisations of several categories of composite numbers in quadratic rings. It will be assumed that we already have a list of primes with norm of lesser modulus than the number to be factorised.

(a) Numbers of form $(x, 0)$. If x is prime in $\mathbb{Z}$, then either $(x, 0)$ is prime in $\mathbb{Z}[\sqrt{k}]$ or it is the product of two prime factors, $(x, 0) = (c, d) \times (c, -d)$ where $c^2 - d^2 k = x$ (see the proof of Theorem 8.4.5). If x is composite, its prime factorisation in $\mathbb{Z}$ will correspond to a factorisation of $(x, 0)$ into factors of form $(p, 0)$ in $\mathbb{Z}[\sqrt{k}]$, each of which may be prime or have two prime factors as described above.

(b) Numbers of form (x, y) where x and y have a common factor. If $x = sd$ and $y = td$ for some $d > 1$, with s and t coprime, then $(x, y) = (s, t) \times (d, 0)$. Then $(d, 0)$ can be factorised as in (a) above, while (s, t) will require one of the methods below, unless it is a unit.

Example 8.5.2. Find the prime factorisation of $(9, 6)$ in $\mathbb{Z}[\sqrt{2}]$.

Solution. Noting the common factor of 3 in 9 and 6, we can write $(9, 6) = (3, 0) \times (3, 2)$. From Exercise 8.4.15 and Corollary 8.4.4, $(3, 0)$ is prime in $\mathbb{Z}[\sqrt{2}]$, while $(3, 2)$ is a unit since $3^2 - 2^2 \times 2 = 1$. So in fact $(9, 6)$ is prime, an associate of (3.0). $\qquad\square$

[xii] This disparity of terminology between elementary writings and specialised ring theory texts is unfortunate. For internal consistency in this book, we shall persist with the use of "prime" to mean that any factorisation must include a unit.

(c) Numbers of form (x, y) where x and y are coprime and there exists a prime in $\mathbb{Z}[\sqrt{k}]$ corresponding to each of the prime factors of $\mathcal{N}(x, y)$. The first step is to find the prime factorisation of $\mathcal{N}(x, y)$ in $\mathbb{Z}$, noting that in real quadratic rings the factors of $\mathcal{N}(x, y)$ may be positive or negative integers. With $\mathcal{N}(x, y) = p_1 \times p_2 \times \ldots \times p_m$, if there are primes in $\mathbb{Z}[\sqrt{k}]$ with norms equal to each of $p_1, p_2, \ldots, p_m$ or their additive inverses, then we expect to find a prime factorisation of (x, y) in terms of such primes; but since there are typically two or more (non-associate) primes in $\mathbb{Z}[\sqrt{k}]$ with a given value of the norm, a trial division process is needed. Suppose that (a_1, b_1) is a prime with norm equal to p_1 (or $-p_1$): then divide (x, y) by (a_1, b_1) according to formula (8.8),

$$(x, y) \div (a_1, b_1) = \frac{(x, y) \times (a_1, -b_1)}{\mathcal{N}(a_1, b_1)}$$

$$= \left(\frac{xa_1 - yb_1 k}{\mathcal{N}(a_1, b_1)}, \frac{-xb_1 + ya_1}{\mathcal{N}(a_1, b_1)} \right);$$

so only if $\mathcal{N}(a_1, b_1)$ is a divisor of both $xa_1 - yb_1 k$ and $-xb_1 + ya_1$ does the division yield a number in $\mathbb{Z}[\sqrt{k}]$ so that (a_1, b_1) is a prime factor of (x, y). Otherwise, (a_1, b_1) is not a factor of (x, y), so we divide (x, y) by any other primes with norm equal to $\pm p_1$, until a prime factor of (x, y) is found. Having divided out this prime factor, we repeat the process, finding prime factors with norms equal to $\pm p_2, \ldots, \pm p_m$. Once all these prime factors have been divided out, we will be left with a unit.

Definition 5.3.22 defines unique prime factorisation in terms of associate classes; but we can find a unique prime factorisation of a *number* in a quadratic ring if we insist on writing it as the product of *fundamental* primes and, if necessary, a unit. Fundamental primes have so far been defined only for real quadratic rings; in $\mathbb{Z}[\sqrt{-1}]$ a prime (a, b) can be regarded as fundamental if $a > 0$ and $b \geq 0$, while in other imaginary quadratic rings a prime (a, b) is fundamental if $a > 0$, or if $a = 0$ and $b > 0$.

Example 8.5.3. Find the prime factorisation of $(6, 7)$ in $\mathbb{Z}[\sqrt{-1}]$.

Solution. The norm in $\mathbb{Z}[\sqrt{-1}]$ is $\mathcal{N}(a, b) = \mathcal{N}(b, a) = a^2 + b^2$, so $\mathcal{N}(6, 7) = 6^2 + 7^2 = 85$, which has prime factorisation $85 = 5 \times 17$. Since $2^2 + 1^2 = 5$, the primes $(2, 1)$ and $(1, 2)$ both have norm equal to 5. First try dividing $(6, 7)$ by $(2, 1)$: using formula (8.8),

$$(6, 7) \div (2, 1) = \frac{(6, 7) \times (2, -1)}{\mathcal{N}(2, 1)}$$

$$= \frac{(6 \times 2 - 7 \times (-1), 6 \times (-1) + 7 \times 2)}{2^2 + 1^2}$$

$$= \frac{(19, 8)}{5},$$

which is clearly not in $\mathbb{Z}[\sqrt{-1}]$, since 5 is not a divisor of either 19 or 8. So try dividing by $(1, 2)$:

$$(6, 7) \div (1, 2) = \frac{(6, 7) \times (1, -2)}{\mathcal{N}(1, 2)}$$

$$= \frac{(6 \times 1 - 7 \times (-2), 6 \times (-2) + 7 \times 1)}{1^2 + 2^2}$$

$$= \frac{(20, -5)}{5}$$

$$= (4, -1)$$

Now $(4, -1)$ is prime (it has norm equal to 17, as indeed it must from our prime factorisation of $\mathcal{N}(6, 7)$), but is not a fundamental prime; it is $(1, 4) \times (0, -1)$ where $(0, -1)$ is a unit in $\mathbb{Z}[\sqrt{-1}]$. So we have found the factorisation in terms of fundamental primes and a unit: $(6, 7) = (1, 2) \times (1, 4) \times (0, -1)$. And this is certainly unique, since $(2, 1)$ and $(1, 2)$ and their associates are the only numbers in $\mathbb{Z}[\sqrt{-1}]$ with norm equal to 5.   □

Exercise 8.5.4. *(a) Find prime factorisations of* $(16, 11)$, $(-4, 23)$, *and* $(-13, -21)$ *in* $\mathbb{Z}[\sqrt{-1}]$. *Express your answers as a product of primes found in Exercise 8.4.14 and, if necessary, a unit.*

(b) Find prime factorisations of $(10, 9)$, $(-17, 15)$ *and* $(1, -29)$ *in* $\mathbb{Z}[\sqrt{2}]$. *Express your results as a product of fundamental primes found in Exercise 8.4.16 and, if necessary, a unit.*

(d) Numbers of form (x, y) where x and y are coprime and there do not exist primes in $\mathbb{Z}[\sqrt{k}]$ corresponding to each of the prime factors of $\mathcal{N}(x, y)$. The number (x, y) is prime if no factor (prime or composite) of $\mathcal{N}(x, y)$ is the norm of any number in $\mathbb{Z}[\sqrt{k}]$, other than (x, y) itself; the simplest such cases are where $\mathcal{N}(x, y)$ has just two prime factors, as in Theorem 8.4.3.

For (x, y) to be composite and to satisfy the definition of this case, $\mathcal{N}(x, y)$ must have at least three prime factors in $\mathbb{Z}$, not necessarily distinct. Suppose $\mathcal{N}(x, y) = p_1 p_2 p_3$ and there is no prime in $\mathbb{Z}[\sqrt{k}]$ with norm equal to $\pm p_1$: for (x, y) to be composite, there must be a prime with norm equal to the product $\pm p_1 p_2$, and another prime with norm equal to $\pm p_3$ (without loss of generality, since the labelling of the p_i is arbitrary). If $\mathcal{N}(x, y)$ has more than three prime factors in $\mathbb{Z}$, there are more possibilities for how the factors of (x, y) in $\mathbb{Z}[\sqrt{k}]$ correspond to factors and products of factors of $\mathcal{N}(x, y)$ in $\mathbb{Z}$.

Example 8.5.5. Find the prime factorisation of $(95, 29)$ in $\mathbb{Z}[\sqrt{10}]$.

Solution. $\mathcal{N}(95, 29) = 95^2 - 29^2 \times 10 = 615$, and 615 has the prime factorisation $3 \times 5 \times 41$. In Exercise 8.4.15(ii) you found that neither ± 3 nor ± 5 can be norms of a number in $\mathbb{Z}[\sqrt{10}]$; but $3 \times 5 = 15$, and in part (iv) of that exercise you should have found the fundamental primes $(5, 1)$ and $(5, 2)$ with norms respectively equal to 15 and -15. Trial division yields $(95, 29) \div (5, 2) = (7, 3)$ whereas $(95, 29)$ is not divisible by $(5, 1)$ in $\mathbb{Z}[\sqrt{10}]$: check this for yourself, and also note that $\mathcal{N}(7, 3) = -41$ as it must be since $\mathcal{N}(5, 2) = -15$ and $\mathcal{N}(95, 29) \div (-15) = -41$; this provides a useful check on the division calculation. It remains to determine whether $(7, 3)$ is a fundamental prime: the fundamental unit in $\mathbb{Z}[\sqrt{10}]$ is $(3, 1)$, and since $(7, 3) \div (3, 1) = (9, -2)$ which has one negative member in the pair, $(7, 3)$ is indeed a fundamental prime. So the required prime factorisation is $(95, 29) = (5, 2) \times (7, 3)$.

If there was any other prime factorisation of $(95, 29)$ in $\mathbb{Z}[\sqrt{10}]$, the prime factors could still only have norms equal to ± 15 and ± 41. The criterion (8.20) can be used (with $k = 10$, $n = 15$ and $(u_1, v_1) = (3, 1)$) to show that no fundamental prime (a, b) with norm equal to ± 15 can have $b > 2$. Hence we have exhausted all possibilities, and the prime factorisation we have found is unique.   □

The next example is our first case of non-unique prime factorisation.

Example 8.5.6. Find the prime factorisation of $(35, 8)$ in $\mathbb{Z}[\sqrt{10}]$.

Solution. $\mathcal{N}(35, 8) = 35^2 - 8^2 \times 10 = 585$, and 585 has the prime factorisation $3^2 \times 5 \times 13$. None of the prime factors can be the norm of a number in $\mathbb{Z}[\sqrt{10}]$, according to the result in Exercise 8.4.15(ii); but this result still allows $9\,(= 3^2)$, $15\,(= 3 \times 5)$, $39\,(= 3 \times 13)$ and $65\,(= 5 \times 13)$ to be norms, and any number in $\mathbb{Z}[\sqrt{10}]$ with a norm equal to one of these values will certainly be prime, according to Theorem 8.4.3 and Corollary 8.4.4. In particular, noting that $585 = 9 \times 65 = 15 \times 39$, we can find the prime factorisations

$$(35, 8) = (1, 1) \times (5, 3) \quad \text{where} \quad \mathcal{N}(1, 1) = -9 \quad \text{and} \quad \mathcal{N}(5, 3) = -65,$$

and

$$(35, 8) = (17, 5) \times (5, 2) \times (-3, 1) \quad \text{where} \quad \mathcal{N}(17, 5) = 39, \quad \mathcal{N}(5, 2) = -15$$
$$\text{and} \quad (-3, 1) \text{ is a unit with } \mathcal{N}(-3, 1) = -1.$$

Since the prime factors in the two factorisations have different norms, they are clearly not associates of each other; so this is definitely an example of non-unique prime factorisation: the Fundamental Theorem of Arithmetic does not apply in the quadratic ring $\mathbb{Z}[\sqrt{10}]$. □

Whereas the last example was in a real quadratic ring, it is also easy to find examples of non-unique prime factorisation in imaginary quadratic rings. The following example is somewhat simpler than the previous one, and is in fact the example that is most often quoted to demonstrate the possibility of non-unique prime factorisation.

Example 8.5.7. Find the prime factorisation of $(6, 0)$ in $\mathbb{Z}[\sqrt{-5}]$.

Solution. $\mathcal{N}(6, 0) = 36$ (in any quadratic ring), and 36 has the prime factorisation $2^2 \times 3^2$. Now, in $\mathbb{Z}[\sqrt{-5}]$, the only numbers (a, b) with norms $a^2 + 5b^2 < 5$ will have $b = 0$: in particular, $(2, 0)$ has norm equal to 4, but there are no numbers in $\mathbb{Z}[\sqrt{-5}]$ with norms equal to the prime factors 2 or 3. So any number with norm equal to $4 (= 2^2), 6 (= 2 \times 3)$ or $9 (= 3^2)$ will be prime. Noting that $36 = 4 \times 9 = 6 \times 6$, we can indeed find the prime factorisations

$$(6, 0) = (2, 0) \times (3, 0) \quad \text{where} \quad \mathcal{N}(2, 0) = 4 \quad \text{and} \quad \mathcal{N}(3, 0) = 9$$

and

$$(6, 0) = (1, 1) \times (1, -1) \quad \text{where} \quad \mathcal{N}(1, 1) = \mathcal{N}(1, -1) = 6. \quad \square$$

The last two examples give the impression that non-unique prime factorisation requires a norm that can be factorised in at least two different ways in $\mathbb{Z}$. However, it is also possible if we have two numbers which are neither associates nor conjugates, but have the same norm:

Exercise 8.5.8. *In* $\mathbb{Z}[\sqrt{-5}]$ *the number* $(9, 0)$ *has the obvious prime factorisation* $(3, 0) \times (3, 0)$. *Find another prime factorisation of* $(9, 0)$ *with factors that have norm equal to 9 but are not associates of* $(3, 0)$. *[Note: to test whether numbers are associates, divide one by the other using formula (8.8); as when doing trial division to determine a factorisation, they are only associates if the result is in the quadratic ring, i.e. with both members of the pair being integers.]*

A superficially similar example may be found in $\mathbb{Z}[\sqrt{-3}]$, where $(4, 0) = (2, 0) \times (2, 0) = (1, 1) \times (1, -1)$; but in this case the situation can be "fixed", although to understand how that works we need to consider quadratic extensions of the Rational Numbers.

8.6   Quadratic Fields: Extensions of $\mathbb{Q}$

Since $\mathbb{Q}$ is a field so that any quadratic extension of $\mathbb{Q}$ is a field (see Theorem 8.1.8), we refer to $\mathbb{Q}[\sqrt{k}]$ as a **real quadratic field** or **imaginary quadratic field** depending on whether k is positive or negative. According to our general prescription for defining a quadratic extension, $\mathbb{Q}[\sqrt{k}]$ exists for any rational number k which does not have a rational square root. However, just as we found that quadratic rings $\mathbb{Z}[\sqrt{k}]$ with non-squarefree k are isomorphic to subrings with squarefree k, we shall now show that quadratic fields $\mathbb{Q}[\sqrt{k}]$ with non-integer k are isomorphic to fields with integer k.

Theorem 8.6.1. *The field $\mathbb{Q}[\sqrt{r/s}]$ is isomorphic to the field $\mathbb{Q}[\sqrt{rs}]$ under both addition and multiplication, where r and s are coprime integers, with s positive.*

Any rational number can be written in lowest terms as r/s where r and s are coprime integers, with s positive. So the theorem claims that the field $\mathbb{Q}[\sqrt{r/s}]$, in which $r/s \in \mathbb{Q}$, has isomorphisms to a field $\mathbb{Q}[\sqrt{rs}]$, in which $rs \in \mathbb{Z}$. Note that if r and s are coprime, then rs is squarefree if r and s individually are squarefree; and if r and/or s is not squarefree, then $\mathbb{Q}[\sqrt{rs}]$ is isomorphic to a subset of $\mathbb{Q}[\sqrt{l}]$, where $rs = lq^2$ with $l, q \in \mathbb{Z}$ and l is squarefree, according to Theorem 8.2.1. The isomorphisms imply that any theorem or calculation in $\mathbb{Q}[\sqrt{r/s}]$ would be equivalent to a theorem or calculation in $\mathbb{Q}[\sqrt{rs}]$ or $\mathbb{Q}[\sqrt{l}]$. Hence it is only necessary to consider quadratic fields $\mathbb{Q}[\sqrt{k}]$ with k taken from the same set of squarefree integers considered in our study of quadratic rings $\mathbb{Z}[\sqrt{k}]$; and in fact the main interest in studying quadratic fields is in shedding light on the behaviour of quadratic rings.

Proof of Theorem 8.6.1. Since the function

$$\tilde{f} : \mathbb{Q} \to \mathbb{Q} \quad \text{defined by} \quad \tilde{f} : x \mapsto \frac{x}{s}$$

(with any $s \in \mathbb{Z}_+$) is a bijection, the function

$$f : \mathbb{Q}[\sqrt{r/s}] \to \mathbb{Q}[\sqrt{rs}] \quad \text{defined by} \quad f : (a, b) \mapsto \left(a, \frac{b}{s}\right)$$

is a bijection.

Let a, b, c, d be any rational numbers. From Definition 8.1.1 for addition,

$$(a, b) + (c, d) = (a + c, b + d) \quad \text{in } \mathbb{Q}[\sqrt{r/s}],$$

while

$$\left(a, \frac{b}{s}\right) + \left(c, \frac{d}{s}\right) = \left(a + c, \frac{b+d}{s}\right) \quad \text{in } \mathbb{Q}[\sqrt{rs}],$$

giving the isomorphism under addition with the bijection f. Then Definition 8.1.2 for multiplication yields

$$(a, b) \times (c, d) = \left(ac + \frac{bdr}{s}, ad + bc\right) \quad \text{in } \mathbb{Q}[\sqrt{r/s}],$$

while in $\mathbb{Q}[\sqrt{rs}]$ we have

$$\left(a, \frac{b}{s}\right) \times \left(c, \frac{d}{s}\right) = \left(ac + \frac{bd}{s^2} rs, \frac{ad}{s} + \frac{bc}{s}\right) = \left(ac + \frac{bdr}{s}, \frac{ad+bc}{s}\right),$$

showing the isomorphism under multiplication with the bijection f. $\quad\square$

8.6.1 Algebraic Numbers in Quadratic Fields

In Section 7.9, we defined algebraic numbers of degree n as being numbers which satisfied equations in which a polynomial of degree n with integer coefficients is equated to zero; but no general means of constructing algebraic numbers was proposed. However, we identified the algebraic numbers of degree 1 as being the Rationals; we shall now show that the set of all algebraic numbers of degree 2 can be identified as the union of quadratic fields $\mathbb{Q}[\sqrt{k}]$ for all squarefree integers k.

Theorem 8.6.2. *(a) Every number in a quadratic field satisfies a quadratic equation with integer coefficients.*

(b) Every quadratic equation with integer coefficients has its solution(s) in a quadratic field $\mathbb{Q}[\sqrt{k}]$ *for some squarefree integer* k.

Proof. (a) For any $c, d \in \mathbb{Q}$ and squarefree $k \in \mathbb{Z}$, we need to find $a_0, a_1, a_2 \in \mathbb{Z}$ such that

$$(a_0, 0) + a_1(c, d) + a_2(c, d)^2 = (0, 0) :$$

this is the general quadratic equation (7.18) in which the solution x is a number $(c, d) \in \mathbb{Q}[\sqrt{k}]$ for some k, the integers a_0, a_1, a_2 in (7.18) correspond to the numbers $(a_0, 0)$, $(a_1, 0)$, $(a_2, 0)$ in $\mathbb{Q}[\sqrt{k}]$, and we have used the scalar multiplication format of Definition 8.1.7.

Now, $(c, d)^2 = (c^2 + d^2 k, 2cd)$, so our quadratic equation requires

$$a_0 + a_1 c + a_2(c^2 + d^2 k) = 0 \quad \text{and} \quad a_1 d + 2 a_2 cd = 0, \tag{8.21}$$

where we have used Definition 8.1.1 for addition, evaluating the sum of numbers in $\mathbb{Q}[\sqrt{k}]$ by separately adding first members and adding second members of ordered pairs. The second equation of (8.21) yields $a_1 = -2 a_2 c$ (unless $d = 0$), and the first equation then gives $a_0 = a_2(c^2 - d^2 k)$. But c and d are rational, whereas we require the coefficients in the quadratic equation to be integers. Writing $c = \frac{r}{s}$ and $d = \frac{u}{v}$ where r, s and u, v are pairs of coprime integers, we can let $a_2 = s^2 v^2 \in \mathbb{Z}$ and we then have $a_1 = -2 r s v^2 \in \mathbb{Z}$ and $a_0 = r^2 v^2 - s^2 u^2 k \in \mathbb{Z}$. In the case where $d = 0$ so that the first equation of (8.21) reduces to $a_0 + a_1 c + a_2 c^2 = 0$, we can again write $c = \frac{r}{s}$, and then let $a_2 = s^2$, $a_1 = s$ and $a_0 = -r - r^2$.

(b) The general quadratic equation $a_0 + a_1 x + a_2 x^2 = 0$ has the formal solutions

$$x = \frac{1}{2 a_2} \left(-a_1 \pm \sqrt{a_1^2 - 4 a_0 a_2} \right),$$

where the form $a + b\sqrt{k}$ corresponds to the behaviour of the number $(a, b) \in \mathbb{Q}[\sqrt{k}]$.

Let $k = a_1^2 - 4 a_0 a_2$. If k is squarefree, then the two solutions are the numbers

$$\left(-\frac{a_1}{2 a_2}, \pm \frac{1}{2 a_2} \right)$$

in $\mathbb{Q}[\sqrt{k}]$. If k is not squarefree but $k = l q^2$ where l is squarefree, then the solutions are the numbers

$$\left(-\frac{a_1}{2 a_2}, \pm \frac{q}{2 a_2} \right)$$

in $\mathbb{Q}[\sqrt{l}]$. Finally if k is an exact square, $k = q^2$, then the solutions are

$$\left(\frac{-a_1 \pm q}{2 a_2}, 0 \right)$$

in an arbitrary quadratic field, corresponding to the ordinary rational numbers $(-a_1 \pm q)/2 a_2$. $\qquad\square$

8.6.2   Quadratic Integers

We have already noted that the Rational Numbers are the algebraic numbers of degree 1, solutions of linear algebraic equations, i.e. of the form

$$a_0 + a_1 x = 0,$$

where a_0 and a_1 are integers. Within the Rationals, the Integers are solutions of equations of the form

$$a_0 + x = 0,$$

in which the coefficient of x is $a_1 = 1$. More generally, we can define an **algebraic integer** of degree n to be a number x satisfying an algebraic equation of degree n in which the coefficient of the highest power of x is 1, i.e.

$$a_0 + a_1 x + \cdots + a_{n-1} x^{n-1} + x^n = 0,$$

with $a_0, a_1, \ldots, a_{n-1} \in \mathbb{Z}$. Such an equation is described as **monic**. In particular, a **quadratic integer** satisfies a monic quadratic equation, i.e. of the form

$$a_0 + a_1 x + x^2 = 0. \tag{8.22}$$

Just as we have found that the solutions of all quadratic algebraic equations are just the numbers in the union of all quadratic extensions of $\mathbb{Q}$, we may expect (by analogy with what happens with linear algebraic equations) that the solutions of all *monic* quadratic algebraic equations should be the numbers in the union of the quadratic extensions of $\mathbb{Z}$. But this is not the case. On the one hand, it is easy to show that a theorem analogous to part (a) of Theorem 8.6.2 does apply to quadratic extensions of $\mathbb{Z}$:

Exercise 8.6.3. *Show that any number (c, d) in any quadratic ring $\mathbb{Z}[\sqrt{k}]$ (i.e. with $c, d \in \mathbb{Z}$) satisfies a quadratic equation of the form (8.22).*

But consider the converse, analogous to part (b) of Theorem 8.6.2, that equations of form (8.22) have solutions in $\mathbb{Z}[\sqrt{k}]$ for some $k \in \mathbb{Z}$: we now show that this is false. Setting $a_2 = 1$ in the solution of the general quadratic equation, we have

$$x = -\frac{a_1}{2} \pm \frac{\sqrt{a_1^2 - 4a_0}}{2}. \tag{8.23}$$

There are two cases to consider.
(i) If a_1 is even, $a_1/2$ is an integer and $a_1^2 - 4a_0 \equiv 0 \,(\mathrm{mod}\ 4)$ so we can write $a_1^2 - 4a_0 = 4k$ for some $k \in \mathbb{Z}$ and hence the solutions (8.23) become $x = -a_1/2 \pm \sqrt{k}$. If k is squarefree, we can identify these solutions as the numbers $(-a_1/2, \pm 1)$ in the quadratic ring $\mathbb{Z}[\sqrt{k}]$; if $k = lq^2$ where l is squarefree, the solutions are the numbers $(-a_1/2, \pm q)$ in $\mathbb{Z}[\sqrt{l}]$; and if $k = q^2$ for some $q \in \mathbb{Z}$, the solutions are $(-a_1/2 \pm q, 0)$ in an arbitrary quadratic ring.
(ii) If a_1 is odd, let $h = a_1^2 - 4a_0$, so $h \equiv 1 \,(\mathrm{mod}\ 4)$ and the solutions of (8.22) are

$$x = (-a_1 \pm \sqrt{h})/2.$$

We shall now show that these solutions can always be written

$$x = c + d \left(\frac{-1 + \sqrt{k}}{2} \right) \text{xiii}$$

---

xiii Why $-1 + \sqrt{k}$ rather than $1 + \sqrt{k}$ in the numerator? It is a matter of convention, and the theory would carry through with some changes to formulae if we had chosen $1 + \sqrt{k}$.

where c, d, k are integers and $k \equiv 1 \pmod 4$. If h is squarefree, this is achieved by letting $k = h$, $d = \pm 1$ and $c = (-a_1 \pm 1)/2$ (so both solutions for c are integers since a_1 is odd). If $h = q^2 l$ where l is squarefree, then since $h \equiv 1 \pmod 4$, q must be odd and $l \equiv 1 \pmod 4$; so let $k = l$ and then $d = \pm q$ and $c = (-a_1 \pm q)/2$, which again are integers. Finally if $h = q^2$, where q is odd since h is odd, we can let k be an arbitrary integer congruent to $1 \pmod 4$, with $d = 0$ and $c = (-a_1 \pm q)/2$.

Let us now take stock of what we have found. Algebraic numbers of degree 2 were found to fit into quadratic extensions of $\mathbb{Q}$, for which the arithmetic operations were defined and their properties derived in Section 8.1. But for algebraic *integers* of degree 2, some fit into quadratic extensions of $\mathbb{Z}$, i.e. they behave like $c + d\sqrt{k}$; but others appear to behave like $c + d(-1 + \sqrt{k})/2$, with $k \equiv 1 \pmod 4$. So to accommodate all quadratic integers, we need to define new number systems $\mathbb{Z}[\omega_k]$, where

$$\omega_k := \frac{-1 + \sqrt{k}}{2},$$

in which numbers are ordered pairs $(c, d) \in \mathbb{Z} \times \mathbb{Z}$ and their arithmetic is defined as "behaving like $c + d\omega_k$". Just as in $\mathbb{Z}[\sqrt{k}]$, where "behaving like $c + d\sqrt{k}$" means that the symbol $\sqrt{k}$ satisfies the equation $(\sqrt{k})^2 = k$, we now need to find an equation satisfied by ω_k to define what "behaving like $c + d\omega_k$" means. Squaring our definition of ω_k, we find

$$\omega_k^2 = -\omega_k + \frac{k-1}{4} \tag{8.24}$$

in which $(k-1)/4$ is an integer because $k \equiv 1 \pmod 4$. So supposing that the usual rules of arithmetic apply to ω_k as well as to integers a, b, c, d, we have

$$(a + b\omega_k) + (c + d\omega_k) = (a + c) + (b + d)\omega_k$$

and

$$
\begin{aligned}
(a + b\omega_k) \times (c + d\omega_k) &= ac + (ad + bc)\omega_k + bd\omega_k^2 \\
&= ac + bd\frac{k-1}{4} + (ad + bc - bd)\omega_k,
\end{aligned}
$$

where we have used equation (8.24) to obtain the final result. Thus the formal definitions of addition and multiplication operations in $\mathbb{Z}[\omega_k]$ are:

Definition 8.6.4.

$$(a, b) \oplus (c, d) := (a + c, b + d).$$

$$(a, b) \otimes (c, d) := \left(ac + bd\frac{k-1}{4}, ad + bc - bd\right).$$

As with quadratic extensions, there are isomorphisms between the set

$$\mathbb{Z}[\omega_k]_0 := \{(a, b) \in \mathbb{Z}[\omega_k] : b = 0\}$$

and $\mathbb{Z}$ under their respectively defined addition and multiplication operations, so we use the usual symbols $+$ and $\times$ for arithmetic operations in $\mathbb{Z}[\omega_k]$ from now on. The usual verifications that Definitions 8.6.4 satisfy all the requirements of a commutative ring may be done, so we can say that a new class of quadratic rings has been defined. To summarise: for each squarefree $k \in \mathbb{Z}$ there is a quadratic ring $\mathbb{Z}[\sqrt{k}]$ which is simply a quadratic extension of $\mathbb{Z}$, but when $k \equiv 1 \pmod 4$ there is also a quadratic ring $\mathbb{Z}[\omega_k]$ as defined above; and all algebraic integers of degree 2 are members of a quadratic ring of one of these types. In the case where $k \equiv 1 \,(\mathrm{mod}\ 4)$ the ring $\mathbb{Z}[\sqrt{k}]$ is actually a subring of $\mathbb{Z}[\omega_k]$: to see this, observe that

$$a + b\sqrt{k} = (a+b) + (2b)\frac{-1+\sqrt{k}}{2} \tag{8.25}$$

so that a number $(a,b) \in \mathbb{Z}[\sqrt{k}]$ behaves exactly like the number $(a+b, 2b) \in \mathbb{Z}[\omega_k]$. This does not invalidate $\mathbb{Z}[\sqrt{k}]$ as a self-contained number system when $k \equiv 1 \,(\mathrm{mod}\ 4)$; but when considering the quadratic integers within the quadratic field $\mathbb{Q}[\sqrt{k}]$, or in the broader context of algebraic integers, we need the ring $\mathbb{Z}[\omega_k]$. The complete set of algebraic integers of degree 2 is the union of the quadratic rings $\mathbb{Z}[\omega_k]$ for $k \equiv 1 \,(\mathrm{mod}\ 4)$ and the rings $\mathbb{Z}[\sqrt{k}]$ for $k \equiv 2$ or $3 \,(\mathrm{mod}\ 4)$, for all squarefree $k \in \mathbb{Z}$.

Exercise 8.6.5. *Show that adding and multiplying numbers $(a+b, 2b)$ and $(c+d, 2d)$ in $\mathbb{Z}[\omega_k]$ according to Definition 8.6.4 yields results corresponding to the addition and multiplication of (a,b) and (c,d) in $\mathbb{Z}[\sqrt{k}]$ according to Definitions 8.1.1 and 8.1.2. This confirms the isomorphisms between $\mathbb{Z}[\sqrt{k}]$ and a subring of $\mathbb{Z}[\omega_k]$.*

We require a little further theory of our new breed of quadratic rings: how to define conjugates and norms in $\mathbb{Z}[\omega_k]$, with similar properties to those in $\mathbb{Z}[\sqrt{k}]$. Now that we understand that numbers in quadratic rings and fields are solutions of algebraic equations of degree 2, we can define conjugates in a more general way than in Definition 8.1.9:

Definition 8.6.6. Algebraic numbers are **conjugates** of each other if they satisfy the same algebraic equation.

For the monic quadratic equation, the two solutions in (8.23) differ by the plus or minus sign in front of the square root; so in $\mathbb{Z}[\sqrt{k}]$, where (a,b) behaves like $a + b\sqrt{k}$, the conjugate $\overline{(a,b)}$ must behave like $a - b\sqrt{k}$, which justifies our earlier definition of $\overline{(a,b)}$ as $(a,-b)$. But in $\mathbb{Z}[\omega_k]$ where (a,b) behaves like $a + b(-1+\sqrt{k})/2$, the conjugate must behave like $a + b(-1-\sqrt{k})/2$. Since

$$a + b\frac{-1-\sqrt{k}}{2} = (a-b) - b\frac{-1+\sqrt{k}}{2} = (a-b) - b\omega_k,$$

we have:

Definition 8.6.7. In $\mathbb{Z}[\omega_k]$ the conjugate is defined by

$$\overline{(a,b)} := (a-b, -b).$$

It is then a matter of routine calculation to establish that this definition yields the expected properties of a conjugate, using Definitions 8.6.4 of the arithmetic operations:

Exercise 8.6.8. *(a) Verify that the conjugate in Definition 8.6.7 satisfies the automorphism properties in Theorem 8.1.10. [Hint: according to Definition 8.6.7, the left-hand sides of the equations in Theorem 8.1.10 are the sum and product of $(a-b,-b)$ and $(c-d,-d)$; on the right-hand sides you need to evaluate a sum and product and then take the conjugates according to Definition 8.6.7.]*
(b) Show that

$$(a,b) \times \overline{(a,b)} = \left(a^2 - ab + b^2\frac{1-k}{4}, 0\right)$$

in $\mathbb{Z}[\omega_k]$.

The result in part (b) of the exercise gives us our definition of the norm:

Definition 8.6.9. In $\mathbb{Z}[\omega_k]$ the norm is defined as

$$\mathcal{N}(a, b) = a^2 - ab + b^2 \frac{1 - k}{4}.$$

The important property that norms preserve multiplication follows from the multiplication automorphism for conjugates: see Exercise 8.1.17. [You are advised not to attempt a "brute force" proof of this property in $\mathbb{Z}[\omega_k]$, unless you really enjoy messy algebra.]

We saw in $\mathbb{Z}[\sqrt{k}]$ that norms are the key to finding units and primes. The same applies in $\mathbb{Z}[\omega_k]$; in particular, because norms still preserve multiplication, units have norms equal to ± 1, and a norm that is prime in $\mathbb{Z}$ implies a number that is prime in $\mathbb{Z}[\omega_k]$. So the calculations are similar to those in $\mathbb{Z}[\sqrt{k}]$, but more messy, and we shall not go into any detail; but it is worth pointing out a few interesting features of certain rings of type $\mathbb{Z}[\omega_k]$.

A particularly important quadratic ring is with $k = -3$, so that $(1 - k)/4 = 1$ and

$$\mathcal{N}(a, b) = a^2 - ab + b^2. \tag{8.26}$$

The ring $\mathbb{Z}[\omega_{-3}]$ is called the **Eisenstein integers**, and like the Gaussian integers it is unique in its number of units. Formula (8.26) shows that $\mathcal{N}(a, b) = 1$ if $a = 0$ and $b = \pm 1$, or if $b = 0$ and $a = \pm 1$, as in $\mathbb{Z}[\sqrt{-1}]$; but unlike the Gaussian integers, (8.26) also gives $\mathcal{N}(a, b) = 1$ with $a = b = 1$ and with $a = b = -1$. Are there any more units? Since $a^2 - ab + b^2 = (a - b)^2 + ab$, we would require either $a - b = 0$ and $ab = \pm 1$, which can only be achieved with the last two cases mentioned, or $a - b = \pm 1$ and $ab = 0$ which only applies to the four cases that are similar to those in $\mathbb{Z}[\sqrt{-1}]$. So there are six units in $\mathbb{Z}[\omega_{-3}]$, namely $(1, 0)$, $(-1, 0)$, $(0, 1)$, $(0, -1)$, $(1, 1)$ and $(-1, -1)$.

With $k = -3$, the quadratic equation (8.24) satisfied by ω_k is

$$\omega_{-3}^2 + \omega_{-3} + 1 = 0.$$

Multiplying this equation by ω_{-3} to obtain

$$\omega_{-3}^3 + \omega_{-3}^2 + \omega_{-3} = 0,$$

and then subtracting the first of these equations from the second, we find

$$\omega_{-3}^3 = 1.$$

So it can be said that the Eisenstein integers are a **cubic extension** of $\mathbb{Z}$. Related to this are some important observations:

Exercise 8.6.10. *Noting that in the Eisenstein integers, the definition 8.6.4 of multiplication becomes*

$$(a, b) \otimes (c, d) := (ac - bd, ad + bc - bd),$$

and that positive integer powers are defined in terms of multiplication in the usual way, verify the following facts about the units in $\mathbb{Z}[\omega_{-3}]$*:*
(a) (i) $\overline{(0, 1)} = (-1, -1)$*;*
    (ii) $(0, 1)^3 = (1, 0)$ *and also* $(-1, -1)^3 = (1, 0)$*;*
    (iii) The six units are the first six powers of $(1, 1)$*, i.e.* $(1, 1)^j$ *for* $j = 1, 2, 3, 4, 5, 6$*.*
(b) Also write each unit (u, v) *in the form* $u + v\frac{-1 + \sqrt{-3}}{2}$*, and verify (iii) above by using standard rules of arithmetic, with* $\sqrt{-3}$ *defined as an object whose square is* -3*.*

Like the quadratic rings $\mathbb{Z}[\sqrt{k}]$ with $k < -1$, the rings $\mathbb{Z}[\omega_k]$ with $k \equiv 1 \pmod 4$ and $k < -3$ only have two units, $(1, 0)$ and $(-1, 0)$.

Exercise 8.6.11. *Show that if $k \equiv 1 \pmod 4$ and $k < -3$, the norm as given in Definition 8.6.9 can only equal ± 1 if $b = 0$ and $a = \pm 1$.*

For positive $k \equiv 1 \pmod 4$, the rings $\mathbb{Z}[\omega_k]$ have groups of units with similar structure to those in real quadratic rings $\mathbb{Z}[\sqrt{k}]$: there are the trivial units $(1, 0)$ and $(-1, 0)$, together with a fundamental unit such that all its integer powers and their additive inverses are also units. The group of units in $\mathbb{Z}[\omega_5]$ is of particular interest because of its connection to the famous Fibonacci sequence: see Investigation 7 at the end of this chapter. From the correspondence in equation (8.25), if (u, v) is a unit in $\mathbb{Z}[\sqrt{k}]$ then $(u + v, 2v)$ will be a unit in $\mathbb{Z}[\omega_k]$ with the same k; but $(u + v, 2v)$ may not be the *fundamental* unit in $\mathbb{Z}[\omega_k]$ when (u, v) is fundamental in its subring $\mathbb{Z}[\sqrt{k}]$.

Exercise 8.6.12. *(a) Find the fundamental units in $\mathbb{Z}[\sqrt{17}]$ and $\mathbb{Z}[\omega_{17}]$ and show that they do correspond according to equation (8.25).*

(b) Find the fundamental units in $\mathbb{Z}[\sqrt{13}]$ and $\mathbb{Z}[\omega_{13}]$. Which power of the fundamental unit in $\mathbb{Z}[\omega_{13}]$ corresponds to the fundamental unit in $\mathbb{Z}[\sqrt{13}]$?

8.7   Norm-Euclidean Rings and Unique Prime Factorisation

The first major theorem in Chapter 5 was the Division Theorem, and everything else in that chapter depended on that theorem. In contrast, we have discussed primes and factorisations in quadratic rings without mentioning a division theorem; indeed, it is not clear how such a theorem could be formulated in a quadratic ring, since its original statement as Theorem 5.1.5 involves dividing by a *positive* integer and placing constraints involving *order relations* on the remainder, whereas order and positivity are not defined in quadratic rings. Nevertheless, we shall now formulate a division property for quadratic rings using ordering of norms rather than of the integers themselves, and we shall see that as in $\mathbb{Z}$, unique prime factorisation follows from this division property.

Definition 8.7.1. A quadratic ring $\mathbb{Z}[\omega]$[xiv] is **norm-Euclidean** if, given (a, b) and $(c, d) \neq (0, 0)$ in $\mathbb{Z}[\omega]$, there exists a quotient $(m, n) \in \mathbb{Z}[\omega]$ and a remainder $(r, s) \in \mathbb{Z}[\omega]$ such that

$$(a, b) = (c, d) \times (m, n) + (r, s) \quad \text{with} \quad |\mathcal{N}(r, s)| < |\mathcal{N}(c, d)|.\text{[xv]} \qquad (8.27)$$

In this definition we can divide by any non-zero quadratic integer (rather than a positive integer), and we require the *modulus* of the norm of the remainder to be smaller than that of the number we are dividing by, since norms can be negative as well as positive. Observe that we have not stated a theorem here: the norm-Euclidean property is something that a quadratic ring may or may not possess. Calculations to determine whether a particular quadratic ring is norm-Euclidean are fairly simple in some cases but, in the absence of the well-ordering property that was used in proving the Division Theorem for ordinary integers, we need to appeal to the concept of *exact division*; this had not been defined when we proved the Division Theorem for $\mathbb{Z}$, but we can use it now because the quadratic integers

---

[xiv]Here we use $\mathbb{Z}[\omega]$ to denote a general quadratic ring, i.e. of either the form $\mathbb{Z}[\sqrt{k}]$ or $\mathbb{Z}[\omega_k]$.

[xv]The designation "norm-Euclidean" derives from the use of the norm in this condition and from a simple algorithm using the division property which Euclid devised to determine the greatest common divisor of two integers; although we have not used this algorithm in the development of theory in this book, you can find it in any textbook on Number Theory.

in $\mathbb{Z}[\sqrt{k}]$ or $\mathbb{Z}[\omega_k]$ exist within the field of quadratic rationals $\mathbb{Q}[\sqrt{k}]$. We start by writing the requirements of (8.27) as

$$|\mathcal{N}(r,s)| \div |\mathcal{N}(c,d)| < 1 \quad \text{with} \quad (r,s) \div (c,d) = (a,b) \div (c,d) - (m,n). \qquad (8.28)$$

Now, since norms preserve multiplication and exact division, we can write

$$|\mathcal{N}(r,s)| \div |\mathcal{N}(c,d)| = |\mathcal{N}\left((r,s) \div (c,d)\right)|; \qquad (8.29)$$

but we must be careful here. On the left of (8.29) we take norms as defined in the ring $\mathbb{Z}[\sqrt{k}]$ or $\mathbb{Z}[\omega_k]$, and divide to obtain an ordinary rational number; whereas on the right of (8.29) we do an exact division of quadratic integers to yield a result which will in general only exist in the field $\mathbb{Q}[\sqrt{k}]$; but the norm that we then take must be the same norm as on the left, i.e. the norm as defined in the *ring*. To emphasise this, we shall temporarily use the notation $\mathcal{N}_{\mathbb{Z}}$ for this norm, which differs from the norm in $\mathbb{Q}[\sqrt{k}]$ in the case of rings $\mathbb{Z}[\omega_k]$, although it has the same form in $\mathbb{Z}[\sqrt{k}]$. In any case, let us define

$$(e,f) := (a,b) \div (c,d),$$

and substitute (8.29) into (8.28), which then becomes

$$|\mathcal{N}_{\mathbb{Z}}((e,f) - (m,n))| < 1.$$

For the ring to be norm-Euclidean, we must be able to satisfy this for *every* choice of (a,b) and non-zero (c,d) in the ring. So for all rational e,f we must be able to find integers m,n such that

$$|\mathcal{N}_{\mathbb{Z}}(e-m, f-n)| < 1. \qquad (8.30)$$

Thus, for any ring of type $\mathbb{Z}[\sqrt{k}]$, for which the norm is defined as

$$\mathcal{N}_{\mathbb{Z}}(a,b) = a^2 - kb^2,$$

the requirement (8.30) is

$$|(e-m)^2 - k(f-n)^2| < 1. \qquad (8.31)$$

For rings of type $\mathbb{Z}[\omega_k]$, the norm is given in Definition 8.6.9 but can more usefully be written as

$$\mathcal{N}_{\mathbb{Z}}(a,b) = \left(a - \frac{b}{2}\right)^2 - k\left(\frac{b}{2}\right)^2,$$

so the requirement (8.30) is

$$\left|\left((e-m) - \frac{f-n}{2}\right)^2 - k\left(\frac{f-n}{2}\right)^2\right| < 1. \qquad (8.32)$$

Now, for any $e,f \in \mathbb{Q}$ there exist $m,n \in \mathbb{Z}$ such that

$$-\frac{1}{2} \leq e - m \leq \frac{1}{2} \quad \text{and} \quad -\frac{1}{2} \leq f - n \leq \frac{1}{2}. \qquad (8.33)$$

Thus we can always find m,n such that the following bounds (required in consideration of the conditions (8.31) and (8.32)) are satisfied:

$$0 \leq (e-m)^2 \leq \frac{1}{4}; \quad 0 \leq (f-n)^2 \leq \frac{1}{4};$$

$$0 \leq \left((e-m) - \frac{f-n}{2}\right)^2 \leq \frac{1}{4}; \quad 0 \leq \left(\frac{f-n}{2}\right)^2 \leq \frac{1}{16}. \qquad (8.34)$$

Exercise 8.7.2. *Using the bounds (8.34) in the condition (8.31), show that the rings $\mathbb{Z}[\sqrt{-1}]$ (the Gaussian integers) and $\mathbb{Z}[\sqrt{-2}]$ are norm-Euclidean.*

Using the bounds (8.34) in the condition (8.32) show that the rings $\mathbb{Z}[\omega_{-3}]$ (the Eisenstein integers), $\mathbb{Z}[\omega_{-7}]$, and $\mathbb{Z}[\omega_{-11}]$ are norm-Euclidean.

For negative k, the conditions (8.31) and (8.32) involve the sum of two non-negative terms: the first term in each condition can be as big as $1/4$, so if the second term can be greater or equal to $3/4$, then the condition can be violated. Hence we can conclude that the *only* norm-Euclidean imaginary rings are the five which were found in Exercise 8.7.2. In particular, whereas the ring $\mathbb{Z}[\omega_{-3}]$ which contains all the quadratic integers within the field $\mathbb{Q}[\sqrt{-3}]$ *is* norm-Euclidean, its subring $\mathbb{Z}[\sqrt{-3}]$ is not.

The situation with real quadratic rings is more complicated, although there are still a few cases for which it is easy to show that such a ring is norm-Euclidean.

Example 8.7.3. Show that (a) $\mathbb{Z}[\sqrt{2}]$, and (b) $\mathbb{Z}[\omega_5]$ are norm-Euclidean.

Solution. (a) We need to satisfy (8.31) with $k = 2$. The bounds in (8.34) imply that

$$(e - m)^2 - 2(f - n)^2 \geq 0 - 2 \times \frac{1}{4} = -\frac{1}{2}$$

and that

$$(e - m)^2 - 2(f - n)^2 \leq \frac{1}{4} - 2 \times 0 = \frac{1}{4}.$$

Hence

$$|(e - m)^2 - 2(f - n)^2| < 1$$

as required.

(b) We need to satisfy (8.32) with $k = 5$. The bounds in (8.34) imply that

$$\left((e - m) - \frac{f - n}{2}\right)^2 - 5\left(\frac{f - n}{2}\right)^2 \geq 0 - 5 \times \frac{1}{16} = -\frac{5}{16}$$

and that

$$\left((e - m) - \frac{f - n}{2}\right)^2 - 5\left(\frac{f - n}{2}\right)^2 \leq \frac{1}{4} - 5 \times 0 = \frac{1}{4}.$$

Hence

$$\left|\left((e - m) - \frac{f - n}{2}\right)^2 - 5\left(\frac{f - n}{2}\right)^2\right| < 1$$

as required. $\square$

Exercise 8.7.4. *Show by similar calculations that the real quadratic rings $\mathbb{Z}[\sqrt{3}]$, and $\mathbb{Z}[\omega_{13}]$ are norm-Euclidean.*

The difficulty with real quadratic rings is that, because of the subtraction in (8.31) and (8.32) when k is positive, we may be able to satisfy these conditions without the individual terms being within the bounds (8.34). So more of these rings may be norm-Euclidean than those found in the above Example and Exercise. It is an intricate problem in Number Theory to determine the full list of real quadratic rings in which it is always possible to find a quotient and remainder satisfying (8.27). The issue was finally settled in 1952 by Eric Barnes and Peter Swinnerton-Dyer, who concluded that $\mathbb{Z}[\sqrt{k}]$ with $k = 2, 3, 6, 7, 11, 19$, and $\mathbb{Z}[\omega_k]$ with $k = 5, 13, 17, 21, 29, 33, 37, 41, 57, 73$ are all the norm-Euclidean real quadratic rings.

We now start to consider the implications of the norm-Euclidean property in quadratic rings. In the ordinary integers we had Theorem 5.6.1, which was framed in terms of modular arithmetic but can be rephrased as: "Given non-zero integers r and d, there exist integers m, n such that $mr + nd = 1$ if and only if r and d are coprime". We want a similar theorem to apply to norm-Euclidean quadratic rings; so we want to prove:

Theorem 8.7.5. *Given non-zero quadratic integers (a, b) and (c, d) in a norm-Euclidean quadratic ring, there exist quadratic integers (m, n) and (p, q) in the ring such that $(a, b) \times (m, n) + (c, d) \times (p, q)$ is equal to a unit if and only if no non-unit quadratic integer is a divisor of both (a, b) and (c, d).*

We have needed to be rather careful in formulating this theorem: the meaning of "coprime" is not clear when there is the possibility of non-unique prime factorisation, hence our use of the phrase, "no non-unit quadratic integer is a divisor of both (a, b) and (c, d)". Whereas with ordinary integers we had already proved uniqueness of prime factorisation before Theorem 5.6.1, we now need a different approach. To make the notation less cumbersome (and to emphasise the generality of the method) we shall use Greek letters rather than ordered pairs for general quadratic integers: α for (a, b), etc. So we want to show that there exist μ, ν such that $\alpha\mu + \beta\nu$ is equal to a unit if and only if the given quadratic integers α and β have no common non-unit divisor. We use the symbol $i_\times$ for $(1, 0)$, the multiplicative identity element in a quadratic ring (to distinguish it from the ordinary integer 1).

Proof of Theorem 8.7.5. Denote our norm-Euclidean ring as $\mathbb{Z}[\omega]$. The quadratic integers α, β are fixed, so the set of all possible values of $\alpha\mu + \beta\nu$ in $\mathbb{Z}[\omega]$ is

$$T := \{\alpha\mu + \beta\nu \in \mathbb{Z}[\omega] : \mu, \nu \in \mathbb{Z}[\omega]\}.$$

To say that "$\alpha\mu + \beta\nu$ is equal to a unit" means that $|\mathcal{N}(\alpha\mu + \beta\nu)| = 1$. Now, the norm of a quadratic integer is an ordinary integer, and its modulus is either zero or a positive integer. Well-ordering applies to the latter: among all values of μ and ν that yield non-zero values of $\alpha\mu + \beta\nu$, there exist μ_0 and ν_0 such that $|\mathcal{N}(\alpha\mu_0 + \beta\nu_0)|$ takes its least non-zero value. [We have used the fact that the norm of a number can only be zero if the number is zero.] Let us denote

$$\delta := \alpha\mu_0 + \beta\nu_0.$$

Now divide α by δ, to obtain a quotient γ and remainder ρ, which is

$$
\begin{aligned}
\rho &= \alpha - \gamma\delta \\
&= \alpha - \gamma(\alpha\mu_0 + \beta\nu_0) \\
&= \alpha(i_\times - \gamma\mu_0) + \beta(-\gamma\nu_0),
\end{aligned}
$$

so ρ is a member of the set T. But since our ring is norm-Euclidean, $|\mathcal{N}(\rho)|$ is less than $|\mathcal{N}(\delta)|$, which (by definition of δ) is the least non-zero value of a norm for any member of T. So $|\mathcal{N}(\rho)| = 0$ and hence the remainder ρ must be zero, which means that δ is a divisor of α.

Similarly, we can show that δ is a divisor of β, simply by swapping the roles of α and β, μ and ν, in the above argument. Given the condition that α and β have no common non-unit divisor, δ must be a unit. Looking at the definition of δ, we see that μ_0 and ν_0 are the desired values of μ and ν that make $\alpha\mu + \beta\nu$ equal to a unit.

The converse is much easier to prove. Suppose that η is a divisor of both α and β, so there are numbers σ and τ such that $\alpha = \sigma\eta$ and $\beta = \tau\eta$. Then

$$\alpha\mu_0 + \beta\nu_0 = (\sigma\mu_0 + \tau\nu_0)\eta.$$

If $\alpha\mu_0 + \beta\nu_0$ is a unit, this means that η is a divisor of a unit, which can only be true if η itself is a unit. So the only divisors that α and β have in common are units. □

Corollary 8.7.6. *Under the conditions of Theorem 8.7.5, there exist quadratic integers (m^*, n^*) and (p^*, q^*) in the ring such that $(a, b) \times (m^*, n^*) + (c, d) \times (p^*, q^*) = (1, 0)$.*

Proof. Having found μ, ν such that $\alpha\mu + \beta\nu = \upsilon$ where υ is a unit, we can multiply through by υ^*, the multiplicative inverse of υ, to get $\alpha\mu^* + \beta\nu^* = i_\times$, where $\mu^* = \mu\upsilon^*$ and $\nu^* = \nu\upsilon^*$. The numbers μ^*, ν^* are the quadratic integers (m^*, n^*) and (p^*, q^*) in the original ordered-pair notation of the corollary. □

Our next step is to prove a generalisation of Theorem 5.3.10 to quadratic rings:

Theorem 8.7.7. *In a norm-Euclidean ring, let ϕ be a number whose only divisors are its associates and the units of the ring. Then if ϕ is a divisor of the product $\alpha\beta$, it must be a divisor of either α or β (or both).*[xvi]

Proof. We need to show that if the conditions of the theorem are satisfied and $\phi \nmid \alpha$, then $\phi | \beta$, where we are using the notation for divisors from Chapter 5.

Consider first a number γ that is a divisor of both ϕ and α. As a divisor of ϕ, γ can only be a unit or an associate of ϕ; if the latter, we have $\phi | \gamma$, and since $\gamma | \alpha$, we would have $\phi | \alpha$. So if $\phi \nmid \alpha$, then γ can only be a unit: we have satisfied the condition of Theorem 8.7.5, that the only common divisors of ϕ and α are units in our norm-Euclidean ring.

Corollary 8.7.6 then tells us that there exist μ^*, ν^* in the ring such that $\alpha\mu^* + \phi\nu^* = i_\times$. Multiply through by β:

$$\beta\alpha\mu^* + \beta\phi\nu^* = \beta.$$

Now if $\phi | (\alpha\beta)$ (a condition of the theorem), then ϕ is a divisor of both terms on the left-hand side, and hence of β on the right, as required. □

Corollary 8.7.8. *In a norm-Euclidean ring, let ϕ be a number whose only divisors are its associates and the units of the ring. Then if ϕ is a divisor of a product of n numbers, $\alpha_1\alpha_2 \ldots \alpha_n$, it must be a divisor of at least one of the α_i $(i = 1, 2, \ldots, n)$.*

Exercise 8.7.9. *Prove Corollary 8.7.8 by induction on n. [Theorem 8.7.7 provides the anchor.]*

The relevance of this theorem and corollary to unique prime factorisation is clear: if a number γ has two different prime factorisations, $\gamma = \phi_1\phi_2$ and $\gamma = \psi_1\psi_2$, then ϕ_1 is a divisor of the product $\psi_1\psi_2$ without being a divisor of either ψ_1 or ψ_2. According to the theorem, this cannot happen in a norm-Euclidean ring. The contrapositive of this is that if a ring *is* norm-Euclidean, it must have unique prime factorisation; but a rigorous proof of this requires us to consider more general prime factorisations.

In the Natural Numbers, the proof of uniqueness of prime factorisation (Theorem 5.3.2) relied on the well-ordering of $\mathbb{N}$, and cannot simply be amended to work for quadratic rings by taking the modulus of norms (try it!). Nevertheless, well-ordering for the modulus of norms of non-zero quadratic integers will be useful in our proof of the theorem:

Theorem 8.7.10. *Prime factorisation is unique in norm-Euclidean rings, in the sense of Definition 5.3.22.*

[xvi]The property of having no divisors other than its associates and the units is what is generally referred to as being *prime* when discussing integers, but is called *irreducible* when considering more general rings; although we shall persist with the designation *prime*.

Proof. Suppose there do exist quadratic integers with more than one prime factorisation in some ring. Of all such numbers, let α be one with the least value of the modulus of its norm: so if β also has non-unique factorisation, then $|\mathcal{N}(\beta)| \geq |\mathcal{N}(\alpha)|$.[xvii] We can write two factorisations of α as

$$\alpha = \phi_1 \phi_2 \ldots \phi_m = \psi_1 \psi_2 \ldots \psi_n,$$

where the ϕ_i $(i = 1, 2, \ldots, m)$ and ψ_j $(j = 1, 2, \ldots, n)$ are prime.

Since the ring is norm-Euclidean, Corollary 8.7.8 implies that ϕ_1 is a divisor of one of the ψ_j $(j = 1, 2, \ldots, n)$; we can label this ψ_1. But the only divisors of ψ_1 are its associates and the units of the ring; and ϕ_1 is not a unit, so it must be an associate of ψ_1. Thus we can do exact division, $\psi_1 \div \phi_1 = \upsilon_1$, a unit, and

$$\alpha \div \phi_1 = \phi_2 \phi_3 \ldots \phi_m = \upsilon_1 \psi_2 \psi_2 \ldots \psi_n.$$

Now,

$$|\mathcal{N}(\alpha \div \phi_1)| = |\mathcal{N}(\alpha) \div \mathcal{N}(\phi_1)|$$
$$< |\mathcal{N}(\alpha)|$$

since ϕ_1 is not a unit. So we have found a number, $\alpha \div \phi_1$, which has two different prime factorisations and for which the modulus of the norm is less than that of α. This contradicts the definition of α; so no number with two distinct prime factorisations can exist in our norm-Euclidean ring. $\square$

The sequence of theorems in this section has shown that if a quadratic ring is norm-Euclidean, then it is a **Unique Factorisation Domain**, or UFD: prime factorisations in the ring are always unique. For example, the ring $\mathbb{Z}[\sqrt{-3}]$ was shown at the end of Section 8.5 to *not* be a UFD, and this has been explained by the finding following Exercise 8.7.2 that the ring is not norm-Euclidean. But $\mathbb{Z}[\sqrt{-3}]$ is a subring of $\mathbb{Z}[\omega_{-3}]$, and Exercise 8.7.2 also showed that the latter ring is norm-Euclidean, and so definitely is a UFD. Obviously the problem of non-uniqueness cannot be fixed in this way when it occurs in rings $\mathbb{Z}[\sqrt{k}]$ which are not subrings of some $\mathbb{Z}[\omega_k]$, such as some of the other examples seen in Section 8.5.

Although a norm-Euclidean ring must be a UFD, the converse is not true: a ring can be a UFD without being norm-Euclidean. Among imaginary quadratic rings, the norm-Euclidean ones are $\mathbb{Z}[\sqrt{-1}]$, $\mathbb{Z}[\sqrt{-2}]$, and $\mathbb{Z}[\omega_k]$ with $k = -3, -7, -11$; but the rings $\mathbb{Z}[\omega_k]$ with $k = -19, -43, -67, -163$ are also UFD's. It was shown by Kurt Heegner that no other imaginary rings are UFD's; the nine negative values of k for which $\mathbb{Z}[\sqrt{k}]$ or $\mathbb{Z}[\omega_k]$ is a UFD are often referred to as **Heegner numbers**, and there is a curious connection between these numbers and primes in $\mathbb{N}$. Recall from Section 5.3 that the polynomial $41 - n + n^2$ yields a prime for every $n \in \mathbb{N}$ such that $1 \leq n \leq 40$: the number 41 in the polynomial relates to the Heegner number of largest modulus by $-163 = 1 - 4 \times 41$, and a similar property applies more generally. If k is a Heegner number apart from $-1, -2, -3$, and a_0 is the integer satisfying $k = 1 - 4a_0$, then $a_0 - n + n^2$ is prime for each n such that $1 \leq n \leq a_0 - 1$.

Among real quadratic rings, we noted earlier that Barnes and Swinnerton-Dyer had determined a list of sixteen positive values of k for which $\mathbb{Z}[\sqrt{k}]$ or $\mathbb{Z}[\omega_k]$ is norm-Euclidean. A much longer list of UFD's with positive k has been obtained, but it is not definitive; in fact, it has not even been resolved whether there are finitely many UFD's among the real quadratic rings.

[xvii]We can allow two or more of these numbers to have the same norm (without being associates). This does not invalidate the proof.

Investigations

1. (a) Find a formula for the fundamental unit in $\mathbb{Z}[\sqrt{k}]$ in the cases where:

   (i) $k = n^2 + 1$ (ii) $k = n^2 - 1$ (iii) $k = n^2 + 2$ (iv) $k = n^2 - 2$

   for some $n \in \mathbb{N}$.

   [For the first two cases, the answer should be fairly obvious. For (iii) and (iv), try some example cases with various values of n, to look for the pattern.]

   (b) Let (a_1, b_1) be the fundamental unit in $\mathbb{Z}[\sqrt{k}]$ for some $k > 1$. Show that in $\mathbb{Z}[\sqrt{4k}]$ there is a unit (a, b) in which $b = a_1 b_1$, and find a formula for the value of a in this unit in terms of a_1 and b_1. [Obviously $4k$ is not squarefree; but calculations in $\mathbb{Z}[\sqrt{4k}]$ are still valid as long as k has no square root in $\mathbb{Z}$.]

2. Find primes in $\mathbb{Z}[\sqrt{-2}]$ as follows.

   (a) Show that $a^2 + 2b^2$ cannot be congruent to 5 or 7 (mod 8).

   (b) Show that $(0, 1)$ is prime in $\mathbb{Z}[\sqrt{-2}]$. Show that every number of form $(2c, d)$ (i.e. with the first member even) is divisible by $(0, 1)$ in $\mathbb{Z}[\sqrt{-2}]$. Thus $(0, 1)$ is the only prime in $\mathbb{Z}[\sqrt{-2}]$ with its first member even.

   (c) Find all primes $(a, b) \in \mathbb{Z}[\sqrt{-2}]$ with norms less than 120, and with $a \geq 0$. [Recall that to capture all prime classes in $\mathbb{Z}[\sqrt{k}]$ with $k \leq -2$, it is necessary to find primes (a, b) with $a \geq 0$ and $b \geq 0$, and then to include their conjugates $(a, -b)$.]

   (d) Find prime factorisations of $(7, 1)$, $(1, 7)$ and $(11, -10)$ in $\mathbb{Z}[\sqrt{-2}]$.

3. In this investigation we consider the determination of primes in $\mathbb{Z}[\sqrt{5}]$ and $\mathbb{Z}[\sqrt{3}]$, where the situation is not as simple as in $\mathbb{Z}[\sqrt{2}]$. Keep in mind that norms can be positive or negative in these real quadratic rings.

   (a) In $\mathbb{Z}[\sqrt{2}]$ and $\mathbb{Z}[\sqrt{-2}]$, modular arithmetic in $\mathbb{Z}_8$ has been useful to identify which primes $p \in \mathbb{N}$ allow solutions of $c^2 - d^2 k = \pm p$. We now consider $\mathbb{Z}[\sqrt{5}]$, where this is not the case.

      (i) Show that $c^2 - 5d^2$ cannot be congruent to 2 or 6 (mod 8). This is not very helpful, since 2 is the only prime which *is* congruent to 2 or 6 (mod 8). However, where $\mathbb{Z}_8$ is not useful, we often find that modular arithmetic in $\mathbb{Z}_k$ can be used to determine which numbers $\pm p$ can be norms of numbers in $\mathbb{Z}[\sqrt{k}]$:

      (ii) Show that $c^2 - 5d^2$ cannot be congruent to 2 or 3 (mod 5). Deduce that $c^2 - 5d^2$ cannot be congruent to -2 or -3 (mod 5).

      (iii) For each prime $p \in \mathbb{N}$ with $p < 50$, either state that $(p, 0)$ is prime in $\mathbb{Z}[\sqrt{5}]$, or else find $c, d \in \mathbb{Z}_+$ such that $(p, 0) = (c, d) \times (c, -d)$.

      (iv) Find all other fundamental primes with norm between -50 and 50 in $\mathbb{Z}[\sqrt{5}]$.

      (v) You should have found $(4, 1)$ in this list of fundamental primes. Writing $(a_0, b_0) = (4, 1)$, find all its associates (a_n, b_n) with $-3 \leq n \leq 3$. Which fundamental prime is an associate of $(4, -1)$?

   (b) (i) Show that $c^2 - 3d^2$ cannot be congruent to 2, 3 or 7 (mod 8), but *can* be congruent to -2, -3 or -7 (mod 8). Show also that $c^2 - 3d^2$ cannot be congruent to 2 (mod 3), but *can* be congruent to -2 (mod 3).

    (ii) Show that $(5,0)$ is prime in $\mathbb{Z}[\sqrt{3}]$, using both congruence (mod 8) and congruence (mod 3). Deduce that $(p,0)$ is prime in $\mathbb{Z}[\sqrt{3}]$ when p is a prime with $p \equiv 5 \,(\text{mod } 24)$.

    (iii) Find all other primes of form $(p,0)$ in $\mathbb{Z}[\sqrt{3}]$, where p is prime in $\mathbb{N}$.

    (iv) Find all other fundamental primes with norm between -50 and 50 in $\mathbb{Z}[\sqrt{3}]$.

    (v) Find prime factorisations of $(0,3)$, $(5,3)$, and $(-18,-11)$ in $\mathbb{Z}[\sqrt{3}]$. Express your results as a product of fundamental primes and, if necessary, a unit.

4. Investigate the prime factorisation of the number $(0,1)$ in $\mathbb{Z}[\sqrt{k}]$ for positive k, as follows.

  (a) Show that if k is prime in $\mathbb{Z}$, then $(0,1)$ is prime in $\mathbb{Z}[\sqrt{k}]$.

  (b) If k is composite in $\mathbb{Z}$, then $(0,1)$ may be composite in $\mathbb{Z}[\sqrt{k}]$ if there exist numbers in $\mathbb{Z}[\sqrt{k}]$ whose norms are the prime factors of k. [Why?]

    (i) Show that $(0,1)$ is prime in $\mathbb{Z}[\sqrt{10}]$, $\mathbb{Z}[\sqrt{15}]$, $\mathbb{Z}[\sqrt{35}]$ and $\mathbb{Z}[\sqrt{26}]$. [Hint: modular arithmetic in $\mathbb{Z}_5$ will help in the first three cases. What modulus d might be useful for a similar calculation for the last case?]

    (ii) Find prime factorisations of $(0,1)$ in $\mathbb{Z}[\sqrt{6}]$, $\mathbb{Z}[\sqrt{14}]$, $\mathbb{Z}[\sqrt{21}]$, $\mathbb{Z}[\sqrt{22}]$ and $\mathbb{Z}[\sqrt{55}]$.

    (iii) Describe the general procedure for determining prime factorisations of $(0,1)$ in $\mathbb{Z}[\sqrt{k}]$ when k is the product of just two prime factors (as in all the examples above). Can you extend this to cases where the prime factorisation of k in $\mathbb{Z}$ contains more than two prime factors, or powers of one or more primes?

5. Find examples of numbers which have two different prime factorisations in each of the quadratic rings $\mathbb{Z}[\sqrt{15}]$, $\mathbb{Z}[\sqrt{-6}]$ and $\mathbb{Z}[\sqrt{-7}]$. It is relatively easy to find numbers of the form $(x,0)$ with this property; try to also find examples (x,y) with $y \neq 0$ in each of these quadratic rings.

6. (a) Find all the associates of an Eisenstein integer (a,b). [Since there are six units in $\mathbb{Z}[\omega_{-3}]$, the associate class consists of six numbers; but you can simplify the calculation by noting that three of them are the additive inverses of the other three.] Hence show that every associate class in $\mathbb{Z}[\omega_{-3}]$ includes a unique member (a,b) with $a > 0$ and $0 \leq b < a$. This means that when searching for primes, we will only need to consider numbers with these restrictions.

  (b) The norm in $\mathbb{Z}[\omega_{-3}]$ is $\mathcal{N}(a,b) = a^2 - ab + b^2 = (a-b)^2 + ab$. Show that if $a > 0$ and $0 \leq b < a$, then $\mathcal{N}(a,b) > 0$. Also show that if $\mathcal{N}(a,b) \leq 50$ with $a > 0$ and $0 \leq b < a$, then $a \leq 8$.

  (c) Show that the norm in $\mathbb{Z}[\omega_{-3}]$ cannot be congruent to 2 (mod 3).

  (d) For each Eisenstein integer (a,b) with $a > 0$ and $0 \leq b < a$ and $\mathcal{N}(a,b) \leq 50$, determine whether it is prime. If not, find its prime factorisation.

7. (a) The Fibonacci sequence is defined by

$$F_1 = 1, \quad F_2 = 1, \quad F_{n+1} = F_{n-1} + F_n \quad \text{for } n > 2.$$

Write down the values of F_n for $n = 3,4,5,6,7$.

  (b) In $\mathbb{Z}[\omega_5]$ the multiplication formula in Definition 8.6.4 is

$$(a,b) \otimes (c,d) := (ac + bd, ad + bc - bd),$$

and Definition 8.6.9 for the norm becomes

$$\mathcal{N}(a, b) = a^2 - ab - b^2.$$

Verify that the fundamental unit in $\mathbb{Z}[\omega_5]$ is $(u_1, v_1) = (1, 1)$ and evaluate the units $(1, 1)^n$ for $n = 2, 3, 4, 5$. Note how your results relate to the members of the Fibonacci sequence.

(c) Prove that $(1, 1)^n = (F_{n+1}, F_n)$ in $\mathbb{Z}[\omega_5]$ for all $n \in \mathbb{N}$.

(d) Show that $(1, 1)^{-1} = (0, 1)$ in $\mathbb{Z}[\omega_5]$, and then prove that

$$(1, 1)^{-n} = (-1)^n \times (F_{n-1}, -F_n)$$

for all $n \in \mathbb{N} \backslash \{1\}$.

(e) Using the last two results, show that $(1, 1)^n - (0, -1)^n = (F_n, 2F_n)$ in $\mathbb{Z}[\omega_5]$.

(f) The arithmetic of numbers $(a, b) \in \mathbb{Z}[\omega_5]$ is defined to behave like real numbers $a + b\frac{-1+\sqrt{5}}{2}$. By writing down the real numbers corresponding to the numbers $(1, 1)$, $(0, -1)$ and $(F_n, 2F_n)$ in $\mathbb{Z}[\omega_5]$, derive the **Binet formula** for the Fibonacci numbers,

$$F_n = \frac{1}{\sqrt{5}} \left(\left(\frac{1 + \sqrt{5}}{2} \right)^n - \left(\frac{1 - \sqrt{5}}{2} \right)^n \right).$$

Chapter 9

Quadratic Extensions II: Complex Numbers, $\mathbb{C}$

9.1 Complex Numbers as a Quadratic Extension

According to our general prescription for quadratic extensions, we could make an extension $\mathbb{R}[\sqrt{k}]$ of the Real Numbers with any $k \in \mathbb{R}$ that does not have a square root in $\mathbb{R}$; which means any negative real k. But the following theorem shows that we only need to consider one such extension.

Theorem 9.1.1. *The quadratic extension $\mathbb{R}[\sqrt{k}]$ with any $k \in \mathbb{R}_-$ is isomorphic to $\mathbb{R}[\sqrt{-1}]$ under both addition and multiplication.*

Proof. Let $l = -k$, so $l > 0$ and $\sqrt{l} \in \mathbb{R}$. Since the function

$$\tilde{f} : \mathbb{R} \to \mathbb{R} \quad \text{defined by} \quad \tilde{f} : x \mapsto x\sqrt{l}$$

(with any $l \in \mathbb{R}_+$) is a bijection, we have a bijection

$$f : \mathbb{R}[\sqrt{-l}] \to \mathbb{R}[\sqrt{-1}] \quad \text{defined by} \quad f : (a, b) \mapsto \left(a, b\sqrt{l} \right).$$

The addition formula, Definition 8.1.1 gives

$$(a, b) \oplus (c, d) = (a + c, b + d) \quad \text{in } \mathbb{R}[\sqrt{-l}]$$

and

$$\left(a, b\sqrt{l} \right) \oplus \left(c, d\sqrt{l} \right) = \left(a + c, (b + d)\sqrt{l} \right) \quad \text{in } \mathbb{R}[\sqrt{-1}],$$

giving the isomorphism under addition with the bijection f.

Multiplication in $\mathbb{R}[\sqrt{-l}]$ is

$$(a, b) \otimes (c, d) := (ac - bdl, ad + bc),$$

while in $\mathbb{R}[\sqrt{-1}]$ we have

$$\left(a, b\sqrt{l} \right) \otimes \left(c, d\sqrt{l} \right) = \left(ac - bdl, (ad + bc)\sqrt{l} \right),$$

showing the isomorphism under multiplication with the bijection f. $\qquad \square$

The isomorphisms mean that any theorems or calculations in $\mathbb{R}[\sqrt{-l}]$ would be equivalent to those in $\mathbb{R}[\sqrt{-1}]$. Thus we only need to consider $\mathbb{R}[\sqrt{-1}]$; and this number system is of such importance that it has its own symbol and name: $\mathbb{C}$, the Complex Numbers. These numbers are "complex" not in the sense of being complicated, but because they consist of more than one part (so all quadratic extensions could be considered "complex" in this sense). The symbol i is commonly used for $\sqrt{-1}$, so a complex number formally written as an ordered pair (x, y) can be informally written $x + yi$, because its arithmetic is defined to behave like $x + yi$ where $i^2 = -1$. As when dealing with quadratic extensions of $\mathbb{Z}$ and $\mathbb{Q}$, we prefer the ordered-pair form, but where this becomes cumbersome we shall use the more common "$x + yi$" notation. There is also a convention of using a single letter to represent a complex number, in particular z for (x, y), and w for (u, v),[i] and there is some associated terminology and notation:

Definition 9.1.2. If the complex number (x, y) is represented as z, then x is the **real part** of z and y is the **imaginary part** of z. We write $x = \text{Re}[z]$ and $y = \text{Im}[z]$.

The use of terms such as "real" and "imaginary", and the letter i standing for "imaginary", should not be taken to mean that Complex Numbers have any less reality than so-called Real Numbers; and note that the imaginary part of z is the *real* number y.

The basic arithmetic of Complex Numbers is that of general quadratic extensions, covered in Section 8.1. In the remainder of this section we shall briefly discuss some features of this arithmetic that are specific to $\mathbb{C}$. In the next section we start to explore exponentiation, which could not be done in quadratic extensions of $\mathbb{Z}$ or $\mathbb{Q}$ (except for integer powers); recall that our main motivation for defining $\mathbb{C}$ was to obtain a number system in which exponentiation is closed.

We have already used Definitions 8.1.1 and 8.1.2 of addition and multiplication in the proof of Theorem 9.1.1. For complex numbers (x, y) and (u, v) we can write these definitions as:

Definition 9.1.3.
$$(x, y) + (u, v) := (x + u, y + v),$$

Definition 9.1.4.
$$(x, y) \times (u, v) := (xu - yv, xv + yu).$$

Here we have used the standard symbols $+$ and $\times$ rather than $\oplus$ and $\otimes$, for the usual reason: isomorphisms have been established between a subset of our new number system and the previous number system. Specifically, $\{(x, y) \in \mathbb{C} : y = 0\}$ is isomorphic to $\mathbb{R}$ under addition and multiplication; this is just the isomorphism between $\mathbb{S}[\sqrt{k}]_0$ and $\mathbb{S}$ that we derived for general quadratic extensions. It is common practice to refer to members of the subset $\{(x, y) \in \mathbb{C} : y = 0\}$ as "real numbers", and simply write x rather than $(x, 0)$, for example writing 1 for $(1, 0)$. In particular, for scalar multiplication or scalar division of a complex number $w = (u, v)$ by a real number x, we may write xw or w/x which mean that we have multiplied or divided (u, v) by the complex number $(x, 0)$: see Definition 8.1.7 and the explanation around that definition. This blurring of the distinction between Real Numbers (members of $\mathbb{R}$) and "real numbers" which are members of a subset of $\mathbb{C}$ is reflected in the common practice of considering the five principal number systems of mathematics to be a hierarchy of subsets,
$$\mathbb{N} \subset \mathbb{Z} \subset \mathbb{Q} \subset \mathbb{R} \subset \mathbb{C},$$

ignoring the pedantic insistence that each system is actually *isomorphic* to a subset of the next system under their respective operations of addition and multiplication.

Equation (8.2) can be written with a change of notation as
$$(x, y) = (x, 0) + (y, 0) \times (0, 1).$$

Writing x for $(x, 0)$, y for $(y, 0)$ and identifying the symbol i with the complex number $(0, 1)$, this formally identifies (x, y) with $x + yi$. A number with zero real part, $(0, y) = (y, 0) \times (0, 1)$, is informally written as yi and is called a **pure imaginary** number.

---

[i]In this chapter the notation (u, v) does not have the connotation of "unit" which it had in Chapter 8.

The conjugate of a complex number $z = (x, y)$ is

$$\overline{z} = \overline{(x, y)} := (x, -y),$$

so that

$$z + \overline{z} = (2x, 0) \quad \text{and} \quad z - \overline{z} = (0, 2y): \tag{9.1}$$

the sum of a complex number and its conjugate is real, and their difference is pure imaginary. The product of a complex number with its conjugate is

$$z \times \overline{z} = (x^2 + y^2, 0),$$

so we define the norm of $z = (x, y)$ as

$$\mathcal{N}(x, y) = x^2 + y^2 \in \mathbb{R}.$$

Since the norm is a non-negative real number, we can take its (non-negative) square root in $\mathbb{R}$:

Definition 9.1.5. The **modulus** of a complex number is

$$|(x, y)| := \sqrt{x^2 + y^2}.$$

Since $|(x, 0)| = \sqrt{x^2}$ where the square root is non-negative, this is consistent with our earlier definition of modulus, which was originally given for $\mathbb{Z}$ in Definition 4.2.12 and can be written in all our *ordered* number systems ($\mathbb{Z}$, $\mathbb{Q}$, and $\mathbb{R}$) in the form:

$$|x| := \begin{cases} x & \text{if } x \geq 0 \\ -x & \text{if } x < 0. \end{cases}$$

The modulus inherits some of the important properties of the norm: $|(x, y)| = 0$ only if $(x, y) = (0, 0)$, and multiplication is preserved:

$$|(x, y) \times (u, v)| = |(x, y)| \times |(u, v)|.$$

Taking the modulus of a conjugate,

$$|\overline{z}| = |(x, -y)| = \sqrt{x^2 + (-y)^2} = \sqrt{x^2 + y^2} = |z|.$$

Since $\sqrt{x^2 + y^2} \geq x$ and $\sqrt{x^2 + y^2} \geq y$ for all $x, y \in \mathbb{R}$, we can say that

$$|z| \geq \text{Re}[z] \quad \text{and} \quad |z| \geq \text{Im}[z] \quad \forall z \in \mathbb{C}.$$

This will be useful in proving a very important property of the modulus:

Theorem 9.1.6 (The Triangle Inequality).

$$|z + w| \leq |z| + |w| \quad \forall z, w \in \mathbb{C}.$$

Proof. We shall show that $|z + w|^2 \leq (|z| + |w|)^2$; since all quantities involved are positive, we can take square roots according to Corollary 4.4.31 to obtain the required inequality. With $z = (x, y)$ and $w = (u, v)$,

$$\begin{aligned} |z + w|^2 &= (x + u)^2 + (y + v)^2 \\ &= x^2 + y^2 + u^2 + v^2 + 2(xu + yv) \\ &= |z|^2 + |w|^2 + 2\text{Re}[z\overline{w}] \\ &\leq |z|^2 + |w|^2 + 2|z\overline{w}| \\ &= |z|^2 + |w|^2 + 2|z| \times |\overline{w}| \\ &= |z|^2 + |w|^2 + 2|z| \times |w| \\ &= (|z| + |w|)^2. \end{aligned}$$

$\square$

We can extend this to the sum of more than two complex numbers:

Exercise 9.1.7. *Let* $z_1, z_2, \ldots, z_n$ *be* n *complex numbers. Using the triangle inequality, prove by induction that*

$$\left| \sum_{j=1}^{n} z_j \right| \leq \sum_{j=1}^{n} |z_j|.$$

The complex number system is a field, and the division formula (8.8) becomes

$$(x, y) \div (u, v) = \frac{(x, y) \times (u, -v)}{u^2 + v^2} = \left(\frac{xu + yv}{u^2 + v^2}, \frac{-xv + yu}{u^2 + v^2} \right), \tag{9.2}$$

more easily remembered as

$$z \div w = \frac{z \times \overline{w}}{|w|^2}. \tag{9.3}$$

Order cannot be defined in $\mathbb{C}$: it was shown in Section 8.1 that no quadratic extension $\mathbb{S}[\sqrt{k}]$ with negative k can be ordered.

9.2   Exponentiation by Real Powers in $\mathbb{C}$: A First Approach

Our path to fully defining exponentiation in $\mathbb{C}$ will be a long and winding one, but this will allow us to develop much of the theory of complex numbers on the way. Most textbooks use geometry to define some of the concepts in complex numbers, and they also assume some familiarity with the Analysis of functions with domains in $\mathbb{R}$, such as exponential and trigonometric functions. We assume no knowledge of geometry beyond a few axiomatic concepts, and the only Analysis we use relates to sequences, which are functions whose domain is $\mathbb{N}$; but we shall develop some geometric concepts and some knowledge of functions in $\mathbb{R}$ through our exploration of exponentiation in $\mathbb{C}$.[ii] Sequences are an important tool because, whereas in $\mathbb{R}$ the Axiom of Completeness allowed us to use the concept of supremum in many definitions, we do not have that in the unordered system $\mathbb{C}$.

A principle that must be followed in defining exponentiation in any number system is that the operation must preserve the fundamental properties known as the Laws of Indices, which were first found when exponentiation was defined in the Natural Numbers: given $\kappa, \lambda, \mu, \nu$ in any number system, we require

$$1^\mu = 1, \tag{9.4}$$

$$\kappa^{\mu+\nu} = \kappa^\mu \times \kappa^\nu, \tag{9.5}$$

$$(\kappa \times \lambda)^\mu = \kappa^\mu \times \lambda^\mu \tag{9.6}$$

$$\kappa^{\mu \times \nu} = (\kappa^\mu)^\nu, \tag{9.7}$$

where "1" in the first of these requirements stands for the multiplicative identity in the number system.

We take the definition of exponentiation in several stages (as we did in $\mathbb{R}$), raising a complex number to a power in each of the successive number systems in our hierarchy.

[ii]It is said that the entrance to Plato's Academy in ancient Athens bore the inscription, "Let no one ignorant of geometry enter". At this point in our book we should consider ourselves ineligible to enter Plato's Academy, but by the end of this chapter we may just be allowed in.

Natural number powers are defined in the usual inductive way in terms of multiplication, and then non-positive integer powers are defined as usual by

$$(x,y)^0 = (1,0), \qquad (x,y)^{-n} = (1,0) \div (x,y)^n.$$

These definitions have already been used in quadratic extensions of $\mathbb{Z}$ and $\mathbb{Q}$. We start venturing beyond what was done in those number systems by considering the square root. To define $(x,y)^{1/2}$ we need to find $(u,v) \in \mathbb{C}$ such that $(u,v)^2 = (x,y)$. Using Definition 9.1.4 for multiplication, this requirement becomes

$$(u^2 - v^2, 2uv) = (x,y).$$

So we need to solve the simultaneous equations

$$u^2 - v^2 = x, \qquad 2uv = y \tag{9.8}$$

for real numbers u, v. Substituting $v = y/2u$ from the second equation into the first, we obtain the quadratic equation for u^2,

$$(u^2)^2 - xu^2 - \frac{y^2}{4} = 0.$$

This has solutions

$$u^2 = \frac{1}{2}\left(x \pm \sqrt{x^2 + y^2}\right);$$

then from the first equation of (9.8),

$$v^2 = u^2 - x = \frac{1}{2}\left(-x \pm \sqrt{x^2 + y^2}\right).$$

To obtain real values for u and v we require u^2 and v^2 to be non-negative, so we take the $+$ option in the $\pm$ signs in these solutions (since $\sqrt{x^2 + y^2} \geq x$ and $\sqrt{x^2 + y^2} \geq -x$ for all $x, y \in \mathbb{R}$). This yields two solutions for each of u and v:

$$u = \pm\frac{1}{\sqrt{2}}\sqrt{x + \sqrt{x^2 + y^2}}, \qquad v = \pm\frac{1}{\sqrt{2}}\sqrt{-x + \sqrt{x^2 + y^2}}, \tag{9.9}$$

which apparently means that there are four possible combinations (u,v) which are square roots of the given (x,y). We can check which of the solutions (9.9) actually satisfy the requirements (9.8): they all satisfy $u^2 - v^2 = x$ but yield $2uv = \pm\sqrt{y^2}$ rather than the required $2uv = y$. To satisfy the latter, we need $2uv \geq 0$ if $y \geq 0$, which means taking either the plus sign in $\pm$ for both u and v, or the minus sign for both u and v, in (9.9); on the other hand, if $y < 0$ we need $2uv < 0$ so we take opposite signs for u and v in (9.9). In both cases, we obtain two solutions satisfying the requirement of a square root of (x,y), as expected; so if we want a single, unambiguous square root (in the same way that in $\mathbb{Q}$ and $\mathbb{R}$ we conventionally take $\sqrt[2]{x}$ to be the *non-negative* number u satisfying $u^2 = x$), we can define the **principal square root** of a complex number to be the one with positive real part ($u > 0$), except that if the real part is zero we take the root with non-negative imaginary part ($v \geq 0$). Thus:

Definition 9.2.1. The **principal square root** of a complex number (x,y) is given by:
    If $y \geq 0$,

$$\sqrt{(x,y)} = \left(\frac{1}{\sqrt{2}}\sqrt{x + \sqrt{x^2 + y^2}}, \; \frac{1}{\sqrt{2}}\sqrt{-x + \sqrt{x^2 + y^2}}\right).$$

If $y < 0$,

$$\sqrt{(x,y)} = \left(\frac{1}{\sqrt{2}} \sqrt{x + \sqrt{x^2 + y^2}}, \ -\frac{1}{\sqrt{2}} \sqrt{-x + \sqrt{x^2 + y^2}} \right).$$

Square roots of real quantities are *non-negative* in these formulae.

Whenever we take a square root in the calculations below, it will be a principal square root unless specified otherwise. The non-principal square root is simply the additive inverse of the principal square root in all cases.

It is important to note that conjugation preserves the square root, i.e.

$$\overline{\sqrt{(x,y)}} = \sqrt{\overline{(x,y)}}.$$

To see this, observe that replacing y with $-y$ leaves the formulae in Definition 9.2.1 unchanged, but swaps between the conditions $y \geq 0$ and $y < 0$ (except when $y = 0$, in which case $\overline{(x,y)} = (x,y)$). It then follows from the identities

$$|(x,y)|^2 = \mathcal{N}(x,y) = (x,y) \times \overline{(x,y)}$$

that the norm and the modulus preserve the square root:

$$\mathcal{N}\left(\sqrt{(x,y)} \right) = \sqrt{\mathcal{N}(x,y)}, \qquad \left| \sqrt{(x,y)} \right| = \sqrt{|(x,y)|}.$$

Exercise 9.2.2. *Find the principal square root of $(0,16)$, and check that $\left| \sqrt{(0,16)} \right| = \sqrt{|(0,16)|}$.*

By taking the square root twice more, find the principal 4'th root, $(0,16)^{1/4}$, and the principal 8'th root, $(0,16)^{1/8}$. Again check that the modulus preserves these roots.

Finding the square root is the key to a method for finding any real power of a complex number. We can take the 2^k'th root of a number for any $k \in \mathbb{N}$ by repeating the square root operation k times; then for any $q \in \mathbb{Z}$ we have $(x,y)^{q/2^k} = \left((x,y)^{1/2^k} \right)^q$, where integer powers have already been defined (and of course all $(q/2^k)$'th powers are then preserved by conjugation, norm and modulus). Next, for any real number α, there is a sequence of base-2 approximations,

$$Q_\alpha : \mathbb{N} \to \mathbb{R}, \qquad Q_\alpha : k \mapsto \frac{q_k}{2^k},$$

which converges to α: see Theorem 7.7.3. So we can calculate a sequence of terms,

$$P_\alpha : \mathbb{N} \to \mathbb{C}, \qquad P_\alpha : k \mapsto (x,y)^{q_k/2^k}$$

which we expect to converge; we can then define $(x,y)^\alpha$ as the complex number to which the sequence P_α converges. But this definition only makes sense if the sequence P_α does actually converge whenever Q_α converges; and to confirm that, we first need to know what it means for a sequence of complex numbers to converge. Although we shall see that Definition 6.8.4 of convergence can be applied in $\mathbb{C}$, we shall not pursue this line of reasoning with the sequence P_α, taking instead a more insightful route.

This starts with the observation that if we divide any non-zero complex number w by its own modulus, the result $w/|w|$ is a complex number with modulus equal to 1. This is because the modulus preserves exact division, so that

$$\left| \frac{w}{|w|} \right| = \frac{|w|}{|(|w|,0)|} = \frac{|w|}{|w|} = 1,$$

where in the first equality we are acknowledging that the scalar division by $|w|$ is actually a shorthand for division by the complex number $(|w|, 0)$. Thus we can write any complex number w as

$$w = |w| \times \frac{w}{|w|}, \tag{9.10}$$

the scalar multiplication of its modulus by a complex number with modulus 1. The requirement that our definition of exponentiation in $\mathbb{C}$ must satisfy (9.6) means that

$$w^{\mu} = |w|^{\mu} \times \left(\frac{w}{|w|} \right)^{\mu}, \tag{9.11}$$

for any power μ. If μ is real, then $|w|^{\mu}$ is already defined as a real power of a real number, so we then need to examine real powers of complex numbers $w/|w|$ with modulus 1. As we do this by the procedure of repeated square root operations, we shall discover that complex numbers with modulus equal to 1 can be characterised by a single parameter called the **argument**, which will then be the key to a convenient way of finding all powers, real and complex, of a complex number. But all of this will be much easier to understand if we consider the geometrical representation of complex numbers.

9.3 Geometry of $\mathbb{C}$; the Principal Value of the Argument, and the Number π

Although it would be possible to fully define exponentiation in $\mathbb{C}$ by pure analysis, it would not make much intuitive sense without some geometrical context. So far in this book, the only references to geometry have been occasional mentions of a "number line": for any ordered number system, we can imagine the numbers arranged along a straight line, with greater numbers to the right of lesser numbers. There is a reference point representing the number 0, and a unit of length: the point representing a number x on the line is at a distance $|x|$ units from the point 0, to the right or left according to whether x is positive or negative. The distance between points representing numbers x_1 and x_2 on the line is $|x_2 - x_1|$ units. The complex numbers are not an ordered number system, so cannot be represented on a number line. However, each complex number is an ordered pair of real numbers, which suggests that complex numbers can be represented using two independent number lines, one for the real part and one for the imaginary part. A general complex number (x, y) is then represented by a point in a two-dimensional space, or plane.[iii] But how does the location of a number in this **complex plane** relate to the two number lines, and how do we measure distance between numbers in the plane? The clue is that the formula for the norm in $\mathbb{C}$, $\mathcal{N}(x, y) = x^2 + y^2$, reminds us of probably the most famous theorem in geometry: Pythagoras' theorem. According to this theorem, we should place our number lines for the real and imaginary parts so that they cross each other at right angles at the point $(0, 0)$ which we call the **origin**; then the point representing (x, y) is found by measuring distances x units from the origin along the real number line and y units parallel to the imaginary number line, and the distance along the straight line from $(0, 0)$ to (x, y) is $\sqrt{x^2 + y^2}$, the modulus of (x, y): see Figure 9.1. We take this as an *axiom* defining the structure of the complex plane, rather than a theorem that needs to be proved.[iv] The **distance** between

---

[iii] The concepts of space, dimension, and independence, which we take to be understood, all have precise definitions in Linear Algebra, which is the area of mathematics that deals with abstractions based on our intuitive notion of physical space in 2 or 3 dimensions; but mathematical spaces can have any countable (finite or infinite) number of dimensions.

[iv] There are probably more proofs of Pythagoras' theorem than any other theorem in the whole of mathematics. Since this is not a geometry textbook, we leave it to interested readers to seek out proof(s) for themselves.

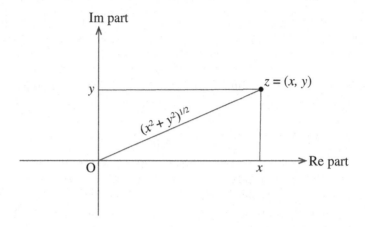

FIGURE 9.1: The complex plane, with origin O at the intersection of two number lines for the Real and Imaginary parts, and a point representing the number $z = (x, y)$.

the points representing the complex numbers $w = (u, v)$ and $z = (x, y)$ is then the length of the straight line between those points in the complex plane,

$$|z - w| = |(x, y) - (u, v)| = |(x - u, y - v)| = \sqrt{(x - u)^2 + (y - v)^2} : \qquad (9.12)$$

see Figure 9.2. Conveniently, this means that we can define convergence of sequences of complex numbers by using Definition 6.8.4 *verbatim*: the condition $|f(n) - L| < \epsilon$, which for rational or real number sequences meant that the *difference* between sequence terms and the limit L had to be less than any desired positive ϵ, now means that the *distance* in the complex plane between sequence terms and the limit must be less than ϵ.

Exercise 9.3.1. *Plot points $z = (x, y)$ and $w = (u, v)$ at randomly chosen positions in a plane. Then plot the point $z + w = (x + u, y + v)$. Draw straight lines from the origin to the points z and $z + w$. Draw the straight line from z to $z + w$, and explain why the length of this line is $|w|$. Satisfy yourself that $|z + w| \leq |z| + |w|$, and explain how the Triangle Inequality (Theorem 9.1.6) got its name.*

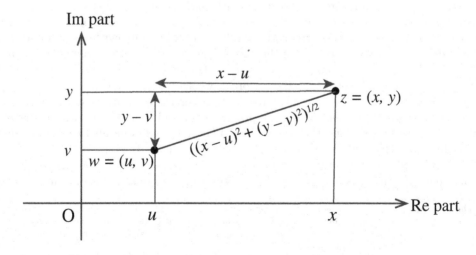

FIGURE 9.2: Distance between points representing complex numbers w and z is $|z - w| = \sqrt{(x - u)^2 + (y - v)^2}$.

9.3.1 The Unit Circle and the Principal Value of the Argument of a Complex Number

Having set up our complex plane to represent complex numbers in a two-dimensional space, any further geometrical concepts must be formally defined. In particular:

Definition 9.3.2. A **circle** is the set of points at a fixed distance from some reference point. The distance from the reference point is called the **radius** of the circle.

Whereas we have taken the concept of length along a straight line to be understood, we have yet to define lengths along a circle (or any other curve). Note our use of the term *length* along a curve, whereas the term *distance* will be reserved for straight-line distance as defined in (9.12).

Of particular interest is the **unit circle**, which is the set of points at a distance of 1 unit from the origin; so it represents the set of complex numbers with modulus equal to 1. It is the exponentiation of these "numbers on the unit circle" that we will now concentrate on, since (9.11) will then enable us to exponentiate all complex numbers.

Suppose that t and w are complex numbers on the unit circle, so $|t| = 1$ and $|w| = 1$. Then

$$|tw| = |t| \times |w| = 1 \times 1 = 1,$$

so tw is on the unit circle. The straight-line distance between the points representing t and tw is

$$\begin{aligned} |tw - t| &= |t(w - (1,0))| \\ &= |t| \times |w - (1,0)| \\ &= |w - (1,0)| \end{aligned} \tag{9.13}$$

since $|t| = 1$; so the distance between t and tw is equal to the distance between $(1,0)$ and w.

Now apply this to taking the square root of a number z on the unit circle. Letting $t = w = z^{1/2}$ so that $tw = z$, the distance between the points representing $z^{1/2}$ and z is equal to that between $(1,0)$ and $z^{1/2}$, as shown by the solid lines of equal length in Figure 9.3. Taking the square root again, with $t = w = z^{1/4}$ so that $tw = z^{1/2}$, the distance between $z^{1/4}$ and $z^{1/2}$ is equal to that between $(1,0)$ and $z^{1/4}$; and it is also equal to the distance between $z^{1/2}$ and $z^{3/4}$, and between $z^{3/4}$ and z (verify this by letting $w = z^{1/4}$, with $t = z^{1/2}$ and $t = z^{3/4}$ in turn). This is shown by the 4 dashed lines of equal length in Figure 9.3. Repeating this "square-root procedure" n times, we obtain 2^n lines, all of equal length: we denote this length at the n'th iteration as $C(n; z)$. So $C(n; z)$ is the length of the j'th line, which connects points at $z^{(j-1)/2^n}$ and $z^{j/2^n}$ for $j = 1, 2, \ldots, 2^n$, and in particular of the first line, connecting $(1,0)$ to $z^{1/2^n}$:

$$\begin{aligned} (C(n; z))^2 &:= \left| z^{j/2^n} - z^{(j-1)/2^n} \right|^2 \\ &= \left| z^{1/2^n} - (1,0) \right|^2 = (x_n - 1)^2 + (y_n - 0)^2 \end{aligned} \tag{9.14}$$

where we have defined

$$x_n := \text{Re}[z^{1/2^n}], \qquad y_n := \text{Im}[z^{1/2^n}].$$

For any given z, the iterations of the square-root procedure yield a *sequence* of line lengths $C(n; z)$, for all $n \in \mathbb{N} \cup \{0\}$ (with $C(0; z) = |z - (1,0)|$). On the unit circle, where $x_n^2 + y_n^2 = 1$, (9.14) simplifies to

$$(C(n; z))^2 = 2 - 2x_n \tag{9.15}$$

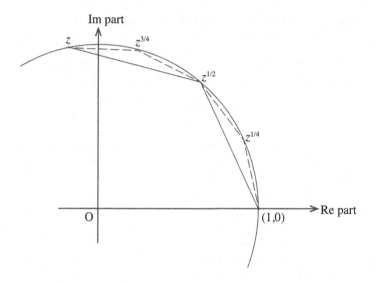

FIGURE 9.3: Part of the unit circle: the lines from $(1,0)$ to $z^{1/2}$ and from $z^{1/2}$ to z are of length $C(1;z)$ (as defined in (9.14)), and the four dashed lines connecting $(1,0)$, $z^{1/4}$, $z^{1/2}$, $z^{3/4}$, and z are of length $C(2;z)$.

where x_n can be evaluated inductively for a given $z = (x_0, y_0)$.[v] Since

$$(x_{S(n)}, y_{S(n)}) = \sqrt{(x_n, y_n)}$$

(where S is the Successor function), we use the principal square-root formulae in Definition 9.2.1: on the unit circle where $x_n^2 + y_n^2 = 1$, these simplify to

$$x_{S(n)} = \sqrt{\frac{x_n + 1}{2}}, \qquad y_{S(n)} = \pm\sqrt{\frac{1 - x_n}{2}}, \tag{9.16}$$

with Definition 9.2.1 keeping $y_{S(n)}$ positive or negative according to whether y_n is positive or negative.

Intuitively, each iteration of the square-root procedure brings the lines closer to the circle: see Figure 9.3. So the total length of the 2^n lines,

$$D(n;z) := 2^n C(n;z), \tag{9.17}$$

is an approximation to the total length along the circle from $(1,0)$ to z, and the approximation becomes better as n increases.[vi] In fact, since we do not have any previous definition of length along a curve, we shall *define* the length along the unit circle as the limit of the sequence $D(n;z)$, if that sequence converges. We also adopt the convention that $D(n;z)$ is the *shortest* length along 2^n lines from $(1,0)$ to z, and is positive going anticlockwise around the circle to any point z with $\operatorname{Im}[z] \geq 0$, and negative going clockwise to any point with $\operatorname{Im}[z] < 0$; which implies that we have taken the positive or negative square root to obtain $C(n;z)$ from (9.15) according to whether $\operatorname{Im}[z]$ is positive or negative, before substituting in (9.17). Now consider the conjugate of z: we have $\operatorname{Im}[\bar{z}] < 0$ if $\operatorname{Im}[z] > 0$, and

---

[v]The induction is anchored at $n = 0$. In this instance it is convenient to use a definition of $\mathbb{N}$ including zero.

[vi]Why do we want to measure lengths from $(1,0)$? Because $(1,0)$ is the multiplicative identity in $\mathbb{C}$, and so is expected to have the special property in relation to exponentiation, that $(1,0)^w = (1,0)$ for all $w \in \mathbb{C}$.

$|\bar{z} - (1,0)| = |z - (1,0)|$ (since $(x-1)^2 + (-y)^2 = (x-1)^2 + y^2$). Hence with our convention, if $\text{Im}[z] > 0$ then

$$C(0; \bar{z}) = -|\bar{z} - (1,0)| = -|z - (1,0)| = -C(0; z),$$

and then the square-root procedure yields

$$D(n; \bar{z}) = -D(n; z). \tag{9.18}$$

So if we calculate $D(n; z)$ for points on the unit circle with $\text{Im}[z] \geq 0$, then (9.18) yields its value for points with $\text{Im}[z] < 0$. For any fixed n, $D(n; z)$ is an injective function from $\{z \in \mathbb{C} : |z| = 1\}$ to $\mathbb{R}$: each point on the unit circle has a unique value of $D(n; z)$, which is not shared by any other point.

We still need to prove that the sequence $D(n; z)$ converges. Our method of proof is inspired by an idea used by Archimedes to estimate the length along a complete circle. As in our procedure, he considered the lengths of straight lines between equally spaced points on the circle. He realised that the sum of line lengths was *less than* the length along the circle, but that doubling the number of lines (as in our square-root procedure) would *increase* their combined length, which would nevertheless remain less than the length along the circle. He also considered lines outside the circle, with each line touching the circle at one point: see Figure 9.4: their combined length is *greater than* the length along the circle, and will *decrease* if the number of lines is doubled. So Archimedes knew that the length along the circle lay between the combined length of the inner lines and that of the outer lines. By starting with 6 lines (for which the length calculation is particularly easy), and successively doubling to 96 ($= 2^4 \times 6$) lines, he showed that the length around a complete unit circle lay between $\frac{446}{71}$ and $\frac{44}{7}$ units; thus he was able to estimate it within an interval of width only $\frac{2}{497}$.[vii] He probably knew intuitively that he could make the interval as small as he liked by continuing to double the number of lines; but whereas the ancient Greeks had well-developed and rigorous theory of geometry, they had none of our modern concepts of convergence of sequences. So we now provide the definitions and theorems that will enable us to determine lengths along a circle as limits of sequences, translating Archimedes' geometrical idea into the language of Analysis.

Definition 9.3.3. A sequence $f : \mathbb{N} \to \mathbb{R}$ is **increasing** if

$$f(S(n)) > f(n) \; \forall \, n \in \mathbb{N},$$

where S is the Successor function.

A sequence $f : \mathbb{N} \to \mathbb{R}$ is **decreasing** if

$$f(S(n)) < f(n) \; \forall \, n \in \mathbb{N}.$$

A sequence $f : \mathbb{N} \to \mathbb{R}$ is **non-decreasing** if

$$f(S(n)) \geq f(n) \; \forall \, n \in \mathbb{N}.$$

[vii] Archimedes' lower and upper bounds, 446/71 and 44/7, are rational, whereas from the formulae developed below it will be obvious that the upper and lower bounds should involve irrational square roots. In particular, Archimedes' calculation starting with the regular hexagon involves $\sqrt{3}$ (see Investigation 1 at the end of this chapter); but he used rational lower and upper approximations to $\sqrt{3}$, respectively 265/153 and 1351/780. If you did Exercise 8.3.11, you should have found these approximations from calculations in $\mathbb{Z}[\sqrt{3}]$, with $(265, 153) = (1,1) \times (2,1)^4$ and $(1351, 780) = (2,1)^6$ where $(2,1)$ is the fundamental unit and $(1,1)$ has norm equal to -2. Archimedes did not say how he found his approximations, and he had no formal theory of quadratic rings, but maybe he did some equivalent calculations?

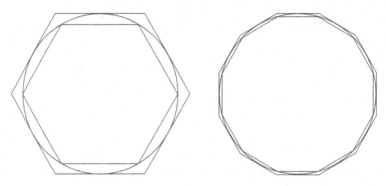

FIGURE 9.4: Archimedes' method for finding the length around a complete circle. On the left, the total lengths of the 6 sides of the inner and outer hexagons are respectively less and greater than that of the circle; the same is true for the 12-sided polygons on the right, but their lengths approximate that of the circle more closely.

A sequence $f : \mathbb{N} \to \mathbb{R}$ is **non-increasing** if

$$f(S(n)) \leq f(n) \; \forall n \in \mathbb{N}.^{\text{viii}}$$

It is important to recognise that all increasing sequences are non-decreasing (since if it is true that $f(S(n)) > f(n)$, then it is certainly true that $f(S(n)) \geq f(n)$), and all decreasing sequences are non-increasing.

Exercise 9.3.4. *Prove that if $f : \mathbb{N} \to \mathbb{R}$ is a non-decreasing sequence, then $f(n) \geq f(N)$ whenever $n > N$. [Hint: use induction on n, anchored at $n = S(N)$.]*

Theorem 9.3.5. *A sequence that is non-decreasing and bounded above converges. A sequence that is non-increasing and bounded below converges.*

The proof of the theorem gives a nice demonstration of the importance of the Axiom of Completeness in the Real Numbers, that every set bounded above has a supremum.

Proof of Theorem 9.3.5. Suppose the sequence $f : \mathbb{N} \to \mathbb{R}$ is non-decreasing, so $f(S(n)) \geq f(n) \; \forall n \in \mathbb{N}$, and bounded above, so that the set $\{f(n) : n \in \mathbb{N}\}$ must have a supremum $L \in \mathbb{R}$ [*Axiom of Completeness*]. We shall prove that the sequence f converges to L.

Choose any $\epsilon > 0$. Suppose that there is no $n \in \mathbb{N}$ for which $f(n) > L - \epsilon$. Then $L - \epsilon$ is an upper bound for $\{f(n) : n \in \mathbb{N}\}$; but $L - \epsilon < L$, so this would contradict the definition of L as the *least* upper bound (i.e. supremum) of the set.

Thus there must exist $N \in \mathbb{N}$ for which $f(N) > L - \epsilon$. Then since f is non-decreasing, $f(n) \geq f(N) \; \forall n > N$ [*Exercise 9.3.4*], and so $f(n) > L - \epsilon \; \forall n > N$. But $f(n) \leq L \; \forall n \in \mathbb{N}$, since L is an upper bound. We have satisfied Definition 6.8.4 for the sequence f to converge to L, by showing that $L - \epsilon < f(n) \leq L \; \forall n > N$.

The proof for a sequence that is non-increasing and bounded below is similar. □

[viii]Many authors use the terms **strictly increasing** and **strictly decreasing** for what we have called "increasing" and "decreasing", and they use "increasing" and "decreasing" for what we have called "non-decreasing" and "non-increasing". The latter usage would mean that a constant sequence (in which all terms are equal) is both increasing and decreasing, which seems a violation of common sense! Our definition is more in line with the way the words are used in non-mathematical English, albeit the terms "non-decreasing" and "non-increasing" are rather inelegant.

Corollary 9.3.6. *Suppose $f(n)$ is a non-decreasing sequence and $g(n)$ is a non-increasing sequence, with $g(n) > f(n)$ for all n. Then both $f(n)$ and $g(n)$ converge.*

This corollary generalises Archimedes' idea of approximating the length of the circle from below by a sequence of inner polygons of increasing length, and from above by a sequence of outer polygons of decreasing length.

Exercise 9.3.7. *Show how Corollary 9.3.6 follows from Theorem 9.3.5.*

Returning to the task of showing that a length along a circle can be found as the limit of a sequence $D(n; z)$, we take $\text{Im}[z] \geq 0$ so that $C(n; z)$ and $D(n; z)$ are positive for all n (we exclude the trivial case $z = (1, 0)$, for which $x_n = 1$, $C(n; z) = 0$ and $D(n; z) = 0$ for all n; thus we take $x_n < 1$ in all calculations below). We will show that $D(n; z)$ is non-decreasing, and then find a non-increasing sequence of numbers which are each greater than the corresponding terms in $D(n; z)$, so that we can use Corollary 9.3.6 to confirm that $D(n; z)$ converges.

Proof that $D(n; z)$ ***converges.*** The sequence $D(n; z)$ is non-decreasing if

$$D(S(n); z) \geq D(n; z).$$

Since

$$D(S(n); z) = 2^{S(n)} C(S(n); z),$$

(from (9.17)), this requires

$$2C(S(n); z) \geq C(n; z), \tag{9.19}$$

which can be verified quite simply:

Exercise 9.3.8. *Use the triangle inequality to verify (9.19): Figure 9.3 provides a clue if you need one.*

However, it is also instructive to verify (9.19) from the formulae we have derived; we shall actually verify the stricter condition that

$$2C(S(n); z) > C(n; z). \tag{9.20}$$

Re-writing (9.15) for the $S(n)$'th iteration, we have

$$(C(S(n); z))^2 = 2 - 2x_{S(n)}$$

$$= 2 - 2\sqrt{\frac{1}{2}(x_n + 1)}. \qquad [\textit{from (9.16)}]$$

Since $C(n; z) \geq 0$ for all n, Corollary 4.4.31 allows us to "square the inequality": (9.20) is true if and only if

$$4(C(S(n); z))^2 > (C(n; z))^2.$$

Substituting the formulae we have obtained, this is

$$4\left(2 - 2\sqrt{\frac{1}{2}(x_n + 1)}\right) > 2 - 2x_n.$$

Using standard manipulations of inequalities (as demonstrated in Section 4.4), this is true if and only if

$$6 + 2x_n > 8\sqrt{\frac{1}{2}(x_n + 1)}.$$

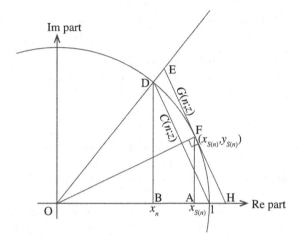

FIGURE 9.5: The required length $G(n;z)$ is twice the length FH, where the point F has coordinates $(x_{S(n)}, y_{S(n)})$: compare with Figure 9.3. Triangles OFA and FHA are similar, so the length ratios FH/FA and OF/OA are equal. Now OF = 1 unit, OA = $x_{S(n)}$ and FA = $y_{S(n)}$. Thus FH = FA × OF/OA = $y_{S(n)}/x_{S(n)}$, and substitution from (9.16) into $G(n;z) = 2$FH then yields the result (9.21).

Both sides of this inequality are non-negative (since $-1 \leq x_n \leq 1$ on the unit circle), so we can square again: after some further manipulation, we obtain

$$4(x_n - 1)^2 > 0.$$

Since this is true for all $x_n < 1$, and every step in the proof is "if and only if", the original inequality (9.20) is true for all $x_n < 1$, and so $D(n;z)$ is increasing.

Next, we find a decreasing (and hence non-increasing) sequence by returning to Archimedes' idea, as illustrated in Figure 9.4. Whereas our increasing sequence involved $C(n;z)$, the length of the lines inside the unit circle, our decreasing sequence will be based on lines outside the circle, of length $G(n;z)$ as shown in Figure 9.5. Using a little geometrical theory, as explained in the caption to Figure 9.5,[ix] we find that

$$G(n;z) = 2\sqrt{\frac{1 - x_n}{1 + x_n}}, \tag{9.21}$$

where we exclude the case $x_n = -1$, which is covered in Subsection 9.3.2 below. The total length of 2^n outer lines is

$$H(n;z) = 2^n G(n;z) = 2^{n+1}\sqrt{\frac{1 - x_n}{1 + x_n}}, \tag{9.22}$$

and to complete our proof we need to show that $H(n;z) > D(n;z)$ and that $H(n;z)$ is decreasing.

Now, from (9.15), (9.17), and (9.22) we obtain

$$\frac{H(n;z)}{D(n;z)} = \sqrt{\frac{2}{1 + x_n}},$$

and $\sqrt{2/(1 + x_n)} > 1$ since $x_n < 1$. Hence $H(n;z) > D(n;z)$.

[ix]Note that geometry will play no part in the proof that $D(n;z)$ converges; it is simply a means to find a suitable sequence which can then be proved to have terms greater than $D(n;z)$ and to be decreasing.

To verify that $H(n; z)$ is decreasing, the calculation is similar to that used for $D(n; z)$ increasing. We start by noting that $H(S(n); z) < H(n; z)$ if

$$2G(S(n); z) < G(n; z),$$

and since $G(n; z)$ is positive for all n, this is true if and only if

$$4(G(S(n); z))^2 < (G(n; z))^2. \tag{9.23}$$

The rest of the calculation is left to the reader:

Exercise 9.3.9. *Re-writing (9.21) for the $S(n)$'th iteration and using (9.16), obtain a formula for $(G(S(n); z))^2$ and hence show that inequality (9.23) becomes*

$$4\left(\frac{1 - \sqrt{\frac{1}{2}(1 + x_n)}}{1 + \sqrt{\frac{1}{2}(1 + x_n)}}\right) < \frac{1 - x_n}{1 + x_n}.$$

By standard manipulations of inequalities, show that this is true if and only if

$$3 + 5x_n < (5 + 3x_n)\sqrt{\frac{1}{2}(1 + x_n)}.$$

This is certainly true if $-1 \leq x_n < -3/5$, when the left side is negative and the right side is positive or zero. If $x_n \geq -3/5$ the inequality can be squared: show that it is true if and only if

$$(x_n - 1)^2(9x_n + 7) > 0.$$

Hence deduce that (9.23) is true whenever $-1 < x_n < 1$, so that $H(n; z)$ is decreasing.

We have now met stricter conditions than those required in Corollary 9.3.6: when $\text{Im}[z] \geq 0$, $D(n; z)$ is an increasing sequence with terms less than the decreasing sequence $H(n; z)$, so that $D(n; z)$ does converge. Equation (9.18) then covers the case where $\text{Im}[z] < 0$.   □

The limit of the sequence $D(n; z)$ is an important quantity in the theory of Complex Numbers: it is called the **principal value of the argument** of z, and denoted $\text{Arg}\, z$; we shall also use the abbreviation PVA for principal value of the argument. It is represented geometrically by the length round the unit circle from $(1, 0)$ to z, anticlockwise to give a positive value of $\text{Arg}\, z$ if $\text{Im}[z] \geq 0$, and clockwise for a negative value of $\text{Arg}\, z$ if $\text{Im}[z] < 0$. Later we shall relax these restrictions on how the length round the unit circle is measured, to yield countably infinitely many other values of the argument of a given z on the unit circle. For numbers w not on the unit circle, we can broaden the definition of PVA by recalling equation (9.10): any complex number w can be written as $w = |w| \times z$ where $z = w/|w|$ is a number on the unit circle. So we define the PVA of w to be equal to the PVA of $w/|w|$:

Definition 9.3.10. Given any non-zero complex number w, let $z = w/|w|$. Then the **principal value of the argument**, or **PVA**, of w is defined as

$$\text{Arg}\, w := \lim_{n \to \infty} \sum_{j=1}^{2^n} \left| z^{j/2^n} - z^{(j-1)/2^n} \right|$$

if $\text{Im}[w] \geq 0$, with $\text{Arg}\, \overline{w} = -\text{Arg}\, w$.

The definition is written in this way to remind you that the PVA is the length along the unit circle, approximated to any desired precision by using the square-root procedure to find 2^n equally spaced points on the circle and then summing the distances between them. The statement, $\operatorname{Arg}\overline{w} = -\operatorname{Arg}w$, follows immediately from (9.18). The "non-zero" proviso is because $|(0,0)| = 0$, so $w/|w|$ is undefined when $w = (0,0)$: the PVA of zero is undefined. Since the distances summed in Definition 9.3.10 are all equal, we could of course write the formula as

$$\operatorname{Arg}w := \lim_{n\to\infty} 2^n \left| z^{1/2^n} - (1,0) \right|. \tag{9.24}$$

Going around the unit circle from $(1,0)$ towards $(-1,0)$, the value of $\operatorname{Re}[z]$ decreases as $|\operatorname{Arg}z|$, the length traversed along either the upper or lower half of the circle, increases. This geometrically obvious fact needs to be proved rigorously, since it will be useful in later proofs.

Theorem 9.3.11. *If $z, t \in \mathbb{C}$ with $|z| = 1$, $|t| = 1$ and $\operatorname{Re}[z] \geq \operatorname{Re}[t]$, then $|\operatorname{Arg}z| \leq |\operatorname{Arg}t|$.*[x]

Proof. Let $x_0 = \operatorname{Re}[z]$ and $r_0 = \operatorname{Re}[t]$, with $x_0 \geq r_0$. If $x_n \geq r_n$, then from the usual theorems on order,

$$\sqrt{\frac{x_n + 1}{2}} \geq \sqrt{\frac{r_n + 1}{2}}$$

and so by induction using the first formula in (9.16),

$$x_n \geq r_n \; \forall\, n \in \mathbb{N}.$$

Hence by (9.15), $|C(n; z)| \leq |C(n; t)|$, and then by (9.17),

$$|D(n; z)| \leq |D(n; t)| \; \forall\, n \in \mathbb{N},$$

where the modulus signs allow for the positive or negative values of $C(n; z)$ and $D(n; z)$ when $\operatorname{Im}[z]$ is positive or negative.

If an order relation applies between the n'th members of two sequences for all n, it seems reasonable that the same order relation will apply between the limits of those sequences. We shall now prove this for the weak ordering $\leq$ (and hence for $\geq$); having studied the proof, think about why it does *not* apply to the strict orderings $<$ and $>$ (which is why Theorem 9.3.11 only involves weak orderings).

Theorem 9.3.12. *If $f : \mathbb{N} \to \mathbb{R}$ and $g : \mathbb{N} \to \mathbb{R}$ are convergent sequences, with $f(n) \leq g(n)$ for each $n \in \mathbb{N}$, then*

$$\lim_{n\to\infty} f(n) \leq \lim_{n\to\infty} g(n).$$

Proof. Suppose the theorem is false: this means that $f(n) \to L$ and $g(n) \to K$ with $L > K$, even though $f(n) \leq g(n)$ for each $n \in \mathbb{N}$. So $L - K = \delta$ for some $\delta > 0$. Because f and g converge, we can find $N \in \mathbb{N}$ such that both $|f(n) - L| < \delta/2$ and $|g(n) - K| < \delta/2$ when $n > N$. Thus

$$g(n) < K + \frac{\delta}{2} = L - \frac{\delta}{2} < f(n)$$

when $n > N$: we have found that $g(n) < f(n)$, which contradicts the specification that $f(n) \leq g(n)$ for each $n \in \mathbb{N}$. Hence it cannot be true that $L > K$. $\qquad\square$

---

[x]The theorem is actually true with strict orderings, $>$ and $<$ rather than $\geq$ and $\leq$, as is obvious from the geometry. Those strict orderings will follow immediately from Theorem 9.4.8, but the weak orderings are sufficient for our purposes now.

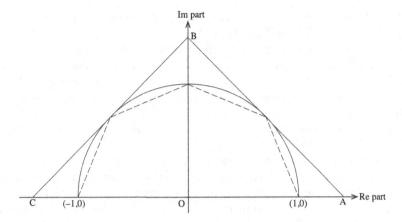

FIGURE 9.6: The length along the semicircle is greater than the straight-line distance between its ends, which is $D(0; z) = 2$ units, and also greater than the total distance along the dashed lines, which is $D(2; z)$ and shown in Exercise 9.3.14 to be greater than 3 units. It is less than the distance along ABC, which is $H(1; z) = 4$ units.

Returning to the proof of Theorem 9.3.11, since $\operatorname{Arg} z$ and $\operatorname{Arg} t$ are the respective limits of the sequences $D(n; z)$ and $D(n; t)$, Theorem 9.3.12 confirms that $|\operatorname{Arg} z| \le |\operatorname{Arg} t|$. $\quad\square$

Although we only have weak orderings in Theorem 9.3.11, there is one case in which we can be strict:

Theorem 9.3.13. *For z on the unit circle,* $\operatorname{Arg} z = 0$ *only if* $z = (1, 0)$

Proof. If $z \ne (1, 0)$ then $|z - (1, 0)| > 0$, so $D(0; z) > 0$ in the case where $\operatorname{Im}[z] > 0$ or $z = (-1, 0)$. Since $D(n; z)$ is an increasing sequence, $D(n; z) > D(0; z) \, \forall n \in \mathbb{N}$ and so by Theorem 9.3.12, $\operatorname{Arg} z \ge D(0; z) > 0$. Similarly if $\operatorname{Im}[z] < 0$, then $\operatorname{Arg} z \le D(0; z) < 0$. $\quad\square$

We shall see later the importance of a zero value of the PVA on the unit circle corresponding only to the multiplicative identity in $\mathbb{C}$.

9.3.2   The Number π

It is perhaps obvious that Definition 9.3.10 does not afford a practical means of calculating the PVA of a given complex number z, nor the equally important task of finding a complex number on the unit circle with a given value of its PVA. However, let us consider $\operatorname{Arg}(-1, 0)$, which is represented geometrically by the length of the upper half of the unit circle, from $(1, 0)$ to $(-1, 0)$. Now, $\operatorname{Arg}(-1, 0)$ is the limit of the sequence $D(n; z)$ with $z = (-1, 0)$, for which we start with $D(0; z) = C(0; z) = |(-1, 0) - (1, 0)| = 2$; so since $D(n; z)$ is increasing we have $\operatorname{Arg}(-1, 0) > 2$. But it is easy to obtain a greater lower bound for $\operatorname{Arg}(-1, 0)$:

Exercise 9.3.14. *By doing two iterations of formula (9.16) starting from $x_0 = -1$, find x_2 and then use (9.15) and (9.17) to obtain $D(2; z)$ for $z = (-1, 0)$. Show without the use of a calculator that $D(2; z) > 3$, and deduce that $\operatorname{Arg}(-1, 0) > 3$.*

An upper bound may be obtained from the decreasing sequence $H(n; z)$. From Exercise 9.3.14 we have $x_1 = 0$ for $z = (-1, 0)$, and then (9.22) yields $H(1; z) = 4$, so

Arg $(-1, 0) < 4.$[xi] Figure 9.6 gives the geometrical interpretation of the various lower and upper bounds that we have found for Arg $(-1, 0)$.

The value of Arg $(-1, 0)$, the limit of the sequence $D(n; z)$ with $z = (-1, 0)$, has been shown to be irrational, and indeed transcendental; it is a number of such importance throughout mathematics that it has a universally recognised symbol, the Greek letter π ("pi"). Its base-ten approximation to 8 digits after the point is $3 \cdot 14159265\ldots$.[xii] It is usually introduced from its geometrical significance, as the length of half a circle of radius 1 unit, but rarely with any acknowledgement that the length of a curve is a concept which needs to be carefully defined.

Since -1 is the least value of Re$[z]$ for any z on the unit circle, Theorem 9.3.11 shows that π is the greatest value that Arg z may attain. Corresponding to each number on the unit circle with Im$[z] > 0$, there exists a number $\overline{z}$ with Im$[z] < 0$ and Arg $\overline{z} = -$Arg z; hence the *infimum* of Arg z is $-\pi$ (but it has no least value). So the principal value of the argument is bounded: for all $z \in \mathbb{C}$,

$$-\pi < \text{Arg } z \leq \pi. \tag{9.25}$$

9.4 Use of the Argument to Define Real Powers in $\mathbb{C}$

We have so far been able to evaluate only integer powers and certain rational powers of complex numbers; and we have defined the principal value of the argument of a complex number, but with no practical means of evaluating it. We also do not have any means of inverting the definition of the PVA, to find a complex number given its modulus and PVA; in fact, we do not even know whether such inversion is possible in general. So we shall now develop some theory of the PVA and introduce its generalisation as the **multiple-valued argument**, which will allow us to:

- Define all real powers of complex numbers;

- Assert the existence of a complex number with any given value of modulus and argument (PVA or multiple-valued);

- Compute the PVA for a given complex number, and *vice versa*, in a limited, but important, selection of cases.

This theory will also prove vital in allowing us to define complex powers in the next section.

9.4.1 The PVA of a Product

We start by recalling some observations made in Section 9.3 about numbers on the unit circle. If $|z| = |w| = 1$, then $|zw| = 1$ and

$$|zw - z| = |w - (1, 0)|.$$

[xi]We have needed to use $H(1; z)$ because $H(0; z)$ is undefined when $x_0 = -1$ according to (9.22).

[xii]By halving Archimedes' bounds for the length along a complete circle, we find that $223/71 < \pi < 22/7$. Use your calculator to check how close these bounds are to the true value of π.

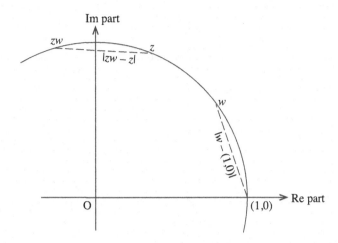

FIGURE 9.7: Points w, z, and zw on the unit circle: $|zw - z| = |w - (1,0)|$, and the length along the circle from z to zw is also equal to that from $(1,0)$ to w.

Geometrically, if z and w are on the unit circle, then so is their product zw; and the straight-line distance between z and zw is the same as that between $(1,0)$ and w: see Figure 9.7. But the length along the unit circle between two points is uniquely determined by the straight-line distance between those points: this seems geometrically obvious, but is also clear from the square-root procedure which begins with the straight-line distance and ultimately generates the length along the circle. So equality between straight-line distances implies equality between lengths along the unit circle. Now, the length along the unit circle from $(1,0)$ to zw is equal to the length from $(1,0)$ to z plus that from z to zw, where the latter has now been shown equal to that from $(1,0)$ to w. Writing these lengths as principal values of arguments, and restricting consideration to cases where all numbers involved have non-negative imaginary parts, we have:

Theorem 9.4.1. *With $|z| = |w| = 1$ and $\operatorname{Arg} z \geq 0$, $\operatorname{Arg} w \geq 0$, and $\operatorname{Arg} zw \geq 0$,*

$$\operatorname{Arg} zw = \operatorname{Arg} z + \operatorname{Arg} w.$$

We have justified this geometrically; but our intention is to use geometry only to illustrate the analysis, so we now need a rigorous proof. This proof is fairly long, but will introduce us to some important elementary properties of sequences.

Proof of Theorem 9.4.1. In the square-root procedure it was seen that if $|z| = 1$, then $|z^{1/2^n}| = 1$ for all $n \in \mathbb{N}$, so

$$\left|(zw)^{1/2^n} - z^{1/2^n}\right| = |z^{1/2^n}| \times |w^{1/2^n} - (1,0)| = |w^{1/2^n} - (1,0)|.$$

Now the definition of Arg in the form (9.24) gives

$$\operatorname{Arg} zw = \lim_{n \to \infty} 2^n \left|(zw)^{1/2^n} - (1,0)\right|.$$

Let us consider the n'th term in the sequence: by the Triangle Inequality,

$$\left|(zw)^{1/2^n} - (1,0)\right| \leq \left|z^{1/2^n} - (1,0)\right| + \left|(zw)^{1/2^n} - z^{1/2^n}\right|$$

$$= \left|z^{1/2^n} - (1,0)\right| + \left|w^{1/2^n} - (1,0)\right|. \tag{9.26}$$

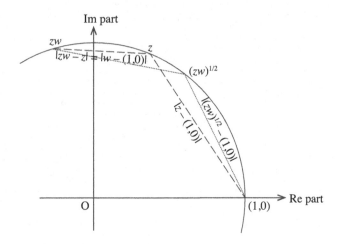

FIGURE 9.8: Lemma 9.4.2 says that the combined length of the two unequal dashed lines is less than the combined length of the two equal dotted lines. [Comparing with inequality (9.27), the annotation in the diagram is for $n = 0$.]

In the case $z = w$,

$$\left|w^{1/2^n} - (1,0)\right| = \left|z^{1/2^n} - (1,0)\right| = \left|(\sqrt{zw})^{1/2^n} - (1,0)\right|$$

so that

$$\left|z^{1/2^n} - (1,0)\right| + \left|w^{1/2^n} - (1,0)\right| = 2\left|(zw)^{1/2^{n+1}} - (1,0)\right|.$$

In fact, this case with $z = w$ yields the greatest value of the expression on the right of (9.26):

Lemma 9.4.2. *With z and w satisfying the conditions of Theorem 9.4.1,*

$$\left|z^{1/2^n} - (1,0)\right| + \left|w^{1/2^n} - (1,0)\right| \leq 2\left|(zw)^{1/2^{n+1}} - (1,0)\right|. \qquad (9.27)$$

This lemma is illustrated in Figure 9.8.

Proof of Lemma 9.4.2. Let $z^{1/2^{n+1}} = (x,y)$, so that $z^{1/2^n} = (x,y)^2$ and hence $\mathrm{Re}[z^{1/2^n}] = x^2 - y^2$. But (x,y) is on the unit circle, so $y^2 = 1 - x^2$ and hence $\mathrm{Re}[z^{1/2^n}] = 2x^2 - 1$. Similarly, let $w^{1/2^{n+1}} = (u,v)$ so that $\mathrm{Re}[w^{1/2^n}] = 2u^2 - 1$.[xiii] Also,

$$\mathrm{Re}[(zw)^{1/2^{n+1}}] = xu - yv = xu - \sqrt{(1-x^2)(1-u^2)}.$$

Now, the distance from $(1,0)$ to a complex number ζ on the unit circle is given by the formula (9.15) (see also (9.14)), which we can write as

$$|\zeta - (1,0)| = \sqrt{2 - 2\mathrm{Re}[\zeta]}.$$

Since $2 - 2\mathrm{Re}[z^{1/2^n}] = 2 - 2(2x^2 - 1) = 4 - 4x^2$, with a similar formula for the term involving w, inequality (9.27) can be written

$$\sqrt{4 - 4x^2} + \sqrt{4 - 4u^2} \leq 2\sqrt{2 - 2\left(xu - \sqrt{(1-x^2)(1-u^2)}\right)}.$$

[xiii]In this proof only, we depart from the usual notation that $(x,y) = z$ and $(u,v) = w$.

Since both sides are positive, this is true if and only if the inequality obtained by squaring both sides is true:

$$4 - 4x^2 + 4 - 4u^2 + 2\sqrt{4 - 4x^2}\sqrt{4 - 4u^2} \le 4\left(2 - 2\left(xu - \sqrt{(1-x^2)(1-u^2)}\right)\right)$$

Many terms cancel between the two sides of this inequality, and we are left with

$$4(x-u)^2 \ge 0$$

which is certainly true. Hence the original inequality (9.27) is true.   □

Returning to the proof of Theorem 9.4.1, we now multiply inequalities (9.26) and (9.27) by 2^n and take the limits of the sequences. We apparently obtain

$$\lim_{n\to\infty} 2^n \left|(zw)^{1/2^n} - (1,0)\right| \le \lim_{n\to\infty} 2^n \left|z^{1/2^n} - (1,0)\right| + \lim_{n\to\infty} 2^n \left|w^{1/2^n} - (1,0)\right|$$

$$\le \lim_{n\to\infty} 2^{n+1} \left|(zw)^{1/2^{n+1}} - (1,0)\right|,$$

which in view of the definition of PVA means that

$$\operatorname{Arg} zw \le \operatorname{Arg} z + \operatorname{Arg} w \le \operatorname{Arg} zw.$$

This can only mean that

$$\operatorname{Arg} zw = \operatorname{Arg} z + \operatorname{Arg} w, \tag{9.28}$$

as required for the theorem. However, in these last steps we made some assumptions about the behaviour of sequences, which may seem obvious, but nevertheless need proving from the definition of convergence. The assumptions were:

- The sequence with terms $2^{n+1} \left|(zw)^{1/2^{n+1}} - (1,0)\right|$ converges to the same limit as the sequence with terms $2^n \left|(zw)^{1/2^n} - (1,0)\right|$; i.e. a shift of 1 in the value of n throughout a sequence $f(n)$ does not affect its limit;

- The limit of the sum of sequences, $\left|z^{1/2^n} - (1,0)\right| + \left|w^{1/2^n} - (1,0)\right|$, is equal to the sum of the limits of the individual sequences;

- The order relation $\le$ is preserved when taking the limit of sequences.

The third assumption is verified by Theorem 9.3.12, but verifying the first two assumptions requires two further theorems; these are proved below, and then the proof of Theorem 9.4.1 will be complete.

Theorem 9.4.3 (The Shift Theorem). *If a sequence $f : \mathbb{N} \to \mathbb{R}$ converges to L, then the sequence $\tilde{f} : \mathbb{N} \to \mathbb{R}$ defined by $\tilde{f}(n) = f(n+M)$ also converges to L, with any $M \in \mathbb{Z}$.*

Proof. We need to show that for any $\epsilon > 0$, we can find $N \in \mathbb{N}$ such that $|f(n+M) - L| < \epsilon$ whenever $n > N$. We know that $|f(n) - L| < \epsilon$ whenever $n > N_0$ for some $N_0 \in \mathbb{N}$ [$f(n)$ *converges to L*].

If $M \ge 0$, then $n + M > N_0$ whenever $n > N_0$, so $|f(n+M) - L| < \epsilon$ whenever $n > N_0$: our required N is simply N_0.

If $M < 0$, let $N = N_0 - M$, so that whenever $n > N$ we have $n + M > N_0$, and hence $|f(n+M) - L| < \epsilon$.   □

A useful implication of this Shift Theorem is that if some other theorem has a condition that must be satisfied by every term in a sequence, then that theorem is still true for a sequence in which every term after the M'th term satisfies the condition, for any $M \in \mathbb{N}$.

Theorem 9.4.4. *Let $f : \mathbb{N} \to \mathbb{R}$ and $g : \mathbb{N} \to \mathbb{R}$ be convergent sequences, with $f(n) \to L$ and $g(n) \to K$ as $n \to \infty$. Then the sequence $f + g$ converges to $L + K$.*

Proof. Because f and g each converge, we can find $N_1, N_2 \in \mathbb{N}$ such that

$$|f(n) - L| < \frac{\epsilon}{2} \quad \text{when} \quad n > N_1$$

and

$$|g(n) - K| < \frac{\epsilon}{2} \quad \text{when} \quad n > N_2,$$

for any choice of $\epsilon > 0$. Then

$$
\begin{aligned}
|(f(n) + g(n)) - (L + K)| &= |(f(n) - L) + (g(n) - K)| \\
&\leq |(f(n) - L)| + |(g(n) - K)| && [\textit{Theorem 9.1.6}] \\
&< \frac{\epsilon}{2} + \frac{\epsilon}{2} \quad \text{when} \quad n > N_1 \quad \text{and} \quad n > N_2.
\end{aligned}
$$

We have shown that $|(f(n) + g(n)) - (L + K)| < \epsilon$ (for any chosen $\epsilon > 0$) when $n > N$, where $N = \max\{N_1, N_2\}$. $\qquad\qquad\square$

Theorem 9.4.1 has now been rigorously established. $\qquad\qquad\square$

We have shown that the property (9.28) of principal values of arguments is true with certain restrictions on the modulus and PVA of the complex numbers involved. Restrictions were placed on the modulus because the calculations in the proof were all done on the unit circle; but if $|z| \neq 1$ and $|w| \neq 1$, we have

$$zw = |zw| \times \left(\frac{z}{|z|} \times \frac{w}{|w|} \right),$$

and then Definition 9.3.10 shows that if (9.28) is true for the numbers $z/|z|$ and $w/|w|$ on the unit circle, it must be true for z and w with any value of modulus. Of more interest is the removal of restrictions on the arguments, first to more general values of PVA and then going beyond principal values to a broader definition of argument.

For negative values of the PVA, suppose that $\operatorname{Arg} z < 0$ and $\operatorname{Arg} w < 0$, so that $\operatorname{Arg} \bar{z} > 0$ and $\operatorname{Arg} \bar{w} > 0$ according to Definition 9.3.10. Then

$$
\begin{aligned}
\operatorname{Arg} z + \operatorname{Arg} w &= (-\operatorname{Arg} \bar{z}) + (-\operatorname{Arg} \bar{w}) && [\textit{Def. 9.3.10}] \\
&= -(\operatorname{Arg} \bar{z} + \operatorname{Arg} \bar{w}) \\
&= -\operatorname{Arg} \overline{zw} && [\textit{Theorem 9.4.1, with } \overline{zw} = \bar{z} \times \bar{w}] \\
&= \operatorname{Arg} zw,
\end{aligned}
$$

so (9.28) applies when $\operatorname{Arg} z$, $\operatorname{Arg} w$ and $\operatorname{Arg} zw$ are all negative. Next, by writing $t = zw$, (9.28) becomes $\operatorname{Arg} t = \operatorname{Arg} z + \operatorname{Arg}[t/z]$, or

$$\operatorname{Arg} \frac{t}{z} = \operatorname{Arg} t - \operatorname{Arg} z \tag{9.29}$$

(with all PVAs positive, or all negative): dividing complex numbers implies subtracting their PVAs. But using (9.3) we have

$$\operatorname{Arg} \frac{t}{z} = \operatorname{Arg} \left[\frac{1}{|z|^2} \times t\bar{z} \right] = \operatorname{Arg} t\bar{z},$$

so (9.29) can be written

$$\operatorname{Arg} t\bar{z} = \operatorname{Arg} t + \operatorname{Arg} \bar{z},$$

which implies that an equation of form (9.28) is true when one of the PVAs being added is positive and the other negative.

9.4.2   The Multiple-Valued Argument and the Definition of Real Powers

The tedious insistence on stating whether the PVAs are positive or negative in the above calculations is because it is possible to have $\operatorname{Arg} z \geq 0$ and $\operatorname{Arg} w \geq 0$ but with $\operatorname{Arg} zw < 0$, in which case (9.28) would certainly not be true. In terms of the real and imaginary parts of $z = (x, y)$ and $w = (u, v)$, this occurs when $y \geq 0, v \geq 0$ and $xv + yu < 0$ (so requires either or both $x < 0$ or $u < 0$). In geometrical terms, it occurs when $z/|z|$ and $w/|w|$ are on the upper half of the unit circle, but adding the lengths along the circle from $(1,0)$ to $z/|z|$ and to $w/|w|$ gives a length greater than π to the point representing $zw/|zw|$ on the lower half of the unit circle. Similarly, we may have $\operatorname{Arg} z < 0$ and $\operatorname{Arg} w < 0$ but with the sum of these negative lengths being $\leq -\pi$, to a point $zw/|zw|$ on the upper half of the unit circle. Since it is certainly desirable to have a theorem of the form (9.28) valid in all cases, and to preserve the identification of argument with length along the unit circle, we introduce a generalised form of argument, denoted arg (with lower-case "a"), initially letting

$$\arg zw := \operatorname{Arg} z + \operatorname{Arg} w. \tag{9.30}$$

In particular, since $\operatorname{Arg}(-1, 0) = \pi$, we have

$$\arg(-z) = \arg(z \times (-1, 0)) = \operatorname{Arg} z + \pi; \tag{9.31}$$

and since $(-1, 0) \times (-1, 0) = (1, 0)$, we have

$$\arg(1, 0) = \pi + \pi = 2\pi,$$

which corresponds to the geometrical fact that traversing a length 2π along the unit circle from $(1, 0)$ takes you back to $(1, 0)$. But so does traversing zero length: $(1, 0) = (1, 0) \times (1, 0)$ so that

$$\arg(1, 0) = \operatorname{Arg}(1, 0) + \operatorname{Arg}(1, 0) = 0 + 0 = 0.$$

Thus whereas Arg is a *function* (there is a unique value of $\operatorname{Arg} z$ corresponding to any given $z \in \mathbb{C}$), arg is a multiple-valued relation.

Since $(1, 0)$ is the multiplicative identity in $\mathbb{C}$, the equations $\arg(1, 0) = 2\pi$ and $z = z \times (1, 0)$ suggest that we should be able to write

$$\arg z = \operatorname{Arg} z + 2\pi. \tag{9.32}$$

Exercise 9.4.5. *Consider the case where* $\operatorname{Arg} z > 0$, $\operatorname{Arg} w > 0$ *and* $\operatorname{Arg} zw < 0$. *Noting that*

$$zw = (-z)(-w), \quad z = (-z) \times (-1, 0) \quad and \quad w = (-w) \times (-1, 0),$$

use (9.28) in cases that have been shown valid for principal values *of arguments to obtain*

$$\operatorname{Arg} z + \operatorname{Arg} w = \operatorname{Arg} zw + 2\pi$$

and hence deduce that

$$\arg zw = \operatorname{Arg} zw + 2\pi.$$

This exercise has shown that (9.32) is consistent with the rigorously proved Theorem 9.4.1 on principal values of arguments. We can then generalise further: since $\arg(1, 0) = 2\pi$, adding or subtracting 2π to the argument is equivalent to multiplying or dividing a complex number by $(1, 0)$ and so leaving it unchanged. So with n additions or subtractions of 2π, we have:

Definition 9.4.6. The **argument** of a complex number z, denoted $\arg z$, has countably infinitely many values,

$$\arg z = \operatorname{Arg} z + 2\pi n \quad \text{for any } n \in \mathbb{Z},$$

where $\operatorname{Arg} z$ is the principal value of its argument.

This has the geometrically pleasing property that the argument can increase or decrease continuously as you travel round the unit circle, whereas the *principal value* of the argument would jump by an amount 2π as you pass through the point $(-1, 0)$.

We can now supersede Theorem 9.4.1 and equation (9.29) with more general forms, which can be formally justified by calculations like that in Exercise 9.4.5:

Corollary 9.4.7. *For any non-zero* $z, w \in \mathbb{C}$,

$$\arg zw = \arg z + \arg w \tag{9.33}$$

and

$$\arg \frac{z}{w} = \arg z - \arg w. \tag{9.34}$$

This is stating that we can take any of the multiple values of $\arg z$ and $\arg w$, and adding or subtracting them will give one of the multiple values of $\arg zw$ or $\arg z/w$, respectively.

If we set $w = z$ in (9.33), we have $\arg z^2 = 2 \arg z$, and then by induction,

$$\arg z^q = q \arg z \ \forall \, q \in \mathbb{N}. \tag{9.35}$$

Since $\arg z^{-1} = \arg \overline{z} = (-1) \times \arg z$ and $z^{-q} = (z^{-1})^q$, equation (9.35) also applies for negative integer powers; so we have

$$\arg z^q = q \arg z \ \forall \, q \in \mathbb{Z}.$$

Next we consider arguments of square roots, returning to the definition of argument as the limit of a sequence. From equation (9.14), noting that $z^{1/2^n} = (z^{1/2})^{1/2^{n-1}}$, we obtain that $C(n; z) = C(n - 1; z^{1/2})$. Then from (9.17),

$$D(n; z) = 2D(n - 1; z^{1/2}). \tag{9.36}$$

By the Shift Theorem, the sequence $D(n-1; z^{1/2})$ converges to the same limit as $D(n; z^{1/2})$; and by Theorem 9.4.4 with $f(n) = g(n)$, the factor of 2 between the sequence terms in (9.36) carries over to the limits of those sequences. Thus

$$\operatorname{Arg} z = 2 \operatorname{Arg} z^{1/2}, \tag{9.37}$$

since $\operatorname{Arg} z$ and $\operatorname{Arg} z^{1/2}$ are the respective limits of the sequences $D(n; z)$ and $D(n; z^{1/2})$. The process can be repeated, so that

$$\operatorname{Arg} z^{1/2^n} = \frac{1}{2^n} \operatorname{Arg} z. \tag{9.38}$$

Combining this with (9.35) we have

$$\arg z^{q/2^n} = \frac{q}{2^n} \arg z, \tag{9.39}$$

which covers all powers of a complex number that we were able to define rigorously in Section 9.2.

We now have enough theory to demonstrate the existence of a unique complex number on the unit circle with any given value of PVA within the bounds (9.25). This is expressed in the following theorem:

Theorem 9.4.8. *The principal value of the argument is a* bijection *from the unit circle to the interval* $(-\pi, \pi]$.

Proof. Definition 9.3.10 provides a unique principal value of the argument for any complex number on the unit circle, so Arg is a *function*; and Theorem 9.3.11 and the definition of π show that its codomain is the interval $(-\pi, \pi]$: see (9.25). To verify that the function Arg is a bijection, we shall first show the *existence* of a number on the unit circle with any given value of PVA in the codomain (so Arg is surjective), and then its *uniqueness* (so Arg is injective).

Any real number in the interval $(-\pi, \pi]$ can be denoted $\alpha\pi$ where $-1 < \alpha \le 1$. Consider the case where $\alpha \ge 0$. For each $\alpha \in [0, 1]$ there is a sequence of base-2 approximations,

$$Q_\alpha : \mathbb{N} \to \mathbb{R}, \qquad Q_\alpha : k \mapsto \frac{q_k}{2^k},$$

that converges to α [*Theorem 7.7.3*]. In Section 9.2 we defined rational powers of the form $z^{q/2^k}$ and speculated that the sequence

$$P_\alpha : \mathbb{N} \to \mathbb{C}, \qquad P_\alpha : k \mapsto z^{q_k/2^k}$$

would converge, in which case we could define its limit as z^α. We now verify this convergence in the case where $z = (-1, 0)$.

Since $\pi = \text{Arg}\,(-1, 0)$, we have

$$\frac{q_k}{2^k}\pi = \text{Arg}\,(-1, 0)^{q_k/2^k}$$

from (9.39), with the argument being the principal value because $0 \le (q_k/2^k)\pi \le \pi$ when $0 \le \alpha \le 1$. But from Theorem 7.7.5, the sequence $Q_\alpha(k)$ is non-decreasing; and Theorem 9.3.11 shows that a non-decreasing sequence of positive PVA's, $Q_\alpha(k)\pi$, can only come from a non-increasing sequence of real parts of complex numbers on the unit circle. Real parts are bounded below by -1 on the unit circle, so by Theorem 9.3.5 the sequence of real parts converges, to some limit u. Thus the sequence of PVA's converging to $\alpha\pi$ belong to a sequence of numbers $(-1, 0)^{q_k/2^k}$ on the unit circle whose real parts converge to u. Then with $v = \sqrt{1 - u^2}$ so that (u, v) is on the unit circle, we have

$$(u, v) = (-1, 0)^\alpha := \lim_{k \to \infty} (-1, 0)^{q_k/2^k}$$

as the number on the unit circle with PVA equal to $\alpha\pi$. For the case where $-1 < \alpha < 0$ a similar proof applies, using the non-increasing sequence of negative base-2 approximations $q_k/2^k$.

To prove uniqueness, suppose that w and t are on the unit circle: we need to show that $w = t$ if and only if $\text{Arg}\,w = \text{Arg}\,t$. If $t = w$, then $\text{Arg}\,[t/w] = \text{Arg}\,(1, 0) = 0$ so from (9.29), $\text{Arg}\,w = \text{Arg}\,t$. Conversely, if $\text{Arg}\,w = \text{Arg}\,t$, then (9.29) shows that $\text{Arg}\,[t/w] = 0$ and then by Theorem 9.3.13, $t = w$. $\qquad\qquad\square$

Although arg is multiple-valued, there is a unique PVA corresponding to any real number found as an argument. Formally:

$$\text{if } n_a = \min\{n \in \mathbb{Z} : \arg z \le (2n + 1)\pi\}, \quad \text{then } \text{Arg}\,z = \arg z - 2\pi n_a.$$

Thus any real value of argument and positive real modulus uniquely define a complex number: the argument corresponds to a unique PVA, which determines a unique number on the unit circle according to Theorem 9.4.8; then multiplying by the modulus yields the required complex number. This result is so important that we should state it as a theorem:

Theorem 9.4.9. *Corresponding to any* $r \in \mathbb{R}_+$ *and* $\theta \in \mathbb{R}$ *there exists a unique complex number* z *with*

$$|z| = r \quad and \quad \arg z = \theta.$$

This theorem is required when using the following definition of real powers of complex numbers:

Definition 9.4.10. For $w \in \mathbb{C}$ and $\mu \in \mathbb{R}$, the exponentiation w^μ is defined by

$$|w^\mu| = |w|^\mu$$

and

$$\arg w^\mu = \mu \arg w. \tag{9.40}$$

This definition is clearly consistent with previous definitions of integer and rational powers: see (9.39). Definition 9.3.10 provides an explicit (but impractical!) formula for $\arg w$, which may be multiplied by μ to yield $\arg w^\mu$; but although Theorem 9.4.8 guarantees the existence of a complex number with this value of argument, we will need to wait until Theorem 9.5.8 for an explicit formula for the number. Note that we cannot use Definition 9.4.10 for complex powers, since an argument can only be real.

Equations (9.33) and (9.40) should remind you of the behaviour of logarithms. Comparing with Theorems 6.4.11 and 6.4.14 (where we have changed the notation from that used in Chapter 6):

$$\arg wz = \arg w + \arg z; \qquad \log_b wz = \log_b w + \log_b z;$$

$$\arg z^\mu = \mu \arg z; \qquad \log_b z^\mu = \mu \log_b z.$$

The correspondence between the behaviour of arguments and logarithms ensures that Definition 9.4.10 of exponentiation in terms of arguments satisfies the Laws of Indices, since the logarithm is defined as the "second inverse" of exponentiation: if $w = \log_b z$, then $z = b^w$. Furthermore, if $\theta = \arg z$, there should be some complex number b such that $z = b^\theta$ or, since $\arg z = \arg(z/|z|)$,

$$z = |z|b^\theta. \tag{9.41}$$

We will need to find the number b that fulfils this, and that calculation will be the key to defining exponentiation for all complex powers in the next section.

But first we consider the implications of there being infinitely many values of the argument for a given non-zero complex number, with values of $\arg w$ differing by an integer multiple of 2π representing the same complex number (if the moduli are the same). Let the principal value be $\mathrm{Arg}\, w = \theta$, so that other values are

$$\arg w = \theta + 2\pi n \quad \text{for } n \in \mathbb{Z}.$$

If μ is an integer, then $\mu \arg w = \mu\theta + 2\pi n\mu$, so that from (9.40) all values of $\arg w^\mu$ differ by integer multiples $(n\mu)$ of 2π: so in fact there is a unique complex number w^μ when $\mu \in \mathbb{Z}$.

Next consider the case where μ is rational, $\mu = q/k$ in lowest terms. Then

$$\arg w^\mu = \mu \arg w = \mu\theta + 2\pi \frac{nq}{k}$$

Consider two values of n which differ by an integer multiple of k: $n_2 = n_1 + mk$ for some $m \in \mathbb{Z}$, so $n_2 \equiv n_1 (\mathrm{mod}\, k)$. Then

$$2\pi \frac{n_2 q}{k} = 2\pi \frac{n_1 q}{k} + 2\pi mq :$$

the values of $\arg w^\mu$ corresponding to $n = n_1$ and $n = n_2$ differ by an integer multiple of

2π, and hence correspond to the same complex number. So there are only as many different values of $w^{q/k}$ as there are congruence classes (mod k). In particular, there are k different k'th roots of a complex number, in accord with our finding by other means of two square roots in Section 9.2.

Finally consider the case where μ is irrational. Then for two different values of n,

$$2\pi n_2 \mu = 2\pi n_1 \mu + (n_2 - n_1)\mu,$$

where $(n_2 - n_1)\mu$ cannot be an integer. So with irrational μ, there are countably infinitely many values of w^μ.

None of the above makes any progress in actually evaluating powers of complex numbers, because we still have no practical means of evaluating arguments. We now start to fill this gap in our knowledge.

9.4.3   Evaluating Rational Powers of Complex Numbers

First consider those complex numbers for which we do know the PVA. We have $\text{Arg}\,(1,0) = 0$ (trivially from Definition 9.3.10), and we have defined $\pi := \text{Arg}\,(-1,0)$. Numbers $(x,0)$ with $x > 0$, which correspond to positive real numbers, have $(x,0) = |x| \times (1,0)$, so positive reals have PVA equal to 0. If $x < 0$, then $(x,0) = |x| \times (-1,0)$, so negative reals have PVA equal to π. With this knowledge, Definition 9.4.10 has solved the problem of defining all real powers of all non-zero real numbers, with $\mathbb{R}$ considered as a subset of $\mathbb{C}$. But although defined, they have not yet been evaluated. Of particular interest are roots of real numbers, $(x,0)^{1/k}$ where $k \in \mathbb{N}$: in $\mathbb{R}$ we could find either two numbers, one number or no numbers u satisfying $u^k = x$, depending on whether x was positive or negative and whether k was even or odd; whereas we now know that in $\mathbb{C}$ we should be able to find k values of $(x,0)^{1/k}$.

Example 9.4.11 (Square roots). Using formula (9.9) to find the two square roots of positive and negative real numbers in $\mathbb{C}$,

$$(x,0)^{1/2} = (\sqrt{x},0) \ \textit{or} \ (-\sqrt{x},0)$$

and

$$(-x,0)^{1/2} = (0,\sqrt{x}) \ \textit{or} \ (0,-\sqrt{x}),$$

for any $x > 0$; in each case we have written the principal square root first. It is immediately clear why in $\mathbb{R}$ we were able to find a positive and a negative square root of a positive number, but no square roots of negative numbers; the latter are pure imaginary numbers in $\mathbb{C}$.

Now use Definition 9.4.10 with $\mu = 1/2$ as an alternative way of finding square roots. For the modulus,

$$\left|(x,0)^{1/2}\right| = |(x,0)|^{1/2} = \sqrt{x} \qquad \text{and} \qquad \left|(-x,0)^{1/2}\right| = |(-x,0)|^{1/2} = \sqrt{x},$$

and for the argument,

$$\arg(x,0) = 0 + 2\pi n, \qquad \arg(-x,0) = \pi + 2\pi n,$$

so (9.40) gives

$$\arg(x,0)^{1/2} = 0 + \pi n, \qquad \arg(-x,0)^{1/2} = \frac{\pi}{2} + \pi n.$$

Take $n = 0$ and $n = 1$ in the result for $(x,0)^{1/2}$ and compare with values of this square root obtained above: this simply yields $\arg(\sqrt{x}, 0) = 0$ and $\arg(-\sqrt{x}, 0) = \pi$ as already known, with $n = 0$ giving the principal square root. Doing a similar comparison for $(-x, 0)^{1/2}$ yields

$$\arg(0, \sqrt{x}) = \frac{\pi}{2}, \qquad \arg(0, -\sqrt{x}) = \frac{3\pi}{2}.$$

Subtracting 2π from the latter result gives the principal value of the argument, $\mathrm{Arg}\,(0, -\sqrt{x}) = -\pi/2$.

Taking square roots again can reveal values of the argument for some further complex numbers; to focus attention on the argument, we consider numbers on the unit circle:

Exercise 9.4.12. *(a) Use both the formula (9.9) and Definition 9.4.10 to take the square roots of $(0, 1)$ and $(0, -1)$; from Example 9.4.11, these will be the 4'th roots of $(-1, 0)$. Hence find the complex numbers on the unit circle with principal values of the argument equal to $\pi/4$, $3\pi/4$, $-\pi/4$ and $-3\pi/4$.*

(b) Find the principal 8'th root of $(-1, 0)$ and hence find the complex number on the unit circle with principal value of the argument equal to $\pi/8$.

The same method can be used to identify complex numbers with arguments which are other simple fractions of π.

Example 9.4.13. Find the complex numbers on the unit circle with arguments equal to $\pi/3$ and $-\pi/3$.

Solution. From (9.40), if $\arg w = \pm\pi/3$, then $\arg w^3 = \pm\pi$. Now $\pi = \mathrm{Arg}\,(-1, 0)$ and also $-\pi = \arg(-1, 0)$ (since $-\pi = \pi - 2\pi$). So we are seeking two of the cube roots of $(-1, 0)$, i.e. numbers $w \in \mathbb{C}$ such that $w^3 = (-1, 0)$. These numbers will be on the unit circle, since $|(-1, 0)^{1/3}| = |(-1, 0)|^{1/3} = 1^{1/3} = 1$.

If $w = (u, v)$, the multiplication formula (9.1.4) can be used to obtain

$$w^3 = (u^3 - 3uv^2, 3u^2v - v^3),$$

to be equated to $(-1, 0)$. Thus we need to solve the simultaneous equations,

$$u^3 - 3uv^2 = -1, \qquad 3u^2v - v^3 = 0 \tag{9.42}$$

for real numbers u, v. The second of these equations can be written

$$(3u^2 - v^2)v = 0,$$

with solutions $v = 0$ and $v^2 = 3u^2$. The case $v = 0$ yields $u^3 = -1$ and hence $u = -1$, so one of the cube roots of $(-1, 0)$ is $(-1, 0)$, with argument equal to π. But if $v^2 = 3u^2$, the first of equations (9.42) becomes $-8u^3 = -1$ and hence $u = (1/8)^{1/3} = 1/2$. Then $v^2 = 3/4$ so that $v = \pm\sqrt{3}/2$: we have the solutions

$$(u, v) = \left(\frac{1}{2}, \frac{\sqrt{3}}{2}\right) \quad \text{and} \quad (u, v) = \left(\frac{1}{2}, -\frac{\sqrt{3}}{2}\right).$$

These must be the cube roots with arguments equal to $\pm\pi/3$. Recalling that the principal value of the argument is positive or negative according to whether the imaginary part is positive or negative, we can identify

$$\mathrm{Arg}\left(\frac{1}{2}, \frac{\sqrt{3}}{2}\right) = \frac{\pi}{3} \quad \text{and} \quad \mathrm{Arg}\left(\frac{1}{2}, -\frac{\sqrt{3}}{2}\right) = -\frac{\pi}{3}. \quad \square$$

It is of interest to compare these results to the units in the Eisenstein integers. In $\mathbb{Z}[\omega_{-3}]$ numbers (a, b) were defined to behave like $a + b\omega_{-3}$, where $\omega_{-3} = (-1 + \sqrt{-3})/2$ and the only sense in which $\sqrt{-3}$ was defined was as an object whose square was -3. But in $\mathbb{C}$ we can identify the integer -3 with the complex number $(-3, 0)$, and evaluate $\sqrt{-3}$ as $(-3, 0)^{1/2}$. From Example 9.4.11, the principal square root is

$$(-3, 0)^{1/2} = (0, \sqrt{3}).$$

One of the units in $\mathbb{Z}[\omega_{-3}]$ is $(1, 1)$, which behaves like

$$1 + 1\frac{-1 + \sqrt{-3}}{2} = \frac{1 + \sqrt{-3}}{2};$$

we can now identify this with the complex number

$$\frac{(1, 0) + (0, \sqrt{3})}{2} = \left(\frac{1}{2}, \frac{\sqrt{3}}{2}\right),$$

the number which we have just found to have argument $\pi/3$.

Exercise 9.4.14. *Show that the other five units in $\mathbb{Z}[\omega_{-3}]$ (which were found in Exercise 8.6.10 to be integer powers of $(1, 1)$) can be identified with complex numbers that have PVAs equal to 0, $2\pi/3$, π, $-\pi/3$ and $-2\pi/3$. Hence the six Eisenstein units can be thought of as complex numbers spaced equally around the unit circle.*

Exercise 9.4.15. *By taking the square root of $(1/2, \sqrt{3}/2)$, found in Example 9.4.13 to have argument $\pi/3$, find complex numbers on the unit circle with arguments equal to $\pi/6$ and $\pi/12$.*

The only other cases where it is reasonably easy to find a complex number with a given argument are when the argument involves fifths of π:

Exercise 9.4.16. *Use the multiplication formula (9.1.4) to obtain a formula for $(u, v)^5$. [Hint: $(u, v)^5 = (u, v)^2 \times (u, v)^3$, and we have already obtained formulae for the square and cube of a complex number.] By equating $(u, v)^5$ to $(-1, 0)$, obtain a pair of simultaneous equations for the 5'th root of $(-1, 0)$. By solving these, find complex numbers with PVAs equal to $\pi/5$ and $3\pi/5$. Also obtain numbers with PVAs equal to $2\pi/5$ and $4\pi/5$.*

[You may find yourself needing to take the fifth root of $176 \pm 80\sqrt{5}$. This seems hopeless without a calculator (which can only give an approximate result). But if you treat these numbers as $(176, \pm 80)$ in the quadratic ring $\mathbb{Z}[\sqrt{5}]$, you will find that their norm is -1024, which is $(-4)^5$. So the fifth roots that you are looking for will be numbers with norm equal to -4 in $\mathbb{Z}[\sqrt{5}]$. It is easy to find some of those, so then you just need to take their 5'th powers to see which ones gives the required $(176, \pm 80)$.]

Even for calculating integer powers of complex numbers, which can be done by multiplication and (for negative powers) division, it is often simpler to use Definition 9.4.10.

Example 9.4.17. Evaluate $(0, -\sqrt{2})^6$.

Solution. We have $|(0, -\sqrt{2})| = \sqrt{2}$ and $\arg(0, -\sqrt{2}) = \arg(0, -1) = -\pi/2$. Hence

$$|(0, -\sqrt{2})^6| = (\sqrt{2})^6 = 2^3 = 8$$

and

$$\arg(0, -\sqrt{2})^6 = 6 \times \left(-\frac{\pi}{2}\right) = -3\pi.$$

We need the number on the unit circle with argument equal to -3π. Since $-3\pi + 2 \times 2\pi = \pi$, this is $(-1, 0)$ (which has PVA equal to π). So from (9.11) we find

$$(0, -\sqrt{2})^6 = 8 \times (-1, 0) = (-8, 0). \qquad \qquad \square$$

9.5   Exponentiation by Complex Powers; the Number e

We now have a complete theory of exponentiation of complex numbers by real powers, but have yet to make any progress in defining complex powers. We defined exponentiation by real powers in terms of modulus and argument, where the argument was derived by considering 2^n equally spaced complex numbers on the unit circle between $(1,0)$ and z. We took the limit of the sum of straight-line distances between pairs of adjacent numbers, where "straight-line distance" is defined as the modulus of the difference between two complex numbers: a real quantity. To define complex powers, we need to consider the complex numbers on the unit circle themselves. Rather than 2^n numbers, we shall consider m equally spaced numbers on the unit circle between $(1,0)$ and z, for any $m \in \mathbb{N}$; so if $\arg z = \theta$, these numbers have arguments equal to $j\theta/m$ for $j = 1, 2, \ldots, m$. For some choice of m, denote the first of these numbers (with $j = 1$) as (x_m, y_m); so $\arg(x_m, y_m) = \theta/m$ and, since $|z| = |(x_m, y_m)| = 1$, we have $z = (x_m, y_m)^m$. But our crucial observation is that when m is large, so that $\arg(x_m, y_m)$ is small and the point (x_m, y_m) is close to $(1,0)$, the point $(1, \theta/m)$ is *very* close to (x_m, y_m): see Figure 9.9. We have used some very vague language here: "large", "small", "close to", "very close to"; and all will be specified in precise mathematical terms in due course; but the key suggestion is that the difference between (x_m, y_m) and $(1, \theta/m)$ is so small that the equation $z = (x_m, y_m)^m$ can be replaced with $z = (1, \theta/m)^m$ when m becomes large enough.

But this still does not seem to help us: we still have a real power of a complex quantity. However, consider the following manipulation, in which we shall use the common notation for complex numbers, $x + yi$, rather than the ordered pair notation (x, y) which becomes rather cumbersome; recall that the arithmetic of (x, y) "behaves like" $x + yi$ where $i^2 = -1$.

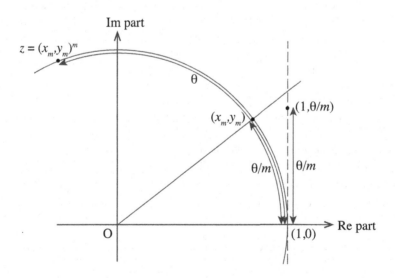

FIGURE 9.9: Part of the unit circle, with a point (x_m, y_m) with argument θ/m, and the nearby point $(1, \theta/m)$. These points are fairly close to $(1,0)$, but much closer to each other.

In particular, $(1, \theta/m)$ behaves like $1 + \theta i/m$. We make the substitution $M = m/\theta i$ so that

$$\left(1 + \frac{\theta i}{m}\right)^m = \left(1 + \frac{1}{M}\right)^{M\theta i} = \left(\left(1 + \frac{1}{M}\right)^M\right)^{\theta i}, \qquad (9.43)$$

where we have assumed that the laws of indices apply, in particular using (9.7) in the second step. Now, we have suggested that a number z on the unit circle with argument θ can be expressed, or at least approximated, as $z = (1 + \theta i/m)^m$ when m is large. In more precise language, we are claiming that z can be written as the limit of a sequence,

$$z = \lim_{m \to \infty} \left(1 + \frac{\theta i}{m}\right)^m, \quad \text{where } \theta = \arg z, \qquad (9.44)$$

a claim to be verified below. If this is true, the result of the manipulation (9.43) suggests that we can write

$$z = \left(\lim_{M \to \infty} \left(1 + \frac{1}{M}\right)^M\right)^{\theta i} \quad \text{where } \theta = \arg z. \qquad (9.45)$$

You have probably noticed the faults in this reasoning: M was defined as a complex number ($= m/\theta i$), but the limit of a sequence as in (9.45) is only defined if $M \in \mathbb{N}$; and we have not justified taking the limit process inside the exponentiation (compare (9.44) with (9.45)). However, what our (illegal!) calculations suggest is that, if the limit in (9.45) exists (with $M \in \mathbb{N}$), then this equation would be a sensible way to *define* a pure imaginary power of a particular real number (the value of the limit). So our plan of action is:

1. Prove that $\lim_{M \to \infty} \left(1 + \frac{1}{M}\right)^M$ does indeed exist; this limit will be a real number, to be denoted e.

2. Show that *real* powers of e can be represented as

$$e^x = \lim_{m \to \infty} \left(1 + \frac{x}{m}\right)^m. \qquad (9.46)$$

   This may seem unnecessary, since real powers of real numbers have already been defined in Section 7.6. However, this formula is clearly related to (9.44), upon which our proposed definition of pure imaginary powers will be based.

3. Prove that (9.44) is true, so that when we define pure imaginary powers of e by

$$e^{\theta i} := \lim_{m \to \infty} \left(1 + \frac{\theta i}{m}\right)^m,$$

   as suggested by (9.43) and (9.45), this will be consistent with our formula for real powers of e and with the laws of indices as used in (9.43).

4. Use the laws of indices and logarithms to derive any complex power of any non-zero complex number from powers of e.

The proofs in steps 1 to 3 are covered in the next subsection: they are rather long, and give some insight into the thinking required in more advanced Analysis, so they are worth detailed study; but maybe not at your first reading.

9.5.1   The Number e and Its Powers

We first prove that the limit in (9.45) exists.

Theorem 9.5.1. *The sequence* $E : \mathbb{N} \to \mathbb{R}$ *defined by*

$$E(M) := \left(1 + \frac{1}{M}\right)^{M}$$

converges, with a limit between 2 and 3.

Proof. We use Corollary 9.3.6, showing that $E(M)$ is an increasing sequence and finding a related but decreasing sequence with terms greater than those of $E(M)$. First, $E(M)$ is increasing if $E(S(M)) > E(M)$ for all $M \in \mathbb{N}$. So we want $E(S(M))/E(M) > 1$. Now,

$$
\begin{aligned}
\frac{E(S(M))}{E(M)} &= \frac{\left(1 + \frac{1}{M+1}\right)^{M+1}}{\left(1 + \frac{1}{M}\right)^{M}} \\
&= \frac{\left(\frac{M+2}{M+1}\right)^{M+1}}{\left(\frac{M+1}{M}\right)^{M}} \\
&= \frac{M+1}{M}\left(\frac{M(M+2)}{(M+1)^2}\right)^{M+1} \\
&= \frac{M+1}{M}\left(1 - \frac{1}{(M+1)^2}\right)^{M+1}.
\end{aligned}
$$

Now from (6.1), the strict form of Bernoulli's inequality, $(1+x)^n > 1+nx$ when $-1 < x < 0$. Since $-1 < -1/(M+1)^2 < 0$ for all $M \in \mathbb{N}$, this gives us

$$\left(1 - \frac{1}{(M+1)^2}\right)^{M+1} > 1 - \frac{M+1}{(M+1)^2} = \frac{M}{M+1}.$$

So we now have

$$\frac{E(S(M))}{E(M)} > \frac{M+1}{M}\,\frac{M}{M+1} = 1,$$

so $E(S(M)) > E(M)\ \forall M \in \mathbb{N}$ as required.

Next consider the sequence $E_+ : \mathbb{N} \to \mathbb{R}$ defined by

$$E_+(M) := \left(1 + \frac{1}{M}\right)^{M+1}.$$

Since $E_+(M) = (1 + 1/M)E(M)$, it is clear that $E_+(M) > E(M)\ \forall M \in \mathbb{N}$.

Exercise 9.5.2. *Show that $E_+(M)$ is a decreasing sequence. [To do this in such a way that you can use Bernoulli's inequality again, you need to show that $E_+(M)/E_+(S(M)) > 1$.]*

With $E(M)$ increasing (and hence non-decreasing), $E_+(M)$ decreasing (so non-increasing), and $E_+(M) > E(M)$, we have satisfied the conditions of Corollary 9.3.6, so $E(M)$ certainly converges. Since $E(1) = 2$ and $E_+(1) = 4$, the limit is clearly between 2 and 4. Numerical calculation of further terms in the two sequences will narrow down the limit further; in particular, $E_+(6) < 3$, so the limit must be less than 3.   □

In fact, the limit of the sequence $E(M)$ is a number of similar importance to π in mathematics, particularly in Analysis; like π, it has been proved to be transcendental, and has a universally recognised symbol:

Definition 9.5.3. The number e is defined to be the limit of the sequence E:

$$e := \lim_{M \to \infty} \left(1 + \frac{1}{M}\right)^M.$$

Its base-ten approximation to 9 digits after the point is $2 \cdot 718281828 \ldots$. Do not imagine that it has a repeating expansion in base ten: the repetition of the four digits 1828 is well known as an aid to memorising this approximation to e, but the next few digits are $459045 \ldots$.[xiv]

Our next step concerns real powers of the number e:

Theorem 9.5.4. *For any* $x \in \mathbb{R}_+$,

$$\lim_{m \to \infty} \left(1 + \frac{x}{m}\right)^m = e^x.$$

Recall the manipulations leading to equation (9.45), in which two steps were unjustified: taking the limit process inside the exponentiation, and substituting a complex number for a natural number in the variable that "tends to infinity". Our proof of Theorem 9.5.4 will proceed by showing that both of these steps are rigorously justified when θi is replaced with the real number x. For the second step, which now involves the substitution of a real number for a natural number, we need a lemma:

Lemma 9.5.5. *The expression* $(1 + x/m)^m$ *is an increasing function of* $m \in \mathbb{R}_+$ *for any* $x \in \mathbb{R}_+$; *i.e. if* $m_2 > m_1$ *with* $m_1, m_2 \in \mathbb{R}_+$, *then*

$$\left(1 + \frac{x}{m_2}\right)^{m_2} > \left(1 + \frac{x}{m_1}\right)^{m_1}.$$

Proof. We start by finding an increasing sequence in a natural number variable M:

Exercise 9.5.6. *By a similar calculation to that used to show that* $(1+1/M)^M$ *is increasing in the proof of Theorem 9.5.1, show that the sequence* $E_u : \mathbb{N} \to \mathbb{R}$ *defined by*

$$E_u : M \mapsto \left(1 + \frac{u}{M}\right)^M$$

is increasing for any real $u > 0$.

It then follows by induction that

$$\left(1 + \frac{u}{M_2}\right)^{M_2} > \left(1 + \frac{u}{M_1}\right)^{M_1} \qquad \text{whenever } M_2 > M_1 \quad (M_1, M_2 \in \mathbb{N}). \qquad (9.47)$$

Now let $m_2 = M_2/q$, $m_1 = M_1/q$ and $x = u/q$ for some $q \in \mathbb{N}$, so $m_1, m_2 \in \mathbb{Q}_+$ and $x \in \mathbb{R}_+$. Taking q'th roots on each side of (9.47), where Corollary 7.6.7 confirms that this preserves the ordering, we obtain

$$\left(1 + \frac{x}{m_2}\right)^{m_2} > \left(1 + \frac{x}{m_1}\right)^{m_1} \qquad \text{whenever } m_2 > m_1 > 0. \qquad (9.48)$$

[xiv]The numbers π and e, which are both of fundamental importance in mathematics, have both been defined as limits of increasing sequences, bounded above by decreasing sequences: curiously, the initial term of the increasing sequence was 2 in both cases, and the initial term of the decreasing sequence was 4 in both cases.

But because any two rational numbers can be put over a common denominator q, this result applies for *any* rational m_1, m_2 with $m_2 > m_1 > 0$. Then with real powers defined as suprema of sets of rational powers (see Section 7.6), the ordering (9.48) also applies with $m_1, m_2 \in \mathbb{R}_+$. $\qquad\qquad\qquad\qquad\qquad\qquad\qquad\qquad\qquad\qquad\qquad\qquad\qquad\quad\square$

Proof of Theorem 9.5.4. We need to show that for any $\epsilon > 0$, we can find $N \in \mathbb{N}$ such that
$$\left| e^x - \left(1 + \frac{x}{m}\right)^m \right| < \epsilon \qquad\qquad\qquad (9.49)$$
whenever $m > N$ with $m \in \mathbb{N}$.

From the proof of Theorem 9.5.1, $(1 + 1/M)^M$ is a positive and increasing sequence converging to e. Thus if we define
$$\delta_M := e - \left(1 + \frac{1}{M}\right)^M,$$
then:
(a) $0 \leq \delta_M/e < 1$ for all $M \in \mathbb{N}$;
(b) for any $\alpha > 0$, there exists $N_M \in \mathbb{N}$ such that $0 \leq \delta_M < \alpha$ whenever $M > N_M$.
    Then
$$e^x - \left(1 + \frac{1}{M}\right)^{Mx} = e^x - (e - \delta_M)^x$$
$$= e^x - e^x \left(1 - \frac{\delta_M}{e}\right)^x. \qquad\qquad (9.50)$$

Now, for any $x \in \mathbb{R}$, we can find $n \in \mathbb{N}$ such that $n > x$, and then
$$\left(1 - \frac{\delta_M}{e}\right)^x > \left(1 - \frac{\delta_M}{e}\right)^n \geq 1 - \frac{n\delta_M}{e}$$
where property (a) above has allowed us to use Theorem 7.6.15(c) for the first ordering and Bernoulli's inequality for the second.

Hence (9.50) yields
$$e^x - \left(1 + \frac{1}{M}\right)^{Mx} < e^x - e^x \left(1 - \frac{n\delta_M}{e}\right) = e^{x-1}n\delta_M.$$

Property (b) above then allows us to find $N_M \in \mathbb{N}$ such that
$$0 \leq \delta_M < \frac{\epsilon}{e^{x-1}n} \qquad \text{whenever} \quad M > N_M,$$
so that
$$0 \leq e^x - \left(1 + \frac{1}{M}\right)^{Mx} < \epsilon$$
with $M \in \mathbb{N}$. This has shown that $(1 + 1/M)^{Mx}$ is an increasing sequence that converges to e^x.

With the substitution $M = m/x$, we have
$$0 \leq e^x - \left(1 + \frac{x}{m}\right)^m < \epsilon \qquad\qquad\qquad (9.51)$$
whenever $m = Mx$ where $M \in \mathbb{N}$ with $M > N_M$. Lemma 9.5.5 then ensures that (9.51) is true for every real number $m > N_M x$ (not just natural number multiples of x), and in particular when $m \in \mathbb{N}$ with $m > N$ where N is a natural number greater than $N_M x$. This fulfils the requirement (9.49). $\qquad\qquad\qquad\qquad\qquad\qquad\qquad\qquad\qquad\quad\square$

Exercise 9.5.7. *Given that Theorem 9.5.4 is true for $x > 0$, use the following procedure to verify that it is also true for $x < 0$.*

With $x < 0$, let $u = -x > 0$, so that $e^x = 1/e^u$: according to the definition of limit of a sequence, we need to find $N \in \mathbb{N}$ such that

$$\left| \frac{1}{e^u} - \left(1 - \frac{u}{m}\right)^m \right| < \epsilon$$

whenever $m > N$, for any choice of $\epsilon > 0$. This will be achieved (according to the triangle inequality, which certainly applies in $\mathbb{R}$ if it applies in $\mathbb{C}$) if we can make

$$\left| \frac{1}{e^u} - \frac{1}{\left(1 + \frac{u}{m}\right)^m} \right| < \frac{\epsilon}{2} \tag{9.52}$$

and also

$$\left| \frac{1}{\left(1 + \frac{u}{m}\right)^m} - \left(1 - \frac{u}{m}\right)^m \right| < \frac{\epsilon}{2}. \tag{9.53}$$

First show that

$$\left| \frac{1}{a} - \frac{1}{b} \right| < a - b$$

when $a > b > 1$. Hence, given that $(1 + u/m)^m$ converges to e^u, show that inequality (9.52) can be satisfied. Evaluate the quantity in the modulus signs in (9.53), and show that this inequality can be achieved with sufficiently large m. [Hint: you can use Bernoulli's inequality on $(1 - u^2/m^2)^m$ when $m > u$.]

We can now proceed to verifying the important relation (9.44) between a complex number and its argument:

Theorem 9.5.8. *A complex number z with $|z| = 1$ is related to its argument θ by*

$$z = \lim_{m \to \infty} \left(1 + \frac{\theta i}{m}\right)^m,$$

where $\theta = \arg z$ as given by Definitions 9.4.6 and 9.3.10.

Proof. We first give some intuitive ideas of why the theorem is true, referring to the discussion in the first paragraph of Section 9.5 and to Figure 9.9, before launching into the rigorous proof. We shall revert to the ordered-pair notation for complex numbers.

Since $z = (x_m, y_m)^m$, the theorem requires that as m increases, the numbers $(1, \theta/m)^m$ and $(x_m, y_m)^m$ become closer to each other. Now, whereas on Figure 9.9 the points $(1, \theta/m)$ and (x_m, y_m) are indeed close to each other, we could reasonably expect taking m'th powers to *magnify* their distance apart by at least a factor of m. If, as m increases, with the points $(1, \theta/m)$ and (x_m, y_m) moving towards $(1, 0)$, their distance apart simply decreases proportionally to $1/m$, this would just cancel out the magnification factor m and would not bring the m'th powers closer together. However, because of the curvature of the circle, the distance between $(1, \theta/m)$ and (x_m, y_m) reduces faster than $1/m$ – in fact proportionally to $1/m^2$ – so that their m'th powers will actually come closer together as m increases: the distance between $(1, \theta/m)^m$ and $(x_m, y_m)^m$ will be proportional to $1/m$.

We start the proof by noting the vital, if somewhat hidden, importance of Theorem 9.4.8, which asserts the existence of a complex number on the unit circle with any given value of PVA (and hence any given value of argument). So, given z on the unit circle with argument θ, there does exist a complex number on the unit circle with argument θ/m (for any $m \in \mathbb{N}$), and we are denoting that number as (x_m, y_m). So $(x_m, y_m)^m = z$, and the definition of limit of a sequence requires us to find $N \in \mathbb{N}$ such that

$$|(1, \theta/m)^m - (x_m, y_m)^m| < \epsilon$$

whenever $m > N$, for any chosen $\epsilon > 0$.

We first relate the distance between $(1, \theta/m)^m$ and $(x_m, y_m)^m$ to that between $(1, \theta/m)$ and (x_m, y_m). Let w and t be any two non-zero complex numbers. Then

$$w^m - t^m = (w - t) \sum_{j=1}^{m} w^{m-j} t^{j-1}, \tag{9.54}$$

which can be proved by induction; a simpler form of this identity appears in Investigation 6 of Chapter 5.

Exercise 9.5.9. *Verify the identity (9.54) by induction on m.*

The distance between w^m and t^m is then

$$
\begin{aligned}
|w^m - t^m| &= |w - t| \times \left| \sum_{j=1}^{m} w^{m-j} t^{j-1} \right| \\
&\leq |w - t| \times \sum_{j=1}^{m} |w^{m-j} t^{j-1}| \qquad\qquad [\textit{Exercise 9.1.7}] \\
&= |w - t| \times \sum_{j=1}^{m} |w|^{m-j} |t|^{j-1}. \tag{9.55}
\end{aligned}
$$

If $|t| = 1$ and $|w| \geq 1$, then the largest of the m terms in the sum in the last line is $|w|^{m-1}$ (the term with $j = 1$), so we have

$$|w^m - t^m| \leq |w - t| \times m|w|^{m-1}. \tag{9.56}$$

For our present purposes we have $t = (x_m, y_m)$ so that $|t| = 1$, and $w = (1, \theta/m)$ so that $|w| = \sqrt{1 + \theta^2/m^2} > 1$. But

$$\sqrt{1 + \frac{\theta^2}{m^2}} \leq \sqrt{1 + 2\frac{|\theta|}{m} + \frac{\theta^2}{m^2}} = 1 + \frac{|\theta|}{m},$$

so

$$|w|^{m-1} \leq \left(1 + \frac{|\theta|}{m}\right)^{m-1} < \left(1 + \frac{|\theta|}{m}\right)^{m}.$$

But from (9.51) in the proof of Theorem 9.5.4, $(1 + |\theta|/m)^m < e^{|\theta|}$. So $|w|^{m-1} < e^{|\theta|}$ and inequality (9.56) now becomes

$$|(1, \theta/m)^m - (x_m, y_m)^m| < |(1, \theta/m) - (x_m, y_m)| \times m e^{|\theta|}. \tag{9.57}$$

Next we need to consider the distance between $(1, \theta/m)$ and (x_m, y_m), which is

$$|(1, \theta/m) - (x_m, y_m)| = \sqrt{(1 - x_m)^2 + \left(\frac{\theta}{m} - y_m\right)^2}. \tag{9.58}$$

Although θ may be any real number, for $m > |\theta|/\pi$ the argument θ/m will be within the bounds for a PVA. So from equations (9.15) and (9.21) for the initial terms in the increasing and decreasing sequences that converge to the PVA (see Figure 9.5, with (x_m, y_m) at the point D), we have

$$\sqrt{2 - 2x_m} < \frac{\theta}{m} < 2\sqrt{\frac{1 - x_m}{1 + x_m}}. \tag{9.59}$$

Now

$$y_m = \sqrt{1 - x_m^2} = \sqrt{(1 - x_m)(1 + x_m)} \le \sqrt{2 - 2x_m}$$

since $x_m \le 1$ on the unit circle, so from (9.59), $\theta/m > y_m$ and

$$\frac{\theta}{m} - y_m < 2\sqrt{\frac{1 - x_m}{1 + x_m}} - \sqrt{(1 - x_m)(1 + x_m)}$$

$$= \sqrt{\frac{1 - x_m}{1 + x_m}}(2 - (1 + x_m))$$

$$= \sqrt{\frac{(1 - x_m)^3}{1 + x_m}}.$$

Thus from (9.58),

$$|(1, \theta/m) - (x_m, y_m)| < \sqrt{(1 - x_m)^2 + \frac{(1 - x_m)^3}{1 + x_m}}$$

$$= (1 - x_m)\sqrt{\frac{2}{1 + x_m}}.$$

From the left inequality in (9.59),

$$1 - x_m < \frac{\theta^2}{2m^2},$$

and for $m > 2|\theta|/\pi$ we have $|\text{Arg}(x_m, y_m)| \le \pi/2$ so that $x_m \ge 0$ and hence

$$\sqrt{\frac{2}{1 + x_m}} \le \sqrt{2}.$$

So

$$|(1, \theta/m) - (x_m, y_m)| < \frac{\theta^2}{\sqrt{2}m^2}$$

and then (9.57) gives

$$|(1, \theta/m)^m - (x_m, y_m)^m| < \frac{1}{\sqrt{2}m}\theta^2 e^{|\theta|}.$$

We have confirmed our intuition that the distance between $(1, \theta/m)$ and (x_m, y_m) is proportional to $1/m^2$, so that the distance between the point z with argument θ and the point obtained by taking the m'th power of $(1, \theta/m)$ reduces proportionally to $1/m$ as m increases. More precisely, by making

$$m > \epsilon\theta^2 e^{|\theta|}/\sqrt{2} \tag{9.60}$$

we can make

$$|(1, \theta/m)^m - (x_m, y_m)^m| < \epsilon.$$

In the course of the proof we have also set the condition $m > |\theta|/\pi$ and the stricter condition $m > 2|\theta|/\pi$. So by setting

$$N > \max\left\{\frac{\epsilon\theta^2 e^{|\theta|}}{\sqrt{2}}, \frac{2|\theta|}{\pi}\right\},$$

all the required conditions are satisfied when $m > N$, for any $\theta \in \mathbb{R}$ and $\epsilon > 0$. Hence the limit claimed in Theorem 9.5.8. is verified. □

Now return to equation (9.45) which, according to Definition 9.5.3, can be written $z = e^{\theta i}$ (or $z = e^{(0,\theta)}$ in ordered-pair notation).[xv] In view of the discussion that followed (9.45), Theorem 9.5.8 justifies using this equation as a definition:

Definition 9.5.10. Pure imaginary powers of the number e are given by

$$e^{\theta i} := \lim_{m \to \infty} \left(1 + \frac{\theta i}{m}\right)^m = z,$$

where z is the unique complex number having

$$|z| = 1 \quad \text{and} \quad \arg z = \theta.$$

The argument is given in terms of its principal value by $\arg z = \operatorname{Arg} z + 2\pi n$ for $n \in \mathbb{Z}$, so Definition 9.5.10 can be written as

$$z = e^{(\operatorname{Arg} z + 2\pi n)i} \quad \text{when } |z| = 1.$$

Definition 9.3.10 then shows that for complex numbers with any non-zero modulus, a full characterisation in terms of modulus and argument is

$$z = |z|e^{(\operatorname{Arg} z + 2\pi n)i} \quad \text{for } n \in \mathbb{Z}. \tag{9.61}$$

After defining the argument we found that it behaves like a logarithm to some base b; comparing with (9.41), we now see that the base is e^i, since $e^{\theta i} = (e^i)^\theta$. Just as when defining real powers in terms of the argument, the logarithmic behaviour of the argument ensures that Definition 9.5.10 satisfies the Laws of Indices. We can also use rule (9.5) to combine real and pure imaginary powers and so define general complex powers of e: since z behaves like $x + yi$ where $x = \operatorname{Re}[z]$ and $y = \operatorname{Im}[z]$,

$$e^z = e^{x+yi} = e^x \times e^{yi},$$

where the real power e^x and the pure imaginary power e^{yi} have already been defined.[xvi] Now, $|e^{yi}| = 1$ (see Theorem 9.5.8 and Definition 9.5.10), so

$$|e^z| = |e^x| \times |e^{yi}| = e^x = e^{\operatorname{Re}[z]}, \tag{9.62}$$

and

$$\arg e^z = \arg \left[\frac{e^z}{|e^z|}\right] = \arg e^{yi} = y = \operatorname{Im}[z]. \tag{9.63}$$

Powers of e are of such importance that they are defined as one of the **elementary functions** of Analysis:

[xv] It is when complex numbers are exponents that the ordered-pair notation is particularly cumbersome. Hence the $x + yi$ notation will be used in the remainder of this section.

[xvi] More pedantically, one can define

$$e^z := \lim_{m \to \infty} \left(1 + \frac{z}{m}\right)^m$$

and then *prove* that this is equal to the product of our earlier formulae for e^x and e^{yi}. A hint as to why this works is that

$$\left(1 + \frac{x}{m}\right)^m \times \left(1 + \frac{yi}{m}\right)^m = \left(1 + \frac{x + yi}{m} + \frac{xyi}{m^2}\right)^m$$

and as m increases, the term xyi/m^2 decreases faster than the other terms, leaving $(1 + z/m)^m$.

Definition 9.5.11. The **exponential function**, denoted exp, is defined on $\mathbb{R}$ and on $\mathbb{C}$:

$$\exp : \mathbb{R} \to \mathbb{R}, \quad \exp(x) := e^x;$$

$$\exp : \mathbb{C} \to \mathbb{C}, \quad \exp(z) := e^z.$$

This definition may appear to be doing nothing more than introducing a new notation for the exponentiation of a particular base, e; but it contrasts with the more usual procedure, which is to define the function exp by other means (which requires a substantial body of knowledge in Real or Complex Analysis) and then to prove that this function obeys the laws of indices and so can be written as the exponentiation of a number e.

9.5.2   General Exponentiation and Logarithms in $\mathbb{C}$

Defining complex powers of a single real number, e, is the crucial breakthrough that will allow us to define all complex powers of all non-zero complex numbers. The next step in this process is to consider logarithms to base e, which are known as **natural logarithms** because they arise naturally in many contexts in Analysis. The notation ln is often used in preference to $\log_e$ (and some authors simply write log when they mean a natural logarithm); but we shall write $\log_e$ for logarithms to base e in $\mathbb{R}$, and reserve the notation ln for logarithms to base e in $\mathbb{C}$. By definition, logarithms are the second inverse to exponentiation:

$$y = e^x \Leftrightarrow x = \log_e y, \qquad w = e^z \Leftrightarrow z = \ln w;$$

but we may also think of natural logarithms as the inverse function of exp:

$$y = \exp x \Leftrightarrow x = \log_e y, \qquad w = \exp z \Leftrightarrow z = \ln w.$$

Setting $z = \ln w$ in (9.62) and (9.63), we then have

$$\mathrm{Re}[\ln w] = \log_e |w|$$

and

$$\mathrm{Im}[\ln w] = \arg w.$$

Thus we can write the natural logarithm of a complex number in ordered-pair notation as

$$\ln w = \left(\log_e |w|, \arg w\right),$$

which therefore has countably infinitely many values, corresponding to the values of $\arg w$. By taking the principal value of the argument, we obtain the **principal value of the logarithm**,

$$\mathrm{Ln}\, w = \left(\log_e |w|, \mathrm{Arg}\, w\right).$$

Now that we have defined natural logarithms and complex powers of e, we can use Theorem 6.4.16 to define all complex powers of all non-zero complex numbers:

Definition 9.5.12. For $z \in \mathbb{C}$ and $w \in \mathbb{C}\backslash\{(0,0)\}$,

$$w^z := e^{z \ln w}.$$

Exercise 9.5.13. *Show that if $z = (x, y)$, then*

$$|w^z| = |w|^x e^{-y \arg w} \quad and \quad \arg w^z = y \log_e |w| + x \arg w.$$

Definition 9.5.12 will certainly obey the Laws of Indices, since powers of e do so and Theorem 6.4.16 is derived from those laws. In general it yields countably infinitely many values of w^z, including a principal value, $e^{z \operatorname{Ln} w}$. In the case where $z = (x, 0)$, so that the power corresponds to the real number x, it is consistent with Definition 9.4.10 for real powers: from Exercise 9.5.13 with $y = 0$, we obtain

$$|w^z| = |w|^x \quad \text{and} \quad \arg w^z = x \arg w.$$

Exercise 9.5.14. *Find all the values of i^i (recalling that the symbol i is shorthand for the complex number $(0, 1)$).*

It is useful at this stage to summarise how the arithmetic operations may be computed in $\mathbb{C}$:

Addition can only be done in ordered-pair form, using Definition 9.1.3. Similarly for subtraction, $(x, y) - (u, v) = (x - u, y - v)$.

Multiplication can be done in ordered-pair form, using Definition 9.1.4, or it can be done in terms of the modulus and argument:

$$|z \times w| = |z| \times |w| \quad \text{and} \quad \arg(z \times w) = \arg z + \arg w.$$

Division can be done in ordered-pair form using formula (9.2), or in terms of modulus and argument by

$$|z \div w| = |z| \div |w| \quad \text{and} \quad \arg(z \div w) = \arg z - \arg w$$

(since division is the inverse operation of multiplication).

Exponentiation w^z for non-integer z requires the ordered-pair form of z and the modulus and argument of w, and can only yield the result in terms of modulus and argument, as shown in Exercise 9.5.13.[xvii] The principal value of w^z is obtained by using the principal value of $\arg w$ in these formulae, while adding integer multiples of 2π to $\operatorname{Arg} w$ yields finitely many values of w^z if z is rational, and countably infinitely many values if z is irrational or complex.

Logarithms to any complex base could in principle be calculated: $\log_w t = z$ if $t = w^z$; but they are seldom of any use.

9.5.3   Trigonometric Functions

The difficulty with using modulus and argument for calculations in $\mathbb{C}$ is that we have no elementary means of calculating the argument of a complex number given in ordered-pair form, or of finding the ordered-pair form given the modulus and argument, except in the special cases covered in Subsection 9.4.3. The transition between ordered-pair and modulus-argument forms is usually defined in terms of **trigonometric functions**, and we are used to doing the evaluations with a touch of a button on an electronic calculator; but that is only possible because the usefulness of these functions has led to sophisticated software being developed to allow them to be calculated quickly. The functions are usually defined in relation to geometry, or in more advanced texts using concepts of Analysis; but we shall define them in terms of what we have previously defined for complex numbers:

[xvii]Of course square roots, and indeed 2^n'th roots, can be calculated using formulae (9.9), but a modulus-argument calculation is often simpler. Integer powers are defined in terms of multiplication and division, but again the modulus-argument method is usually simpler.

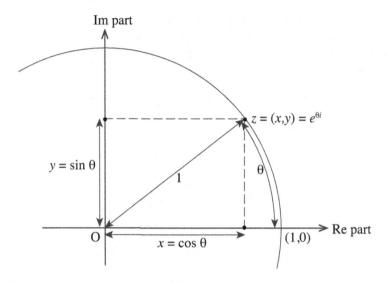

FIGURE 9.10: A complex number z on the unit circle is related to its argument θ by $z = e^{\theta i}$, and its real and imaginary parts are $\cos \theta$ and $\sin \theta$, respectively.

Definition 9.5.15. The **cosine** of a real number θ is the real part of $e^{\theta i}$:

$$\cos \theta := \operatorname{Re}\left[e^{\theta i}\right].$$

The **sine** of a real number θ is the imaginary part of $e^{\theta i}$:

$$\sin \theta := \operatorname{Im}\left[e^{\theta i}\right].$$

Hence we can write

$$e^{\theta i} = \cos \theta + i \sin \theta,$$

and (9.61) becomes

$$z = |z|\left(\cos(\arg z) + i \sin(\arg z)\right)$$

so that

$$\operatorname{Re}[z] = |z| \cos(\arg z), \qquad \operatorname{Im}[z] = |z| \sin(\arg z).$$

Figure 9.10 confirms that Definition 9.5.15 accords with the usual geometric definitions of cosine and sine.

An important property of the cosine and sine functions derives from the fact that adding any integer multiple of 2π to the argument θ leaves the value of the complex number $z = e^{\theta i}$ unchanged:

Definition 9.5.16. A function f is **periodic** with period equal to p if

$$f(x) = f(x + np) \ \forall \, n \in \mathbb{Z}.$$

Thus $e^{\theta i}$, and hence $\cos \theta$ and $\sin \theta$, are periodic functions of θ, with period equal to 2π:

$$\cos(\theta + 2\pi n) = \cos \theta, \qquad \sin(\theta + 2\pi n) = \sin \theta \qquad \text{for all } n \in \mathbb{Z}.$$

By definition, the norm of a complex number is $\mathcal{N}(z) = (\operatorname{Re}[z])^2 + (\operatorname{Im}[z])^2$; but $\mathcal{N}\left(e^{\theta i}\right) = 1$, so from Definition 9.5.15,

$$(\cos \theta)^2 + (\sin \theta)^2 = 1 \ \forall \, \theta \in \mathbb{R}. \tag{9.64}$$

Figure 9.10 shows that this identity is just an expression of Pythagoras' Theorem.

Further useful formulae involving cosine and sine can be derived from the properties of complex numbers. Given $z = (x, y)$ and $w = (u, v)$ with $|z| = 1$ and $|w| = 1$, let $\theta = \arg z$ and $\phi = \arg w$ so that

$$x = \cos\theta, \quad y = \sin\theta, \quad u = \cos\phi, \quad v = \sin\phi. \tag{9.65}$$

Now, $|zw| = |z| \times |w| = 1$, so

$$\mathrm{Re}[zw] = \cos(\arg zw) = \cos(\arg z + \arg w) = \cos(\theta + \phi)$$

and similarly

$$\mathrm{Im}[zw] = \sin(\theta + \phi).$$

But from Definition 9.1.4 and the representations (9.65),

$$\mathrm{Re}[zw] = xu - yv = \cos\theta\cos\phi - \sin\theta\sin\phi$$

and

$$\mathrm{Im}[zw] = xv + yu = \cos\theta\sin\phi + \sin\theta\cos\phi.$$

We have shown that

$$\cos(\theta + \phi) = \cos\theta\cos\phi - \sin\theta\sin\phi$$

and

$$\sin(\theta + \phi) = \cos\theta\sin\phi + \sin\theta\cos\phi,$$

a pair of well-known trigonometric formulae, which can also be proved by purely geometric methods.

We can also find formulae for $\cos m\theta$ and $\sin m\theta$, which are the real and imaginary parts of $e^{m\theta i}$ for any $m \in \mathbb{N}$, by noting that

$$e^{m\theta i} = \left(e^{\theta i}\right)^m$$

so that in ordered-pair form,

$$(\cos m\theta, \sin m\theta) = (\cos\theta, \sin\theta)^m,$$

which is called **De Moivre's Theorem**. Evaluating the m'th power on the right by multiplication in ordered-pair form, the real and imaginary parts will provide formulae for $\cos m\theta$ and $\sin m\theta$ in terms of $\cos\theta$ and $\sin\theta$.

Exercise 9.5.17. *Find formulae for* $\cos 2\theta, \sin 2\theta, \cos 3\theta, \sin 3\theta, \cos 4\theta, \sin 4\theta$ *in terms of* $\cos\theta$ *and* $\sin\theta$.

We can define the cosine and sine of any complex number, but not directly from Definition 9.5.15. If $\arg z = \theta$, then $\arg \overline{z} = -\theta$ and with $|z| = 1$, Definition 9.5.10 gives $z = e^{\theta i}$ and $\overline{z} = e^{-\theta i}$. So, using (9.1),

$$e^{\theta i} + e^{-\theta i} = z + \overline{z} = 2\mathrm{Re}[z] = 2\cos\theta$$

and

$$e^{\theta i} - e^{-\theta i} = z - \overline{z} = 2i\mathrm{Im}[z] = 2i\sin\theta.$$

By replacing the real number θ with a general complex number w in these identities, we obtain:

Definition 9.5.18. For any $w \in \mathbb{C}$:

$$\cos w = \frac{1}{2}\left(e^{wi} + e^{-wi}\right), \qquad \sin w = \frac{1}{2i}\left(e^{wi} - e^{-wi}\right).$$

From its derivation, this is obviously consistent with the earlier definition of cosine and sine of real numbers, and its greater generality is because it does not involve taking real and imaginary parts. These trigonometric functions of complex numbers do not retain the simple geometrical interpretation that the functions in $\mathbb{R}$ have, but they do retain many of the algebraic properties.

Exercise 9.5.19. *Prove that addition formulae similar to those for $\cos(\theta+\phi)$ and $\sin(\theta+\phi)$ do apply to cosines and sines of complex numbers:*

$$\cos(z + w) = \cos z \cos w - \sin z \sin w,$$

$$\sin(z + w) = \sin z \cos w + \cos z \sin w.$$

9.6   The Fundamental Theorem of Algebra

A major theme running through our survey of number systems has been the binary operations. Addition and multiplication have been defined, obeying commutative, associative, and distributive laws in all our number systems. In contrast, exponentiation was defined in $\mathbb{N}$ and found to obey the Laws of Indices, but we were unable to define exponentiation as closed and obeying these laws on any other number system until we defined $\mathbb{C}$.[xviii] However, the requirement for a number system in which exponentiation is closed is not the motivation that is usually quoted for defining the complex numbers; rather, it is the need to solve algebraic equations, as discussed in Section 7.9 of this book. Obviously the two issues are connected: for example, the exponentiation $(-1)^{1/2}$ gives the solutions of the algebraic equation $z^2 + 1 = 0$; and one of the remarkable features of $\mathbb{C}$ is that it does completely resolve both issues. In fact, it goes beyond solving algebraic equations, which we defined as involving polynomials with *integer* coefficients:

Theorem 9.6.1 (The Fundamental Theorem of Algebra). *Every polynomial equation,*

$$a_0 + a_1 z + a_2 z^2 + \ldots + a_n z^n = 0,$$

with $n \in \mathbb{N}$ and $a_0, a_1, a_2, \ldots, a_n \in \mathbb{C}$, has a solution for z in $\mathbb{C}$.

So the coefficients in the polynomial may be complex numbers, which provides a nice closure: any polynomial equation with coefficients in $\mathbb{C}$ has a solution in $\mathbb{C}$.

The truth of the Fundamental Theorem of Algebra [FTA] for equations of degree 2, 3 or 4 is verified by the existence of algorithms to generate their solutions. Methods for solving problems that we would now recognise as constituting quadratic equations (degree 2) go back to the ancient world, and in the 9th century, Abu Ja'far Muhammad ibn Musa al-Khwarizmi provided systematic methods for solving them. However, the need for complex numbers was not apparent, since the only quadratic equations considered were those that arose from practical applications such as geometry, where the solutions would be real. It was

---

[xviii]The exception for powers of zero will remain in any number system, just like division by zero.

not until the 16th century, when algorithms to solve cubic and quartic equations (degree 3 and 4) were devised, that the need to work with square roots of negative numbers became unavoidable, as they would sometimes appear in the course of finding a solution even when the solution ultimately obtained was real. In any case, the existence of solutions in $\mathbb{C}$ for equations of degree up to 4 suggested that equations of any degree may have solutions in $\mathbb{C}$, and attempts to prove this were made by many of the greatest mathematical minds of the 18th century. The first proof of the FTA that would meet modern standards of rigour was by Carl Friedrich Gauss; in fact, he ultimately published four different proofs, and many other proofs have been devised by other mathematicians.

Every proof of the FTA relies on the idea that polynomials are **continuous functions**. A common intuitive idea, for functions whose domain is $\mathbb{R}$ or a subset thereof, is that a function is **continuous** at any point through which you can draw its graph without taking your pencil off the paper. How can we embody this idea in a formal definition? If the graph of a function is a continuous curve through a point $x = c$, then the value of $f(x)$ must be close to $f(c)$ when x is close to c. The definition below says that if f is continuous at c, then we can ensure that $f(x)$ is as close as we like to $f(c)$ by keeping x sufficiently close to c. Since "close to" concerns difference or distance, which is expressed by the modulus in both $\mathbb{R}$ and $\mathbb{C}$, the definition below works in both of these number systems. So the domain D may be a subset or the entirety of either $\mathbb{R}$ or $\mathbb{C}$, and the codomain $\mathbb{S}$ may be either $\mathbb{R}$ or $\mathbb{C}$ (with the notation x in the definition standing for either a real or complex variable).

Definition 9.6.2. A function $f : D \to \mathbb{S}$ is **continuous** at a point $c \in D$ if, for any $\epsilon > 0$, there exists $\delta > 0$ such that $|f(x) - f(c)| < \epsilon$ whenever $|x - c| < \delta$.[xix]

Having given the definition, we need to verify that every polynomial satisfies the definition at every point in $\mathbb{C}$.

Proof that polynomials are continuous everywhere. Given a polynomial

$$P(z) = a_0 + a_1 z + a_2 z^2 + \ldots + a_n z^n \doteq \sum_{k=0}^{n} a_k z^k \tag{9.66}$$

and an arbitrary non-zero point $c \in \mathbb{C}$, Definition 9.6.2 requires us to find $\delta > 0$ such that

$$\left| \sum_{k=0}^{n} a_k z^k - \sum_{k=0}^{n} a_k c^k \right| < \epsilon \tag{9.67}$$

whenever $|z - c| < \delta$, for any chosen $\epsilon > 0$. Now,

$$\left| \sum_{k=0}^{n} a_k z^k - \sum_{k=0}^{n} a_k c^k \right| = \left| \sum_{k=0}^{n} a_k (z^k - c^k) \right|$$

$$\leq \sum_{k=1}^{n} |a_k| \, |z^k - c^k|$$

[xix]This definition would upset many Analysts. The definition that is required for more advanced work with functions defined on any kind of domain specifies that $|f(x) - f(c)| < \epsilon$ whenever $|x - c| < \delta$ *and* $x \in D$. This has counter-intuitive effects: for example, any function whose domain is restricted to isolated points, such as a function only defined on $\mathbb{Z}$, is continuous at each of those isolated points! Our simpler, more intuitive definition, which may be found in some older textbooks on Analysis and many Calculus texts, implicitly requires the function f to be defined within an open interval of the real line containing c, or throughout some region of the complex plane surrounding c. But in any case, when dealing with functions such as polynomials for which the domain is the entirety of $\mathbb{R}$ or $\mathbb{C}$, the two definitions are effectively the same.

by the extended triangle inequality (Exercise 9.1.7), also noting that $z^0 - c^0 = 1 - 1 = 0$. The factors $|z^k - c^k|$ are bounded according to (9.55):

$$|z^k - c^k| \leq |z - c| \sum_{j=1}^{k} |z|^{k-j}|c|^{j-1}.$$

By choosing $\delta \leq 1$ so that $|z - c| \leq 1$, we have

$$|z| = |c + (z - c)| \leq |c| + |z - c| \leq |c| + 1$$

so that

$$|z^k - c^k| \leq B_k|z - c|,$$

where

$$B_k = \sum_{j=1}^{k} (|c| + 1)^{k-j}|c|^{j-1}. \tag{9.68}$$

We now have

$$\left| \sum_{k=0}^{n} a_k z^k - \sum_{k=0}^{n} a_k c^k \right| \leq \sum_{k=1}^{n} |a_k| B_k |z - c|$$

$$= |z - c| \sum_{k=1}^{n} |a_k| B_k.$$

From (9.68), B_k is strictly positive for each k and depends only on c; thus, for any $c \in \mathbb{C}$, we can choose $\epsilon > 0$ and then let

$$\delta = \min \left\{ \frac{\epsilon}{\sum_{k=1}^{n} |a_k| B_k}, 1 \right\}$$

(since we required $\delta \leq 1$ earlier) to fulfil the requirement (9.67) when $|z - c| < \delta$. $\qquad \square$

Exercise 9.6.3. *We excluded the case $c = 0$ in the first sentence of the above proof. What goes wrong if $c = 0$? Write down a proof that any polynomial $P(z)$ is continuous at 0; it is much easier than the general case of a point $c \in \mathbb{C}$!*

Different proofs of the FTA diverge after this step. We shall give a proof which essentially follows that devised by Argand, examining the *modulus* of the polynomial: if we can find z such that $|P(z)| = 0$, then we have $P(z) = 0$ as required by the FTA. The advantage of considering the modulus is that it is real, so we can consider order: in particular, Argand's proof shows that $|P(z)|$ must have a minimum for some $z \in \mathbb{C}$, and then that the only possible value of this minimum is zero. The existence of a minimum relies on Theorem 9.6.4 below, which involves the concept of a **compact set**. To avoid a lengthy digression into more advanced Analysis, we shall only give an informal discussion of compact sets, and will not prove Theorem 9.6.4; in doing so, we are actually following Argand, who assumed the truth of that theorem without proof.

The idea of a compact set is best understood through examples. In $\mathbb{R}$ a closed interval, the set of real numbers x such that $a \leq x \leq b$, is a compact set: the important features are that the set is bounded and includes its boundary points a and b. In $\mathbb{C}$ an example of a compact set would be $\{z \in \mathbb{C} : |z - t| \leq R\}$ for some $t \in \mathbb{C}$ and $R \in \mathbb{R}_+$: geometrically, this is the set of points on and inside a circle of finite radius R in the complex plane, so is again bounded and includes the boundary points. An important property of continuous functions is the following:

Theorem 9.6.4. *A function* $h : D \to \mathbb{R}$ *that is continuous at every point of a compact set* $T \subseteq D$ *has a* minimum *on* T; *i.e.* $\exists s \in T$ *such that* $h(s) \leq h(z) \, \forall z \in T$.

Now, the Axiom of Completeness guarantees that every bounded set of real numbers has an *infimum*; but this theorem, concerning the set of numbers that are the images of all points of a compact set under a continuous function, first guarantees that set to be bounded below, and then says that its infimum will actually be a *minimum*, or least member.[xx]

Our problem with applying Theorem 9.6.4 to $|P(z)|$ is that we want to show that $|P(z)|$ has a minimum for some $z \in \mathbb{C}$, but $\mathbb{C}$ is not a compact set. Note that $|P(z)|$ *does* have an infimum for some $z \in \mathbb{C}$, because the modulus of any function is bounded below by 0; but one could imagine a function $f : \mathbb{C} \to \mathbb{C}$ for which, as $|z|$ increases, $|f(z)|$ decreases towards its infimum but never reaches a minimum value. We need to show that this cannot happen with polynomials $P(z)$, by identifying a compact set within $\mathbb{C}$ such that the infimum of $|P(z)|$ cannot lie outside that compact set. The proof is based on the intuitive idea that the term $a_n z^n$ in a polynomial of form (9.66) becomes larger than all the other terms as $|z|$ becomes very large.

Proof that $|P(z)|$ ***has a minimum in*** $\mathbb{C}$. For a polynomial $P(z)$ of form (9.66), we can write

$$a_n z^n = P(z) - \sum_{k=0}^{n-1} a_k z^k. \tag{9.69}$$

By the extended triangle inequality,

$$\left| \sum_{k=0}^{n-1} a_k z^k \right| \leq \sum_{k=0}^{n-1} |a_k| \, |z^k|.$$

Where $|z| > 1$, so that $|z^k| \leq |z^{n-1}|$ for all $k \leq n - 1$, the last inequality becomes

$$\left| \sum_{k=0}^{n-1} a_k z^k \right| \leq \left(\sum_{k=0}^{n-1} |a_k| \right) |z|^{n-1}. \tag{9.70}$$

If we also impose the condition

$$|z| > 2 \frac{\sum_{k=0}^{n-1} |a_k|}{|a_n|}, \tag{9.71}$$

and then multiply (9.71) by $|a_n z^{n-1}|$ and invoke (9.70), we obtain

$$|a_n z^n| > 2 \left| \sum_{k=0}^{n-1} a_k z^k \right|. \tag{9.72}$$

We have shown that the term $a_n z^n$ is more than twice as big as the sum of all the other terms in $P(z)$ in the region

$$|z| > R \quad \text{where} \quad R = \max \left\{ 2 \frac{\sum_{k=0}^{n-1} |a_k|}{|a_n|}, 1 \right\}.$$

Next we apply the triangle inequality to (9.69) to obtain

$$|a_n z^n| \leq |P(z)| + \left| \sum_{k=0}^{n-1} a_k z^k \right|$$

[xx]Similarly, the set will be bounded above and will have a maximum, or greatest member; but that is not of interest to us here.

which, together with (9.72), yields

$$|P(z)| > \frac{1}{2}|a_n z^n|$$

throughout the region $|z| > R$. Also when $|z| > R$ the condition

$$|z| > \left(\frac{2|a_0|}{|a_n|}\right)^{1/n} \tag{9.73}$$

is satisfied; to see this, you should consider separately the cases where $2|a_0|/|a_n|$ is greater or less than 1. Condition (9.73) ensures that $\frac{1}{2}|a_n z^n| > |a_0|$, and hence

$$|P(z)| > |a_0|. \tag{9.74}$$

We have established that $|P(z)| > |a_0|$ everywhere *outside* the compact set $|z| \leq R$; but $|P(z)| = |a_0|$ at the point $z = 0$ *inside* that compact set. So the infimum of $|P(z)|$ must be at a point inside the compact set, and so according to Theorem 9.6.4 it must be the minimum of $|P(z)|$.   □

The final step in the proof of the FTA is to show that this minimum cannot take any value other than zero. To do this, we show that near any point t where $|P(t)| > 0$, there is a point z where $|P(z)| < |P(t)|$. Notice that this would not be true on the Real Number line: for example, consider the function $P(x) = 1 + x^2$, for which $|P(0)| = 1$ and $|P(x)| \geq 1$ for all $x \in \mathbb{R}$, so that $|P(x)|$ takes a minimum value which is not zero. But if we consider $P(z) = 1 + z^2$ in the complex plane, while it is still true that $|P(z)| \geq |P(0)|$ for all real z (i.e. with $\text{Im}[z] = 0$), if we consider pure imaginary z, i.e. with $z = yi$, then for $0 < y \leq 1$ and $-1 \leq y < 0$ we have $|P(z)| = 1 - y^2 < |P(0)|$. On the real line you can only go in two directions from any point, but from a point t in the complex plane you can go in infinitely many directions: here the **direction** from t to z (a geometrical concept) is defined in terms of complex numbers as $\text{Arg}\,[z-t]$, which can take any value between $-\pi$ and π. Our method is to show that for any polynomial $P(z)$ and any point t at which $P(t) \neq 0$, we can find a point z with direction $\text{Arg}\,[z-t]$ and distance $|z-t|$ from t, such that $|P(z)| < |P(t)|$.

Proof that the minimum of $|P(z)|$ ***is zero.*** Let $w = z - t$, and substitute $z = t + w$ in the polynomial (9.66) to obtain

$$P(z) = P^*(w) = b_0 + b_1 w + b_2 w^2 + \cdots + b_n w^n$$

where the coefficients $b_0, b_1, b_2, \ldots, b_n$ depend on the value of t and the coefficients $a_0, a_1, a_2, \ldots, a_n$ in (9.66), and the point $z = t$ corresponds to $w = 0$. If $b_0 = 0$ then $P(t) = P^*(0) = 0$ and we have obtained our desired solution of $P(z) = 0$ to verify the FTA. Otherwise, we can write

$$P^*(w) = b_0 \left(1 + \frac{b_1}{b_0}w + \frac{b_2}{b_0}w^2 + \cdots + \frac{b_n}{b_0}w^n\right) \tag{9.75}$$

so that $|P^*(0)| = |b_0|$. Suppose now that $b_1 \neq 0$. By choosing

$$\arg w = \pi - \text{Arg}\left[\frac{b_1}{b_0}\right] \tag{9.76}$$

we have

$$\arg\left[\frac{b_1}{b_0}w\right] = \pi$$

so that $(b_1/b_0)w$ is real and negative, and hence

$$\left| 1 + \frac{b_1}{b_0} w \right| = 1 - \left| \frac{b_1}{b_0} w \right| \tag{9.77}$$

as long as $|(b_1/b_0)w| \leq 1$, which can be ensured by choosing $|w|$ sufficiently small: $|w| \leq |b_0/b_1|$.

Now consider the remaining terms in (9.75). The extended triangle inequality gives

$$\left| \frac{b_2}{b_0} w^2 + \cdots + \frac{b_n}{b_0} w^n \right| \leq \sum_{j=2}^{n} \frac{|b_j|}{|b_0|} |w|^j;$$

then if $|w| < 1$ so that $|w|^j < |w|^2$ for all $j \geq 2$, we have

$$\left| \frac{b_2}{b_0} w^2 + \cdots + \frac{b_n}{b_0} w^n \right| \leq \frac{\sum_{j=2}^{n} |b_j|}{|b_0|} |w|^2.$$

If we also require that

$$|w| < \frac{|b_1|}{\sum_{j=2}^{n} |b_j|},$$

we ensure that

$$\left| \frac{b_2}{b_0} w^2 + \cdots + \frac{b_n}{b_0} w^n \right| < \left| \frac{b_1}{b_0} w \right|. \tag{9.78}$$

Next, the triangle inequality gives

$$\left| 1 + \frac{b_1}{b_0} w + \frac{b_2}{b_0} w^2 + \cdots + \frac{b_n}{b_0} w^n \right| \leq \left| 1 + \frac{b_1}{b_0} w \right| + \left| \frac{b_2}{b_0} w^2 + \cdots + \frac{b_n}{b_0} w^n \right|,$$

and substituting (9.77) and (9.78) then gives

$$\left| 1 + \frac{b_1}{b_0} w + \frac{b_2}{b_0} w^2 + \cdots + \frac{b_n}{b_0} w^n \right| < 1$$

so from (9.75),

$$|P^*(w)| < |b_0| = |P^*(0)|.$$

This applies when $\operatorname{Arg} w$ is given by (9.76) and $|w|$ satisfies all the conditions imposed above, which can be combined as

$$|w| < \min \left\{ \left| \frac{b_0}{b_1} \right|, 1, \frac{|b_1|}{\sum_{j=2}^{n} |b_j|} \right\};$$

and Theorem 9.4.9 assures us that a w satisfying these requirements does exist. In terms of the original variable z, the result $|P^*(w)| < |P^*(0)|$ becomes

$$|P(z)| < |P(t)|$$

as required.

There remains the possibility that $b_1 = 0$, and maybe $b_2 = \cdots = b_{k-1} = 0$, with b_k being the first non-zero coefficient in $P^*(w)$, so that

$$P^*(w) = b_0 \left(1 + \frac{b_k}{b_0} w^k + \frac{b_{k+1}}{b_0} w^{k+1} + \cdots + \frac{b_n}{b_0} w^n \right).$$

In this case we can make $\arg[(b_k/b_0)w^k] = \pi$ by choosing

$$\arg w = \frac{1}{k}\left(\pi - \text{Arg}\left[\frac{b_k}{b_0}\right]\right)$$

so that

$$\left|1 + \frac{b_k}{b_0}w^k\right| = 1 - \left|\frac{b_k}{b_0}w^k\right|$$

if $|w| < |b_0/b_k|^{1/k}$. Then we can again choose $|w|$ sufficiently small that the modulus of the sum of the remaining terms (with $(k+1)$'th to n'th powers of w) is smaller than that of the k'th power term (similarly to (9.78)), and it follows that $|P(z)| < |P(t)|$.

Finally, the result that there exists z such that $|P(z)| < |P(t)|$ whenever $|P(t)| \neq 0$ contradicts the proposition that the minimum of $|P(z)|$ is non-zero; so there exists a point $z = z_0$ where $|P(z_0)| = 0$ and hence $P(z_0) = 0$. This concludes our proof of the Fundamental Theorem of Algebra. $\qquad\square$

9.6.1 Factorisation of Polynomials

The theory of polynomials in Section 7.9 remains valid for complex polynomials, the difference now being that we know that there does exist a solution of a polynomial equation

$$P(z) := a_0 + a_1 z + a_2 z^2 + \cdots + a_n z^n = 0, \qquad (9.79)$$

and hence by induction there are in fact n solutions in $\mathbb{C}$ (rather than *at most* n solutions in $\mathbb{R}$ to a polynomial equation with real coefficients). Some of the n solutions in $\mathbb{C}$ may be equal to each other, so there are at most n *distinct* solutions. Then by induction it follows from Corollary 7.9.4 that if the equation $P(z) = 0$ has solutions $\alpha_1, \alpha_2, \ldots, \alpha_n \in \mathbb{C}$, the polynomial $P(z)$ can be factorised as

$$P(z) = (z - \alpha_1)(z - \alpha_2)\ldots(z - \alpha_n) = \prod_{j=1}^{n}(z - \alpha_j).$$

It is of interest to look more closely at polynomial equations with real coefficients:

Theorem 9.6.5. *If $z = \alpha$ is a solution of (9.79) where $a_0, a_1, \ldots, a_n \in \mathbb{R}$, then $z = \overline{\alpha}$ is also a solution.*

Proof. To say that $z = \alpha$ is a solution of (9.79) means that

$$a_0 + a_1\alpha + a_2\alpha^2 \ldots + a_n\alpha^n = 0.$$

Take the conjugate of this equation, noting that $\overline{a_j} = a_j$ for the real coefficients $a_0, a_1, a_2, \ldots, a_n$, to give

$$a_0 + a_1\overline{\alpha} + a_2\overline{\alpha}^2 + \ldots + a_n\overline{\alpha}^n = 0 \qquad (9.80)$$

where we have used the additive and multiplicative properties of conjugates in Theorem 8.1.10, with $\overline{\alpha^j} = \overline{\alpha}^j$ following from the multiplicative property. But equation (9.80) identifies $\overline{\alpha}$ as a solution of (9.79). $\qquad\square$

If α is a real solution, i.e. with $\text{Im}[\alpha] = 0$, then $\overline{\alpha} = \alpha$ so this theorem has not identified a distinct solution. But for any solution with $\text{Im}[\alpha] \neq 0$, it does imply the existence of a distinct solution $\overline{\alpha}$, so that the polynomial $P(z)$ has a pair of factors $(z - \alpha)(z - \overline{\alpha})$. But

$$\begin{aligned} (z - \alpha)(z - \overline{\alpha}) &= z^2 - (\alpha + \overline{\alpha})z + \alpha\overline{\alpha} \\ &= z^2 - 2\text{Re}[z] + |\alpha|^2, \end{aligned}$$

which is a *real* quadratic factor.[xxi] We have therefore verified:

Theorem 9.6.6. *Every polynomial with real coefficients can be factorised into real linear and/or quadratic factors.*

Since the real quadratic factors take up an even number of complex linear factors, we can also deduce:

Corollary 9.6.7. *Every polynomial equation of odd degree with real coefficients has at least one real solution.*

This corollary is usually deduced from the Intermediate Value Theorem, which says that if a real function $f(x)$ is continuous on a closed interval $[a, b]$, with $f(a) < f(b)$, then for any $y \in \mathbb{R}$ with $f(a) < y < f(b)$ there is a solution of the equation $f(x) = y$ with $a < x < b$. This seems obvious from the intuitive idea that the graph of a continuous function is a continuous curve, but needs proving from the definition of a continuous function. We have apparently taken a shortcut to Corollary 9.6.7; but in fact we have reasoned via the FTA, which involved Theorem 9.6.4, another property of continuous functions which we did not prove.

9.7 Cardinality of $\mathbb{C}$

We have seen how the Rational Numbers $\mathbb{Q}$ have the same cardinality as the Integers $\mathbb{Z}$, even though a rational number can in general only be expressed as an ordered pair of integers. So it is reasonable to expect that the Complex Numbers $\mathbb{C}$, which are ordered pairs of Real Numbers, may have the same cardinality as $\mathbb{R}$. This is indeed true, but cannot be proved in the same way as was done with $\mathbb{Q}$, for which the proof used theorems that were specific to countable sets.

When proving that $\mathbb{R}$ is uncountable, we used base-b representations to show that the real numbers in the interval $[0, 1)$ were uncountable, and later bijected this interval to the whole of $\mathbb{R}$. To find a bijection between $\mathbb{R}$ and $\mathbb{C}$ we shall again use base-b representations, starting with real numbers in the interval $[0, 1)$ and complex numbers in $[0, 1) \times [0, 1)$ [xxii] (i.e with both the real and imaginary parts in the interval $[0, 1)$). Georg Cantor proposed that a bijection could be made between these subsets of $\mathbb{R}$ and $\mathbb{C}$ by **interleaving**: given a complex number $w = (u, v)$ whose real and imaginary parts are represented in some base b as

$$\begin{aligned} u &= 0 \cdot r_1 r_2 r_3 \ldots \\ v &= 0 \cdot s_1 s_2 s_3 \ldots, \end{aligned}$$

[xxi] We refer to *factors* as being real when all the *coefficients* are real; although of course the value of $z - \alpha$ or $z^2 + \beta z + \gamma$ is only real if z is real as well as α, β and γ being real.

[xxii] We are using the Cartesian product notation here, since complex numbers are ordered pairs of real numbers.

we can form the base-b representation

$$0 \cdot r_1 s_1 r_2 s_2 r_3 s_3 \ldots$$

which certainly represents a real number $x \in [0, 1)$.

However, there is a problem. It was pointed out during the proof of the uncountability of $\mathbb{R}$ that any number with a terminating base-b representation, which can be considered as having an infinite string of zeroes after the last non-zero digit, also has a representation with an infinite string of the digit $b-1$, for example, a string of 9's in base ten. To maintain uniqueness of representations, we do not allow the latter. In the present context, we could never obtain a disallowed representation for x from allowed representations of u and v, but it is possible for an allowed representation of a real number to correspond to a disallowed representation of either u or v. For example in base ten, $x = 0 \cdot 10909090\ldots$ (which actually represents the rational number $6/55$) is obtained from the interleaving of $u = 0 \cdot 1999\ldots$, a disallowed representation of $1/5$, with $v = 0 \cdot 0000\ldots$, which is allowed. Thus Cantor's interleaving is *injective* from $\mathbb{C}$ to $\mathbb{R}$ (no allowed representation of a real number can be obtained from more than one pair of representations for the two parts of a complex number), but it is not *surjective*: not all real numbers can be formed by interleaving allowed representations of the parts of complex numbers.

Recalling that ordering of cardinalities is defined in terms of injections, Cantor's interleaving implies that $|\mathbb{C}| \leq |\mathbb{R}|$, where we are here using the notation $|\cdot|$ for cardinality, as introduced in Section 3.7, not modulus. The complex numbers are certainly uncountable, so if we believe the Continuum Hypothesis, that there is no cardinality between those of $\mathbb{N}$ and $\mathbb{R}$, then we must conclude that the cardinality of $\mathbb{C}$ is equal to that of $\mathbb{R}$; but since the Continuum Hypothesis has been found to be undecidable, this cannot be used to prove that $|\mathbb{C}| = |\mathbb{R}|$. The most elegant proof uses:

Theorem 9.7.1 (The Cantor-Bernstein-Schröder Theorem).[xxiii] *If A and B are sets with both* $|A| \leq |B|$ *and* $|B| \leq |A|$, *then* $|A| = |B|$.

For finite sets, the theorem is obviously true, since their cardinalities are natural numbers, for which it is a fundamental property of order relations that if $m \leq n$ and $n \leq m$, then $m = n$. But with infinite cardinalities, nothing is obvious, and the proof is not easy. If we accept the theorem, the fact that $|\mathbb{C}| = |\mathbb{R}|$ follows from the interleaving injection from $\mathbb{C}$ to $\mathbb{R}$ combined with the injection that maps any $x \in \mathbb{R}$ to $(x, 0) \in \mathbb{C}$. We shall not prove the Cantor-Bernstein-Schröder Theorem here, but instead demonstrate an amended interleaving algorithm, due to Gyula König, that provides the required bijection.

König's interleaving algorithm proceeds in the same way as Cantor's, *except* when the digit $b - 1$ is encountered in either part of the complex number, in which case a "chunk" of digits is taken. This chunk consists of that digit $b - 1$ and as many further $b - 1$'s that follow it, until a digit not equal to $b - 1$ is encountered; that then becomes the last digit of the chunk.

Example 9.7.2. In base ten, so that $b - 1 = 9$: with $u = 0 \cdot 74905991\ldots$ and $v = 0.01899933\ldots$, König's interleaving yields

$$x = 0 \cdot 7\ 0\ 4\ 1\ 90\ 8\ 5\ 9993\ 991\ 3 \ldots$$

in which spaces have been inserted to separate each digit or chunk taken from either u or v.

The process can be reversed unambiguously by assigning digits or chunks in x alternately to u and v:

[xxiii]Also known in other texts by various permutations of the three names, or of any two of the names.

Example 9.7.3. Given $x = 0 \cdot 9735999912994952\ldots$, we obtain

$$u = 0 \cdot 97\ 5\ 2\ 95\ldots, \qquad v = 0 \cdot 3\ 99991\ 994\ 2\ldots$$

in which spaces are again inserted to indicate where a chunk or a single digit has been taken from the base-ten representation of x.

Returning to our earlier example of a real number that could not be obtained from any $(u, v) \in \mathbb{C}$ by Cantor's interleaving, $x = 0 \cdot 1090909090\ldots$ now corresponds to

$$u = 0 \cdot 19090\ldots, \qquad v = 0 \cdot 09090\ldots,$$

with chunks of "90" in x coming alternately from u and v.

With Kőnig's algorithm, every pair of base-b representations for a complex number $(u, v) \in [0, 1) \times [0, 1)$ is mapped to the representation of a unique real number $x \in [0, 1)$, and the base-b representation of every $x \in [0, 1)$ corresponds to the representation of a unique $(u, v) \in [0, 1) \times [0, 1)$. Disallowed infinite strings of the digit $b - 1$ can never occur in x because they do not appear in u or v; in the reverse process, separating the representation of x into chunks which always end with a digit other than $b - 1$ ensures that neither u nor v have an infinite string of $b - 1$'s.

Exercise 9.7.4. *(a) Use Kőnig's algorithm to obtain the base-2 representation of the real number corresponding to the complex number (u, v) represented in base 2 as*

$$u = 0 \cdot 111001011100\ldots, \qquad v = 0 \cdot 001101010010\ldots.$$

(b) Find the base-5 representation of the complex number corresponding to the real number with base-5 representation

$$x = 0.401324442224144041443\ldots.$$

Having obtained a bijection between the complex numbers in $[0, 1) \times [0, 1)$ and the real numbers in $[0, 1)$, the final step to obtain a bijection from $\mathbb{C}$ to $\mathbb{R}$ is to use bijections between intervals and the whole of $\mathbb{R}$, which were discussed in the proof of Theorem 7.8.4. To biject from $[0, 1) \times [0, 1)$ to $\mathbb{C}$, this needs to be done in two stages, first from $[0, 1) \times [0, 1)$ to $[0, 1) \times \mathbb{R}$, and then to $\mathbb{R} \times \mathbb{R} = \mathbb{C}$.

Investigations

1. Find upper and lower approximations to π by Archimedes' method as follows, without using rational approximations to $\sqrt{3}$.

   Let

   $$z = (x_0, y_0) = \left(\frac{1}{2}, \frac{\sqrt{3}}{2} \right).$$

   It is shown in Example 9.4.13 that this point z has $\mathrm{Arg}\, z = \pi/3$; why does this identify it as one of the points on the inner hexagon in Figure 9.4, which Archimedes used as the starting point for his calculation?

   Do 4 iterations of the square-root procedure using (9.16) to obtain x_4, starting from $x_0 = 1/2$. Then use formulae given in Section 9.3 to evaluate $D(4; z)$ and $H(4; z)$

from x_4; these are your lower and upper approximations to $\pi/3$, corresponding to $1/6$ of the total lengths of 96-sided inner and outer regular polygons. Multiply these approximations by 6, and use a calculator to compare the results with 2π (the length of the circle) and with Archimedes' rational lower and upper approximations, $446/71$ and $44/7$.

2. Corollary 9.3.6 says that the non-decreasing and non-increasing sequences both converge; but they need not converge to the same limit. In the case of the sequences used when defining the PVA, prove that $H(n; z)$ *does* converge to the same limit as $D(n; z)$, by showing that for any $\epsilon > 0$, there exists $n \in \mathbb{N}$ such that

$$0 \leq H(n; z) - D(n; z) < \epsilon \quad \text{whenever} \quad n > N.$$

3. During the first half of the 18th century, several mathematicians gave examples of quartic polynomials with real coefficients which they thought could not be factorised into real linear or quadratic factors, thereby contradicting the FTA (which had not yet been proved). Find the real quadratic factors of the following quartic polynomials; the names of the mathematicians who proposed these examples are given in brackets.

(a) [Gottfried von Leibniz] $P(z) = a^4 + z^4$.

(b) [Christian Goldbach] $P(z) = -20 + 72z^2 + z^4$.

(c) [Nicolaus Bernoulli] $P(z) = 4 + 4z + 2z^2 - 4z^3 + z^4$. [This one is too difficult to do without some help! Show by substitution into $P(z)$ that the equation $P(z) = 0$ is satisfied by

$$z = 1 \pm \sqrt{2 \pm \sqrt{-3}},$$

which gives the four solutions when each of the options in the $\pm$ signs is taken.

Chapter 10

Yet More Number Systems

Complex numbers seem to do everything we could want from a number system: their addition and multiplication operations have all the properties required to constitute a field, the exponentiation operation satisfying the Laws of Indices is closed on $\mathbb{C}\backslash\{0\}$, and any polynomial equation with complex numbers as coefficients has a solution within the complex numbers. But mathematicians always want to go further. Complex numbers are ordered sets of two real numbers, but there are also useful applications for systems of ordered sets of $4, 8, \ldots$ real numbers. The bulk of this chapter concerns such systems, but in the first section we briefly discuss what happens when you pile quadratic extensions upon quadratic extensions.

10.1   Constructible Numbers

One way in which we can generate new number systems is to re-apply the quadratic extension procedure to a quadratic ring or field of the type studied in Chapter 8. Given a number system $\mathbb{S}$, the procedure requires $k \in \mathbb{S}$ such that $\nexists x \in \mathbb{S}$ with $x^2 = k$, and then defines ordered pairs $(a, b) \in \mathbb{S} \times \mathbb{S}$ to behave arithmetically like $a + b\sqrt{k}$, with the usual commutative, associative, and distributive laws and with $\sqrt{k}$ defined as an object satisfying $(\sqrt{k})^2 = k$. This leads to the addition and multiplication rules in Definitions 8.1.1 and 8.1.2.

Now suppose we start with $\mathbb{S}_1 = \mathbb{Q}[\sqrt{k_1}]$ where k_1 is a squarefree integer. We can then take k_2 to be any squarefree integer coprime with k_1 (so that the product $k_1 k_2$ is still squarefree) to create the quadratic extension $\mathbb{S}_1[\sqrt{k_2}] = \mathbb{Q}[\sqrt{k_1}][\sqrt{k_2}]$. Numbers in this system would be ordered pairs of ordered pairs of rationals: the arithmetic of $((a, b), (c, d))$ would behave like $(a + b\sqrt{k_1}) + (c + d\sqrt{k_1})\sqrt{k_2}$; or they could be regarded as ordered quadruples of rationals (a, b, c, d) behaving like $a + b\sqrt{k_1} + c\sqrt{k_2} + d\sqrt{k_1 k_2}$.

There are more options for k_2 in such an extension. It can be any number in $\mathbb{Q}[\sqrt{k_1}]$, for example $(0, 1)$, which we think of more intuitively as $\sqrt{k_1}$; so our quadratic extension could be denoted $\mathbb{Q}[\sqrt{k_1}][\sqrt[4]{k_1}]$, and would consist of ordered quadruples (a, b, c, d) behaving like

$$a + b\sqrt{k_1} + c\sqrt[4]{k_1} + d\sqrt[4]{k_1^3}.$$

More generally, we could take $k_2 = (x, y) \in \mathbb{Q}[\sqrt{k_1}]$ so that $(a, b, c, d) \in \mathbb{Q}[\sqrt{k_1}][\sqrt{k_2}]$ behaves like

$$a + b\sqrt{k_1} + (c + d\sqrt{k_1})\sqrt{x + y\sqrt{k_1}}.$$

Having quadratically extended for a second time in this way, there is nothing to stop us taking some $k_3 \in \mathbb{Q}[\sqrt{k_1}][\sqrt{k_2}]$ to define a third quadratic extension; and the process can be repeated as often as we like, to provide an infinite variety of fields (or rings, if we start with $\mathbb{Z}$ rather than $\mathbb{Q}$). The numbers produced in this way are called **constructible** if each of $k_1, k_2, \ldots$ is positive, so has a square root in $\mathbb{R}$. This terminology arises from the fact that lengths corresponding to such numbers can be constructed geometrically using a straightedge and compass, starting from a line of length 1 unit: essentially, intersections of straight lines can be constructed to yield any rational number, and intersections of lines with circles, or of two circles, yield lengths involving square roots of previously constructed lengths.

All these constructible numbers (and those in which some k_j is negative, so that $\sqrt{k_j}$ cannot be constructed as a real length) are algebraic and hence, according to the Fundamental Theorem of Algebra, they exist within $\mathbb{C}$. It is of more interest to devise number systems beyond $\mathbb{C}$, and we now briefly discuss a method which again proceeds by successively taking ordered pairs from the previous system, but produces completely new numbers.

10.2 Hypercomplex Numbers

William Rowan Hamilton was the first to conceive of complex numbers as ordered pairs of real numbers, to be added or multiplied according to rules as expressed in our Definitions 9.1.3 and 9.1.4. Noting the relevance of complex numbers to the geometry of two-dimensional space, he spent considerable effort trying to devise a system of ordered triples of real numbers to represent three-dimensional space, but found that he was unable to define multiplication of these triples in such a way that it retained the standard properties. Eventually, in 1843, he realised that to extend the system of complex numbers while retaining desirable arithmetical properties would require ordered quadruples, which he named **quaternions**. Very shortly afterwards, John Graves discovered that a system of ordered octuples (sets of 8 numbers) could be devised with useful arithmetical properties; these **octonions** were rediscovered independently by Arthur Cayley, and are sometimes called "Cayley numbers".

An ordered quadruple can be considered as an ordered pair of ordered pairs, identifying

$$(a, b, c, d) := ((a, b), (c, d)). \tag{10.1}$$

Similarly, an ordered octuple can be identified as an ordered pair of ordered quadruples; and by induction we can obtain ordered 2^n-tuples for any $n \in \mathbb{N}$. With $a, b, c, d \in \mathbb{R}$ in (10.1), the pairs (a, b) and (c, d) are complex numbers if their arithmetic (addition, multiplication, conjugates, and norms) is defined as in Section 9.1. But how should we define the arithmetic when constructing quaternions as ordered pairs of complex numbers, or octonions as ordered pairs of quaternions? In 1919 Leonard Dickson realised that the arithmetical rules for complex numbers, quaternions, and octonions could all be considered as examples of an inductive procedure which, given the arithmetic of a system $\mathbb{C}_n$ of ordered 2^n-tuples of real numbers for any $n \in \mathbb{N}$, defines the arithmetic of the successor system $\mathbb{C}_{S(n)} = \mathbb{C}_n \times \mathbb{C}_n$ in such a way as to retain as many as possible of the desirable properties. The procedure is now known as the **Cayley-Dickson construction**. The induction is anchored by the known arithmetic in $\mathbb{C}_1 := \mathbb{C}$: here we have taken advantage of the common initial letter of "complex" and "Cayley" in our notation for the systems of ordered 2^n-tuples. These systems are called **Cayley-Dickson algebras**, and for $n > 1$ the 2^n-tuples of real numbers are called **hypercomplex numbers**. We shall adopt the following notation conventions: a, b, c, d will represent 2^n-tuples in $\mathbb{C}_n$, so $a = (a_1, a_2, \ldots, a_{2^n})$, etc., with $a_1, a_2, \ldots, a_{2^n} \in \mathbb{R}$; $\alpha = (a, b)$ and $\beta = (c, d)$ are ordered pairs of 2^n-tuples, so they are $2^{S(n)}$-tuples in $\mathbb{C}_{S(n)}$. Arithmetic operations in $\mathbb{C}_n$ will be denoted $+$ and $\times$, while those in $\mathbb{C}_{S(n)}$ will be $\oplus$ and $\otimes$, as has been done when defining arithmetic in new number systems in earlier chapters.[i] The first element of a 2^n-tuple will be called its **real part**:

$$\text{Re}[(a_1, a_2, \ldots, a_{2^n})] := a_1;$$

and a 2^n-tuple $(a_1, 0, \ldots, 0)$ in which all but the first element is zero will be called a **real** 2^n-tuple. We shall show that real 2^n-tuples are isomorphic to real numbers under addition and multiplication.

Addition in the Cayley-Dickson construction is easy to define, having exactly the same form as in $\mathbb{C}$:

Definition 10.2.1. Given $a, b, c, d \in \mathbb{C}_n$, the addition of (a, b) and (c, d) is defined in $\mathbb{C}_{S(n)}$ by

$$(a, b) \oplus (c, d) := (a + c, b + d). \tag{10.2}$$

It is also easy to show that addition in $\mathbb{C}_n$ has all the same properties as in $\mathbb{R}$ and $\mathbb{C}$:

Theorem 10.2.2. *The algebraic structure $\langle \mathbb{C}_n, + \rangle$ is an Abelian group.*

The properties that constitute an Abelian group are set out in Theorem 4.3.1: we must show that the addition on $\mathbb{C}_n$ is associative and commutative, that there is an additive identity element, and that every element of $\mathbb{C}_n$ has an additive inverse. These properties are certainly true in $\mathbb{C}$, so we only need to complete the inductive step:

Exercise 10.2.3. *Show that if $\langle \mathbb{C}_n, + \rangle$ is an Abelian group with 0 as the additive identity element[ii] and $-a$ as the additive inverse of a, then:*
$\oplus$ *is commutative and associative on $\mathbb{C}_{S(n)}$;*
$(0, 0)$ *is the identity element for $\oplus$ in $\mathbb{C}_{S(n)}$;*
$(-a, -b)$ *is the inverse of (a, b) under $\oplus$ in $\mathbb{C}_{S(n)}$.*

Thus addition has the Abelian group properties in $\mathbb{C}_n$ for all $n \in \mathbb{N}$.

Noting that a $2^{S(n)}$-tuple (a, b) in $\mathbb{C}_{S(n)}$ consists of the 2^n real numbers in a followed by the 2^n real numbers in b, induction from Definition 10.2.1 also yields the addition formula in $\mathbb{C}_n$ explicitly in terms of 2^n-tuples,

$$(a_1, a_2, \ldots, a_{2^n}) + (c_1, c_2, \ldots, c_{2^n}) = (a_1 + c_1, a_2 + c_2, \ldots, a_{2^n} + c_{2^n}). \tag{10.3}$$

Thus $(a_1, 0, \ldots, 0) + (c_1, 0, \ldots, 0) = (a_1 + c_1, 0, \ldots, 0)$, verifying the isomorphism between real 2^n-tuples and real numbers under addition.

The next definition in the Cayley-Dickson construction is for conjugation:

Definition 10.2.4. The conjugate of $(a, b) \in \mathbb{C}_{S(n)}$ is

$$\overline{(a, b)} := (\overline{a}, -b),$$

where $\overline{a}$ is the conjugate of $a \in \mathbb{C}_n$.

For $(a, b) \in \mathbb{C}$ the numbers a, b are real, which means that $\overline{a} = a$; so Definition 10.2.4 is consistent with the definition of conjugation in $\mathbb{C}$. To check that conjugation has the same properties in $\mathbb{C}_n$ as in $\mathbb{C}$, it is useful to have an explicit formula for the conjugate in terms of 2^n-tuples:

---

[i]It might be useful to also introduce a notation for conjugation and norms in $\mathbb{C}_{S(n)}$ which is different from that in $\mathbb{C}_n$, but there isn't any convenient notation available.

[ii]Here 0 stands for the 2^n-tuple, $(0, 0, \ldots, 0)$.

Exercise 10.2.5. *Prove by induction that*

$$\overline{(a_1, a_2, a_3, \ldots, a_{2^n})} = (a_1, -a_2, -a_3, \ldots, -a_{2^n}). \tag{10.4}$$

The following properties follow immediately from (10.4) and the formula (10.3) for addition of 2^n-tuples:

Theorem 10.2.6. *Conjugation in $\mathbb{C}_n$ has the following properties for all $n \in \mathbb{N}$:*
(a) $\overline{(\overline{a})} = a$: *conjugation of a conjugate returns to the original number.*
(b) $\overline{a + b} = \overline{a} + \overline{b}$: *conjugation preserves addition.*
(c) $a + \overline{a} = (2\text{Re}[a], 0, 0, \ldots, 0)$: *adding a 2^n-tuple to its conjugate gives a real 2^n-tuple.*

The central feature of the Cayley-Dickson construction is the definition of multiplication:

Definition 10.2.7. *Given $a, b, c, d \in \mathbb{C}_n$, the multiplication of (a, b) and (c, d) is defined in $\mathbb{C}_{S(n)}$ by*

$$(a, b) \otimes (c, d) := (a \times c - \overline{d} \times b, d \times a + b \times \overline{c}). \tag{10.5}$$

This definition does not look very intuitive, but it is consistent with Definition 9.1.4 for multiplication in $\mathbb{C}$, since with $a, b, c, d \in \mathbb{R}$ the multiplication of real numbers on the right of (10.5) is commutative and these real numbers are equal to their conjugates. But more generally, with $a, b, c, d \in \mathbb{C}_n$ we cannot assume that the multiplication of hypercomplex numbers is commutative, so we cannot swap the order of quantities being multiplied.

Definition 10.2.7 is designed to retain as many as possible of the properties of multiplication in $\mathbb{C}$. However, we shall see that as n increases in the succession of number systems $\mathbb{C}_n$, useful properties are progressively lost; but one property that does persist in all $\mathbb{C}_n$ is the distributive law:

Theorem 10.2.8. *Multiplication distributes over addition in all $\mathbb{C}_n$.*

Proof. The distributive law applies in $\mathbb{C}$, so the induction is anchored. To complete the proof, we need to show that if it applies in $\mathbb{C}_n$, then it applies in $\mathbb{C}_{S(n)}$.
Given $a, b, c, d, e, f \in \mathbb{C}_n$,

$$((a, b) \oplus (c, d)) \otimes (e, f) = (a + c, b + d) \otimes (e, f) \qquad [Def.\ 10.2.1]$$

$$= ((a + c) \times e - \overline{f} \times (b + d), f \times (a + c) + (b + d) \times \overline{e})$$
$$[Def.\ 10.2.7]$$

$$= (a \times e + c \times e - \overline{f} \times b - \overline{f} \times d, f \times a + f \times c + b \times \overline{e} + d \times \overline{e})$$
$$[Inductive\ hypothesis:\ distributive\ law\ in\ \mathbb{C}_n]$$

$$= (a \times e - \overline{f} \times b, f \times a + b \times \overline{e}) \oplus (c \times e - \overline{f} \times d, f \times c + d \times \overline{e})$$
$$[Def.\ 10.2.1]$$

$$= ((a, b) \otimes (e, f)) \oplus ((c, d) \otimes (e, f)). \qquad [Def.\ 10.2.7]$$

This shows that multiplication distributes over addition from the right; since multiplication may not be commutative, a similar calculation with the multiplication on the left will complete the proof.

Exercise 10.2.9. *Show that with $a, b, c, d, e, f \in \mathbb{C}_n$,*

$$(a, b) \otimes ((c, d) \oplus (e, f)) = ((a, b) \otimes (c, d)) \oplus ((a, b) \otimes (e, f)). \quad \square$$

An important consequence of the distributive law is that

$$a \times 0 = 0 \times a = 0$$

in $\mathbb{C}_n$ for all $n \in \mathbb{N}$. This follows because in the proof of Theorem 4.3.7 the distributive law is the only property of multiplication required to verify that multiplication by the additive identity always yields the additive identity. The rule for multiplication by zero is useful in the proof of the next theorem:

Theorem 10.2.10. *Multiplication by a real 2^n-tuple is scalar multiplication in all $\mathbb{C}_n$:*

$$(a_1, 0, 0, \ldots, 0) \times (c_1, c_2, \ldots, c_{2^n}) = (c_1, c_2, \ldots, c_{2^n}) \times (a_1, 0, 0, \ldots, 0) = (a_1 c_1, a_1 c_2, \ldots, a_1 c_{2^n}).$$
$$(10.6)$$

Proof. Note first that the theorem includes the statement that multiplication by a real 2^n-tuple commutes, even if multiplication is not generally commutative in $\mathbb{C}_n$.

Scalar multiplication applies in $\mathbb{C}$. We assume as our inductive hypothesis that it applies in $\mathbb{C}_n$. Then

$$(a, 0) \otimes (c, d) = (a \times c - \overline{d} \times 0, d \times a + 0 \times \overline{c})$$
$$= (a \times c, d \times a).$$

If a is a real 2^n-tuple $(a_1, 0, 0, \ldots, 0)$, then $(a, 0)$ is the real $2^{S(n)}$-tuple $(a_1, 0, 0, \ldots, 0)$; also by the inductive hypothesis, multiplication by a commutes, so $d \times a = a \times d$. So

$$(a, 0) \otimes (c, d) = (a \times c, a \times d),$$

where $(a \times c, a \times d)$ is the $2^{S(n)}$-tuple

$$(a_1 c_1, a_1 c_2, \ldots, a_1 c_{2^n}, a_1 d_1, a_1 d_2, \ldots, a_1 d_{2^n})$$

which is the required form; if this is not clear, relabel the elements of d as $(d_1, d_2, \ldots, d_{2^n}) = (c_{2^n+1}, c_{2^n+2}, \ldots, c_{2^{S(n)}})$.

A similar calculation with the real $2^{S(n)}$-tuple multiplying on the right completes the proof. $\qquad\square$

An important instance of scalar multiplication is:

Corollary 10.2.11. *The real 2^n-tuple $(1, 0, 0, \ldots)$ is the multiplicative identity element for $\mathbb{C}_n$.*

A scalar multiplication of a real 2^n-tuple by another real 2^n-tuple confirms the isomorphism between real 2^n-tuples and real numbers under multiplication:

$$(a_1, 0, \ldots, 0) \times (c_1, 0, \ldots, 0) = (a_1 c_1, 0, \ldots, 0).$$

The scalar multiple in equation (10.6) may be written as $a_1 c$ where $a_1 \in \mathbb{R}$ and $c \in \mathbb{C}_n$. Similarly, a scalar multiplication

$$\left(\frac{1}{a_1}, 0, 0, \ldots, 0 \right) \times (c_1, c_2, \ldots, c_{2^n})$$

may be written as a scalar division,

$$\frac{c}{a_1}.$$

Next we turn to the combination of conjugation with multiplication:

Theorem 10.2.12. *For $a, c \in \mathbb{C}_n$,*

$$\overline{a \times c} = \overline{c} \times \overline{a}.$$

This differs from the simple "conjugation preserves multiplication" theorem in $\mathbb{C}$, since the conjugation reverses the order of the numbers being multiplied. However, since multiplication is commutative in $\mathbb{C}$, the rule in $\mathbb{C}$ is consistent with the theorem in $\mathbb{C}_n$.

Proof of Theorem 10.2.12. Having established that the theorem is satisfied in $\mathbb{C}$, we take the inductive hypothesis that $\overline{a \times c} = \overline{c} \times \overline{a}$ in $\mathbb{C}_n$. Then in $\mathbb{C}_{S(n)}$ we have

$$
\begin{aligned}
\overline{(a,b) \otimes (c,d)} &= \overline{(a \times c - \overline{d} \times b, d \times a + b \times \overline{c})} \\
&= (\overline{a \times c - \overline{d} \times b}, -(d \times a + b \times \overline{c})) \\
&\qquad [\textit{Def. 10.2.4 \& Theorem 10.2.6(b)}] \\
&= (\overline{c} \times \overline{a} - \overline{b} \times d, -d \times a - b \times \overline{c}) \qquad [\textit{Inductive hypothesis}] \\
&= \left(\overline{c} \times \overline{a} - (\overline{-b}) \times (-d), (-b) \times \overline{c} + (-d) \times \overline{(\overline{a})}\right) \\
&= (\overline{c}, -d) \otimes (\overline{a}, -b) \qquad [\textit{Def. 10.2.7}] \\
&= \overline{(c,d)} \otimes \overline{(a,b)}. \qquad [\textit{Def. 10.2.4}] \ \square
\end{aligned}
$$

In $\mathbb{C}$ the product of a number z with its conjugate is real, which motivates the definition of the norm. In $\mathbb{C}_n$ the norm arises in a similar way:

Theorem 10.2.13. *If $a = (a_1, a_2, \ldots, a_{2^n}) \in \mathbb{C}_n$, then*

$$a \times \overline{a} = \overline{a} \times a = \left(\sum_{j=1}^{2^n} a_j^2, 0, \ldots, 0\right). \tag{10.7}$$

Definition 10.2.14. The norm $\mathcal{N} : \mathbb{C}_n \to \mathbb{R}$ is defined by

$$\mathcal{N}(a) := \mathrm{Re}[a \times \overline{a}] = \sum_{j=1}^{2^n} a_j^2.$$

The theorem states firstly that multiplication of a hypercomplex number a with its conjugate always commutes (even if multiplication is not generally commutative in $\mathbb{C}_n$), and secondly that the result of this multiplication is a real 2^n-tuple, with its real part being the sum of squares of the 2^n elements of a. We then define that real part as the norm of a.

Proof of Theorem 10.2.13. For $a = (a_1, a_2) \in \mathbb{C}$ we know that

$$a \times \overline{a} = \overline{a} \times a = (a_1^2 + a_2^2, 0):^{\text{iii}}$$

this anchors the induction.

Our inductive hypothesis is equation (10.7). Then for any $a, b \in \mathbb{C}_n$, the multiplication in $\mathbb{C}_{S(n)}$ is

$$
\begin{aligned}
(a,b) \otimes \overline{(a,b)} &= (a,b) \otimes (\overline{a}, -b) \\
&= \left(a \times \overline{a} - (\overline{-b}) \times b, (-b) \times a + b \times \overline{(\overline{a})}\right) \qquad [\textit{Def. 10.2.7}] \\
&= (a \times \overline{a} + \overline{b} \times b, (-b + b) \times a) \qquad [\textit{Thms. 10.2.6(a), 10.2.8}] \\
&= (a \times \overline{a} + \overline{b} \times b, 0). \tag{10.8}
\end{aligned}
$$

---

iii To check this against the usual notation in $\mathbb{C}$, write z for a, x for a_1, and y for a_2.

By the inductive hypothesis,

$$a \times \overline{a} + \overline{b} \times b = \sum_{j=1}^{2^n} a_j^2 + \sum_{j=1}^{2^n} b_j^2,$$

which is the sum of squares of all the elements of the $2^{S(n)}$-tuple (a, b).

To complete the proof, it remains to show that the same result is obtained from $\overline{(a, b)} \otimes (a, b)$.

Exercise 10.2.15. *Verify that*

$$\overline{(a, b)} \otimes (a, b) = \sum_{j=1}^{2^n} a_j^2 + \sum_{j=1}^{2^n} b_j^2. \qquad \qquad \square$$

The sum-of-squares formula assures us that the norm of a hypercomplex number can only be zero when the number itself is the zero 2^n-tuple, $(0, 0, \ldots, 0)$.

Theorem 10.2.13 and Definition 10.2.14 allow us to write the multiplicative inverse of a number $a \in \mathbb{C}_n$ as

$$\frac{\overline{a}}{\mathcal{N}(a)} \qquad \qquad (10.9)$$

which is the scalar division of the conjugate by the norm; and when a number has a multiplicative inverse, we may divide by that number according to Theorem 4.2.13, by multiplying by its multiplicative inverse. However, because we have not shown that multiplication is either associative or commutative in $\mathbb{C}_n$ for any $n > 1$, we need to make some provisos. Non-associativity would mean that Theorem 4.2.10 would not apply: we could not guarantee that the multiplicative inverse in (10.9) is unique. Even if multiplication is associative and the multiplicative inverse *element* is unique, non-commutativity would imply that there are two inverse *operations* to multiplication, as discussed in Section 3.8. Given $b, c \in \mathbb{C}_n$, we may seek $a_L, a_R \in \mathbb{C}_n$ such that

$$a_L \times b = c \quad \text{and} \quad b \times a_R = c.$$

If $b \neq 0$, we obtain using the inverse element of form (10.9) that

$$a_L = \frac{c \times \overline{b}}{\mathcal{N}(b)} \quad \text{and} \quad a_R = \frac{\overline{b} \times c}{\mathcal{N}(b)}, \qquad \qquad (10.10)$$

which can be described respectively as the **left division** and **right division** of c by b.[iv] We do not use the symbol $\div$ because it does not distinguish between left division and right division, and the fraction notation is only used for scalar division, for the same reason.

Having so far discussed theory that applies to all Cayley-Dickson algebras, we now consider those properties of multiplication that do not remain true in $\mathbb{C}_n$ for all $n \in \mathbb{N}$. We shall present several theorems, and consider their implications in specific Cayley-Dickson algebras, notably the quaternions and octonions. In the notation we have been using, quaternions are $\mathbb{C}_2$ and octonions are $\mathbb{C}_3$; but the usual notations are respectively $\mathbb{H}$ (in honour of Hamilton) and $\mathbb{O}$, and we shall use these below.

Theorem 10.2.16. *Multiplication is commutative in $\mathbb{C}_{S(n)}$ if it is commutative in $\mathbb{C}_n$ and if $\overline{a} = a \ \forall a \in \mathbb{C}_n$.*

[iv]Note the contrast with the non-commutative operation of exponentiation (in $\mathbb{Q}$ or $\mathbb{R}$), which had no identity element, so that inverse elements could not be defined. We were able to define two inverse operations, root-extraction and logarithms, which have completely different character; whereas the left and right division in $\mathbb{C}_n$ only differ in whether the inverse element is multiplied on the left or right.

Exercise 10.2.17. *Prove Theorem 10.2.16 from formula (10.5): you need to show that $(c,d) \otimes (a,b) = (a,b) \otimes (c,d)$ when the conditions of the theorem are satisfied.*

This theorem seems fairly useless: the condition that $\bar{a} = a$ is not even satisfied in $\mathbb{C}$, so the theorem does not imply that any system of hypercomplex numbers has commutative multiplication. It does not exclude that possibility, but the calculation in Exercise 10.2.17 does suggest that having $\bar{a} = a$ in $\mathbb{C}_n$ is a *necessary* condition for commutativity in $\mathbb{C}_{S(n)}$. In fact, no Cayley-Dickson algebra with $n > 1$ has commutative multiplication. We now verify this for the quaternions by obtaining a formula for multiplying quaternions as ordered quadruples of real numbers. In Definition 10.2.7, suppose that a, b, c, d are complex numbers, so that (a,b) and (c,d) are quaternions and we can write $a = (a_1, a_2)$, etc., with $a_1, a_2 \in \mathbb{R}$. The multiplications on the right of (10.5) become:

$$a \times c = (a_1, a_2) \times (c_1, c_2) = (a_1 c_1 - a_2 c_2, a_1 c_2 + a_2 c_1),$$
$$\tilde{d} \times b = (d_1, -d_2) \times (b_1, b_2) = (d_1 b_1 + d_2 b_2, d_1 b_2 - d_2 b_1),$$
$$d \times a = (d_1, d_2) \times (a_1, a_2) = (d_1 a_1 - d_2 a_2, d_1 a_2 + d_2 a_1),$$
$$b \times \tilde{c} = (b_1, b_2) \times (c_1, -c_2) = (b_1 c_1 + b_2 c_2, -b_1 c_2 + b_2 c_1).$$

The subtraction and addition on the right of (10.5) then yield the multiplication of quaternions written as ordered pairs of ordered pairs of real numbers,

$$(a,b) \otimes (c,d) := ((a_1, a_2), (b_1, b_2)) \otimes ((c_1, c_2), (d_1, d_2))$$
$$= ((a_1 c_1 - a_2 c_2 - d_1 b_1 - d_2 b_2, a_1 c_2 + a_2 c_1 - d_1 b_2 + d_2 b_1),$$
$$(d_1 a_1 - d_2 a_2 + b_1 c_1 + b_2 c_2, d_1 a_2 + d_2 a_1 - b_1 c_2 + b_2 c_1)).$$

This result is clearer if we write the quaternions as ordered quadruples of real numbers, introducing the notation

$$\alpha = (\alpha_1, \alpha_2, \alpha_3, \alpha_4) := (a_1, a_2, b_1, b_2)$$

and

$$\beta = (\beta_1, \beta_2, \beta_3, \beta_4) := (c_1, c_2, d_1, d_2).$$

We then have

$$\alpha \otimes \beta = (\alpha_1 \beta_1 - \alpha_2 \beta_2 - \alpha_3 \beta_3 - \alpha_4 \beta_4, \alpha_1 \beta_2 + \alpha_2 \beta_1 + \alpha_3 \beta_4 - \alpha_4 \beta_3,$$
$$\alpha_1 \beta_3 - \alpha_2 \beta_4 + \alpha_3 \beta_1 + \alpha_4 \beta_2, \alpha_1 \beta_4 + \alpha_2 \beta_3 - \alpha_3 \beta_2 + \alpha_4 \beta_1). \quad (10.11)$$

For a clearer picture of how multiplication works in $\mathbb{H}$, observe that any $\alpha \in \mathbb{H}$ can be written in terms of scalar multiples of its real part α_1 and its three imaginary parts $\alpha_2, \alpha_3, \alpha_4$ with **unit quaternions:**

$$(\alpha_1, \alpha_2, \alpha_3, \alpha_4) = \alpha_1(1,0,0,0) + \alpha_2(0,1,0,0) + \alpha_3(0,0,1,0) + \alpha_4(0,0,0,1).$$

Just as in $\mathbb{C}$ we commonly write i for $(0,1)$, the conventional symbols for the imaginary unit quaternions are

$$i := (0,1,0,0), \quad j := (0,0,1,0), \quad k := (0,0,0,1),$$

while the multiplicative identity $(1,0,0,0)$ may be written as 1, and other real quaternions

$(\alpha_1, 0, 0, 0)$ as α_1. Multiplication of the three imaginary unit quaternions according to (10.11) yields

$$i \times i = -1, \quad j \times j = -1, \quad k \times k = -1,^{\text{v}}$$

$$i \times j = k, \quad j \times k = i, \quad k \times i = j, \qquad j \times i = -k, \quad k \times j = -i, \quad i \times k = -j, \quad (10.12)$$

where we are now using the standard symbol $\times$ for multiplication in $\mathbb{H}$. Many authors use these formulae to *define* multiplication of quaternions, without reference to the Cayley-Dickson construction. It is immediately apparent from (10.12) that multiplication is not commutative in $\mathbb{H}$: $i \times j \neq j \times i$, etc.

It is of interest to consider the multiplication of two quaternions whose real parts are both zero: from (10.11) we obtain

$$(0, \alpha_2, \alpha_3, \alpha_4) \times (0, \beta_2, \beta_3, \beta_4) = (-\alpha_2\beta_2 - \alpha_3\beta_3 - \alpha_4\beta_4,$$
$$\alpha_3\beta_4 - \alpha_4\beta_3, -\alpha_2\beta_4 + \alpha_4\beta_2, \alpha_2\beta_3 - \alpha_3\beta_2). \quad (10.13)$$

Readers who are familiar with vectors in 3-dimensional space will recognise that if we have vectors $\boldsymbol{\alpha} := (\alpha_2, \alpha_3, \alpha_4)$ and $\boldsymbol{\beta} := (\beta_2, \beta_3, \beta_4)$, the first element of the quaternion on the right of (10.13) is the negative of their scalar product, while the second, third and fourth elements constitute their vector product. So by defining a 4-dimensional number system, Hamilton achieved his original aim of devising a formalism that would be useful for studying geometry and physics in 3-dimensional space.

Having considered commutativity of multiplication, we next examine its associativity:

Theorem 10.2.18. *Multiplication is associative in $\mathbb{C}_{S(n)}$ if it is commutative and associative in $\mathbb{C}_n$.*

Exercise 10.2.19. *Show from formula (10.5) that if multiplication in $\mathbb{C}_n$ is associative (but not assumed commutative), then*

$$((a, b) \otimes (c, d)) \otimes (e, f) = (ace - \overline{d}be - \overline{f}da - \overline{f}b\overline{c}, fac - f\overline{d}b + da\overline{e} + b\overline{c}\overline{e})$$

and

$$(a, b) \otimes ((c, d) \otimes (e, f)) = (ace - e\overline{d}b - a\overline{f}d - \overline{c}\overline{f}b, fca - b\overline{d}f + d\overline{e}a + b\overline{e}\overline{c}).$$

[For brevity, we have omitted the $\times$ symbols for multiplication in $\mathbb{C}_n$ on the right of these equations. We are allowed to write the products of three quantities without brackets on the right because of the condition that multiplication in $\mathbb{C}_n$ is associative.]

To complete the proof of Theorem 10.2.18, we only need to observe that the two expressions given in Exercise 10.2.19 are equal if the multiplication on the right is commutative, so that $\overline{d}be = e\overline{d}b$, $\overline{f}da = a\overline{f}d$, etc.

This theorem implies that multiplication is associative in the quaternions, because it is commutative and associative in the complex numbers. However, since multiplication is not commutative in the quaternions, we cannot expect it to be associative in the octonions. We leave it to the reader to investigate the details of the multiplication of octonions, and to confirm that it is not associative: see Investigation 1 at the end of this chapter.

We have already noted that non-associativity allows non-uniqueness of multiplicative inverses. We now examine the implications of associativity or the lack of it for the behaviour of norms.

Theorem 10.2.20. *If multiplication is associative in $\mathbb{C}_n$, then norms preserve multiplication in $\mathbb{C}_{S(n)}$,*

$$\mathcal{N}(\alpha \otimes \beta) = \mathcal{N}(\alpha) \times \mathcal{N}(\beta) \,\forall\, \alpha, \beta \in \mathbb{C}_{S(n)}.$$

---

^{v}Here -1 stands for the real quaternion $(-1, 0, 0, 0)$.

Proof. We first show that if multiplication is associative in $\mathbb{C}_n$, then norms preserve multiplication in the *same* Cayley-Dickson algebra $\mathbb{C}_n$. Given $a, b \in \mathbb{C}_n$,

$$
\begin{aligned}
\mathcal{N}(a \times b) &= \text{Re}[(a \times b) \times \overline{(a \times b)}] && [\textit{Def. 10.2.14}] \\
&= \text{Re}[(a \times b) \times (\overline{b} \times \overline{a})] && [\textit{Theorem 10.2.12}] \\
&= \text{Re}[a \times (b \times \overline{b}) \times \overline{a}]. && [\textit{Associative multiplication}]
\end{aligned}
$$

Now, $b \times \overline{b} = (\mathcal{N}(b), 0, \ldots, 0)$, a real 2^n-tuple for which multiplication does always commute [*Theorem 10.2.10*]. So

$$
\begin{aligned}
\text{Re}[a \times (b \times \overline{b}) \times \overline{a}] &= \text{Re}[(a \times \overline{a}) \times (b \times \overline{b})] \\
&= \text{Re}[(\mathcal{N}(a), 0, \ldots, 0) \times (\mathcal{N}(b), 0, \ldots, 0)] \\
&= \text{Re}[(\mathcal{N}(a) \times \mathcal{N}(b), 0, \ldots, 0)], \\
&= \mathcal{N}(a) \times \mathcal{N}(b); && (10.14)
\end{aligned}
$$

so $\mathcal{N}(a \times b) = \mathcal{N}(a) \times \mathcal{N}(b)$ for $a, b \in \mathbb{C}_n$.

Next, for $\alpha, \beta \in \mathbb{C}_{S(n)}$, we have

$$
\mathcal{N}(\alpha \otimes \beta) = \text{Re}[(\alpha \otimes \beta) \otimes \overline{(\alpha \otimes \beta)}], \qquad (10.15)
$$

and we can write α and β as ordered pairs of 2^n-tuples, $\alpha = (a, b)$ and $\beta = (c, d)$ with $a, b, c, d \in \mathbb{C}_n$. Then (omitting $\times$ symbols for multiplication in $\mathbb{C}_n$)

$$
\alpha \otimes \beta = (ac - \overline{d}b, da + b\overline{c}),
$$

so that

$$
\begin{aligned}
(\alpha \otimes \beta) \otimes \overline{(\alpha \otimes \beta)} &= \left((ac - \overline{d}b)\overline{(ac - \overline{d}b)} + \overline{(da + b\overline{c})}(da + b\overline{c}), 0 \right) && [\textit{Equation (10.8)}] \\
&= \left((ac - \overline{d}b)(\overline{c}\,\overline{a} - \overline{b}d) + (\overline{a}\overline{d} + c\overline{b})(da + b\overline{c}), 0 \right) && [\textit{Theorem 10.2.12}] \\
&= \left(ac\overline{c}\,\overline{a} - \overline{d}b\overline{c}\,\overline{a} - ac\overline{b}d + \overline{d}b\overline{b}d + \overline{a}\overline{d}da + c\overline{b}da + \overline{a}\overline{d}b\overline{c} + c\overline{b}b\overline{c}, 0 \right) && (10.16)
\end{aligned}
$$

where as well as using the distributive law to obtain the last expression, we have used the associative property to allow us to write products of four quantities in $\mathbb{C}_n$ without brackets.

We can identify two sets of four terms in (10.16). First, using the result (10.14) we have

$$
\begin{aligned}
\text{Re}[ac\overline{c}\,\overline{a} + \overline{d}b\overline{b}d + \overline{a}\overline{d}da + c\overline{b}b\overline{c}] &= \mathcal{N}(a)\mathcal{N}(c) + \mathcal{N}(d)\mathcal{N}(b) + \mathcal{N}(a)\mathcal{N}(d) + \mathcal{N}(c)\mathcal{N}(b) \\
&= (\mathcal{N}(a) + \mathcal{N}(b))(\mathcal{N}(c) + \mathcal{N}(d)) \\
&= \mathcal{N}(\alpha)\mathcal{N}(\beta) && (10.17)
\end{aligned}
$$

since the sum-of-squares formula for norms implies that

$$
\mathcal{N}(a) + \mathcal{N}(b) = \mathcal{N}(a, b) = \mathcal{N}(\alpha) \quad \text{and} \quad \mathcal{N}(c) + \mathcal{N}(d) = \mathcal{N}(c, d) = \mathcal{N}(\beta).
$$

The remaining four terms in (10.16) are

$$
-(\overline{d}b\overline{c}\,\overline{a} + ac\overline{b}d) + (c\overline{b}da + \overline{a}\overline{d}b\overline{c}). \qquad (10.18)
$$

Considering the terms in the second parentheses, $\overline{a}\overline{d}b\overline{c}$ is the conjugate of $c\overline{b}da$ according to Theorem 10.2.12, so the sum of the two terms is a *real* 2^n-tuple according to Theorem

10.2.6(c). Hence it multiplies any other 2^n-tuple commutatively (i.e. by scalar multiplication), so that

$$(\bar{c}bda + \bar{a}\bar{d}b\bar{c})a\bar{a} = a(\bar{c}bda + \bar{a}\bar{d}b\bar{c})\bar{a}$$
$$= ac\bar{b}d(a\bar{a}) + (a\bar{a})\bar{d}b\bar{c}a$$

since multiplication in $\mathbb{C}_n$ is associative. But the 2^n-tuple $a\bar{a}$ is real [*Theorem 10.2.13*], so can be scalar-divided out; thus

$$\bar{c}bda + \bar{a}\bar{d}b\bar{c} = ac\bar{b}d + \bar{d}b\bar{c}a,$$

so that the expression (10.18) is zero. This, combined with the results (10.15), (10.16), and (10.17), yields

$$\mathcal{N}(\alpha \otimes \beta) = \mathcal{N}(\alpha) \times \mathcal{N}(\beta). \qquad \qquad \square$$

In the quaternions, associativity of multiplication ensures that multiplicative inverses are unique, so that left division and right division by any non-zero quaternion each produce a unique result. In the octonions we do not have associativity, but according to Theorem 10.2.20 norms preserve multiplication in $\mathbb{O}$, and in fact this is sufficient to ensure uniqueness of multiplicative inverses:

Theorem 10.2.21. *If norms preserve multiplication in a Cayley-Dickson algebra, then multiplicative inverses are unique.*

Proof. Suppose that a non-zero hypercomplex number a has two multiplicative inverses, c and d. Then

$$a \times c = a \times d = (1, 0, \ldots, 0).$$

Then, because the distributive law applies in all $\mathbb{C}_n$,

$$a \times (c - d) = (0, 0, \ldots, 0);$$

and since norms preserve multiplication,

$$\mathcal{N}(a)\mathcal{N}(c - d) = \mathcal{N}(0, 0, \ldots, 0) = 0.$$

Norms are real numbers, so this implies that either $\mathcal{N}(a) = 0$ or $\mathcal{N}(c-d) = 0$. But the norm of a hypercomplex number is only zero if the number itself is zero; and we have specified that a is non-zero, so this means that $c - d = (0, 0, \ldots, 0)$. Hence $c = d$, and the two supposed inverses are the same. $\qquad \qquad \square$

The proof is saying that where norms preserve multiplication, there can be no **zero divisors** (non-zero numbers which yield zero when multiplied together, see Definition 5.6.9), and where there are no zero divisors, multiplicative inverses are unique. This applies in the octonions; but because multiplication is not associative in $\mathbb{O}$, Theorem 10.2.20 does not extend to the next Cayley-Dickson algebra $\mathbb{C}_4$, known as the **sedenions**. It has been shown that zero divisors do exist in $\mathbb{C}_4$ and indeed all Cayley-Dickson algebras $\mathbb{C}_n$ with $n \geq 4$, so that multiplicative inverses are not unique and although left division and right division can be done, they may not yield unique results. Indeed, it has been proved that systems of m-tuples of real numbers with norms that preserve multiplication and in which left division and right division yield unique results can only exist with $m = 1, 2, 4$ or 8.

Finally, we note that just as the Gaussian integers $\mathbb{Z}[\sqrt{-1}]$ are an extension of $\mathbb{Z}$ which can be considered as a subset of $\mathbb{C}$ (because $\mathbb{Z}[\sqrt{-1}]$ consists of ordered pairs of integers with the same rules of arithmetic as the pairs of real numbers that constitute $\mathbb{C}$), there are extensions of $\mathbb{Z}$ into 2^n-tuples which are subsets of $\mathbb{C}_n$. With $n = 2$, the obvious way to do this is with quadruples of integers,

$$(\alpha_1, \alpha_2, \alpha_3, \alpha_4) \in \mathbb{Z} \times \mathbb{Z} \times \mathbb{Z} \times \mathbb{Z}, \tag{10.19}$$

subject to the rules for addition, multiplication, conjugates, and norms given above for the quaternions. However, recall how we introduced rings of quadratic integers $\mathbb{Z}[\omega_k]$ where ω_k satisfied a monic quadratic equation with integer coefficients. If we set $\beta = \alpha$ in equation (10.11), we find that

$$\alpha^2 := \alpha \otimes \alpha = 2\alpha_1 \alpha - (\mathcal{N}(\alpha), 0, 0, 0),$$

so that every quaternion α satisfies a monic quadratic equation with real coefficients. So we can define integer quaternions as those for which the coefficients $2\alpha_1$ and $\mathcal{N}(\alpha)$ are integers. This includes those of type (10.19), known as the **Lipschitz integers**, but also those in which $2\alpha_1$ is an odd integer, for which the condition that $\mathcal{N}(\alpha) \in \mathbb{Z}$ is satisfied by requiring that each of $\alpha_2, \alpha_3, \alpha_4$ is also half of an odd integer. This leads to the definition of **Hurwitz integers**: whereas a Lipschitz integer can be written as

$$\alpha_1(1, 0, 0, 0) + \alpha_2(0, 1, 0, 0) + \alpha_3(0, 0, 1, 0) + \alpha_4(0, 0, 0, 1),$$

a Hurwitz integer is

$$\beta_1 \left(\frac{1}{2}, \frac{1}{2}, \frac{1}{2}, \frac{1}{2} \right) + \beta_2(0, 1, 0, 0) + \beta_3(0, 0, 1, 0) + \beta_4(0, 0, 0, 1)$$

with $\beta_1, \beta_2, \beta_3, \beta_4 \in \mathbb{Z}$. The Lipschitz integers are the subring of the Hurwitz integers in which β_1 is even.

As in quadratic rings, units have norm equal to 1. In the Lipschitz integers there are eight units:

$$(\pm 1, 0, 0, 0), \ (0, \pm 1, 0, 0), \ (0, 0, \pm 1, 0), \ (0, 0, 0, \pm 1);$$

but in the Hurwitz integers there are 16 further units:

$$\left(\pm \frac{1}{2}, \pm \frac{1}{2}, \pm \frac{1}{2}, \pm \frac{1}{2} \right).$$

Theory regarding primes can be developed for Lipschitz and Hurwitz integers in an analogous way to that for quadratic rings in Chapter 8. Similarly to the way that $\mathbb{Z}[\omega_{-3}]$ is norm-Euclidean while its subring $\mathbb{Z}[\sqrt{-3}]$ is not, the Hurwitz integers are norm-Euclidean while the Lipschitz integers are not. The Hurwitz integers are useful in number theory, in particular for formulating an elegant proof of the theorem that every natural number can be expressed as the sum of four squares of integers.

Investigations

1. Investigate the multiplication of octonions, as follows. It would be possible to derive a general formula for the multiplication of two octuples of real numbers, similar to (10.11) for quadruples, by using the definition formula (10.5) with $a, b, c, d \in \mathbb{H}$. However, this would be a very cumbersome calculation, and it might be better to only

consider the multiplication of unit octonions. These are the eight octuples,

$$(1,0,0,0,0,0,0,0), \quad (0,1,0,0,0,0,0,0), \quad (0,0,1,0,0,0,0,0)\ldots,$$
$$(0,0,0,0,0,0,0,1),$$

which we can denote respectively as $1, i_1, i_2, \ldots, i_7$. They can also be represented as ordered pairs of unit quaternions, for example

$$i_2 := (0,0,1,0,0,0,0,0) = ((0,0,1,0),(0,0,0,0)) = (j,0),$$

$$i_7 := (0,0,0,0,0,0,0,1) = ((0,0,0,0),(0,0,0,1)) = (0,k).$$

Using these representations in formula (10.5), generate a table of all 49 products of two imaginary unit octonions $i_1, i_2, \ldots, i_7$. [You do not need to consider products involving the unit octonion 1, since that is known to be the multiplicative identity element.] Confirm that multiplication is not commutative in $\mathbb{O}$. By considering some examples of products of three unit octonions, verify that multiplication is not associative in $\mathbb{O}$.

2. Some authors have used an alternative definition of multiplication in Cayley-Dickson algebras,

$$(a,b) \otimes (c,d) := (a \times c - d \times \overline{b}, \overline{a} \times d + c \times b),$$

rather than that in our Definition 10.2.7. This can be shown to retain all the essential properties of multiplication that we have proved (distributive law, scalar multiplication, behaviour of norms, etc.), but changes some of the multiplication formulae for particular Cayley-Dickson algebras.

Show that when this alternative definition of multiplication is used for the quaternions, it reverses the $+$ and $-$ signs in the products of the imaginary unit quaternions in (10.12), with corresponding changes to $+$ and $-$ signs in the general quaternion multiplication formula (10.11). [If we regard (10.11) as defining a right-handed version of quaternions, the alternative multiplication formula used here gives an equally valid left-handed version.]

What differences does the alternative multiplication formula make to the results you found for multiplication of the imaginary unit octonions in Investigation 1 above?

Chapter 11

Where Do We Go from Here?

In Section 1.3 we suggested that in a logical progression through mathematics, Number Systems would come after Sets, and would open the way to several other major areas of study within mathematics. In this chapter, we shall briefly discuss how the theory of number systems leads into these further topics.

11.1 Number Theory and Abstract Algebra

Number Theory deals mainly, but not exclusively, with the Integers. Some of the fundamental themes of Number Theory were introduced in Chapter 5. Prime numbers and factorisations are simple concepts, and yet remain a topic of continuing research, with applications in cryptography being prominent in recent decades. We also introduced congruence and modular arithmetic, which led to an example of the interaction of Number Theory with Abstract Algebra: that the finite number system $\mathbb{Z}_n$ is a field if and only if n is prime. It can be shown that a finite set with two binary operations satisfying the definition of a field can only exist if the number of elements is p^m where p is prime and $m \in \mathbb{N}$ – although with $m > 1$ it is not possible to represent the elements of such a field simply as numbers.

In Chapter 8 our study of quadratic extensions of $\mathbb{Z}$ and $\mathbb{Q}$ introduced some of the fundamentals of **algebraic number theory**.[i] The algebraic structures of particular interest in that chapter were rings. We saw that some rings may not have unique prime factorisation, but we did not resolve this issue. This can be done by taking the next step in the theory of rings, introducing **ideals** which broaden the concept of the set of multiples of a number. As with finite fields, deeper study of rings is best done by working with polynomials rather than numbers.

Rings and fields are structures with two binary operations; groups have only one operation, so have much wider applicability. All the groups we encountered were Abelian, until we came to the quaternions in which multiplication is not commutative; so the non-zero quaternions with the operation of multiplication constitute a non-Abelian group. Quaternions can be written as matrices, which are structures that also have non-commutative multiplication but commutative addition. Quaternions have also been used to represent rotations in three-dimensional space, which are not commutative; whereas rotations about a point in two-dimensional space do commute. Other geometrical symmetries, such as reflections, can be combined with rotations, and then the group of symmetries will not be Abelian, even in two dimensions. These are just a few of the contexts in which the group structure appears.

---

[i]As remarked by Ian Stewart and David Tall in the preface to their book on the subject, "Algebraic number theory" could mean the theory of algebraic numbers, or the study of number theory through the methods of [abstract] algebra; and in fact it has both meanings.

11.2   Analysis

Of the topics that would typically be taught in a course on Analysis, only Sequences (functions with domain $\mathbb{N}$) have been discussed in this book – and we have only included theorems on sequences that are needed for the development of our theory of number systems. There remains a considerably greater body of theory on sequences and series, in particular tests that can be used to determine whether a sequence or series converges.

We defined Analysis as the study of functions whose domain is infinite. Whereas an infinite set of Natural Numbers must have no upper bound, there is an infinite set of Real Numbers within a distance δ of any real number c, for any positive value of δ, however small. So Analysis in $\mathbb{R}$ is concerned with what happens in the neighbourhood of any number in the domain of a function, as well as what happens when the independent variable becomes unboundedly large (positive or negative). Thus we may want to find the limit of a function as the independent variable x approaches any value c in its domain as well as when $x \to \pm\infty$. Behaviour of a function near a point in its domain relates to its **continuity**, which was discussed briefly in Section 9.6. Analysis also deals with **differentiation**, the mathematical representation of rates of change; and with **integration**, which is the evaluation of quantities that can only be represented as the sum of infinitely many terms, each of infinitesimal size. Our definition of the principal value of the argument, interpreted geometrically as length along a circle, is a rather specialised example of the process of integration.

Beyond $\mathbb{R}$, Analysis reveals many elegant and unexpected results for functions with domains in $\mathbb{C}$. More generally, one may study functions of two or more real or complex variables, which then requires some measure of distance in a multi-dimensional space. The modulus provides this in $\mathbb{R}$ and $\mathbb{C}$, but in more general spaces there is a variety of ways to define a **metric** to fulfil this requirement; the rules that a metric must satisfy are that: it is symmetric (the distance from a to b must be the same as that from b to a); it is non-negative (and zero only when a and b are the same point); and the triangle inequality must be satisfied.

Ultimately, all of Analysis relies on the fundamental properties of number systems, particularly $\mathbb{R}$ and $\mathbb{C}$, which we have defined and derived in this book.

Appendix A

How to Read Proofs: The "Self-Explanation" Strategy

Prepared by Lara Alcock, Mark Hodds, Matthew Inglis,
Mathematics Education Centre, Loughborough University.

The "self-explanation" strategy has been found to enhance problem solving and comprehension in learners across a wide variety of academic subjects. It can help you to better understand mathematical proofs: in one recent research study students who had worked through these materials before reading a proof scored 30% higher than a control group on a subsequent proof comprehension test.

How to Self-Explain

To improve your understanding of a proof, there is a series of techniques you should apply.

After reading each line:

- Try to identify and elaborate the main ideas in the proof.

- Attempt to explain each line in terms of previous ideas. These may be ideas from the information in the proof, ideas from previous theorems/proofs, or ideas from your own prior knowledge of the topic area.

- Consider any questions that arise if new information contradicts your current understanding.

Before proceeding to the next line of the proof you should ask yourself the following:

- Do I understand the ideas used in that line?

- Do I understand why those ideas have been used?

- How do those ideas link to other ideas in the proof, other theorems, or prior knowledge that I may have?

- Does the self-explanation I have generated help to answer the questions that I am asking?

On the next page, you will find an example showing possible self-explanations generated by students when trying to understand a proof (the labels "(L1)" etc. in the proof indicate line numbers). Please read the example carefully in order to understand how to use this strategy in your own learning.

Example Self-Explanations

Theorem
No odd integer can be expressed as the sum of three even integers.

Proof

(L1) Assume, to the contrary, that there is an odd integer x, such that $x = a + b + c$, where a, b, and c are even integers.

(L2) Then $a = 2k, b = 2l$, and $c = 2p$, for some integers k, l, and p.

(L3) Thus $x = a + b + c = 2k + 2l + 2p = 2(k + l + p)$.

(L4) It follows that x is even; a contradiction.

(L5) Thus no odd integer can be expressed as the sum of three even integers.   □

After reading this proof, one reader made the following self-explanations:

- "This proof uses the technique of proof by contradiction."

- "Since a, b and c are even integers, we have to use the definition of an even integer, which is used in L2."

- "The proof then replaces a, b and c with their respective definitions in the formula for x."

- "The formula for x is then simplified and is shown to satisfy the definition of an even integer also; a contradiction."

- "Therefore, no odd integer can be expressed as the sum of three even integers."

Self-Explanation Compared with Other Comments

You must also be aware that the self-explanation strategy is not the same as *monitoring* or *paraphrasing*. These two methods will not help your learning to the same extent as self-explanation.

Paraphrasing

"a, b and c have to be positive or negative, even whole numbers".

There is no self-explanation in this statement. No additional information is added or linked. The reader merely uses different words to describe what is already represented in the text by the words "even integers". You should avoid using such paraphrasing during your own proof comprehension. Paraphrasing will not improve your understanding of the text as much as self-explanation will.

Monitoring

"OK, I understand that $2(k + l + p)$ is an even integer".

This statement simply shows the reader's thought process. It is not the same as self-explanation, because the student does not relate the sentence to additional information in the text or to prior knowledge. Please concentrate on self-explanation rather than monitoring.

A possible self-explanation of the same sentence would be:

"OK, $2(k + l + p)$ is an even integer because the sum of 3 integers is an integer and 2 times an integer is an even integer".

In this example the reader identifies and elaborates the main ideas in the text. They use information that has already been presented to understand the logic of the proof

This is the approach you should take after reading every line of a proof in order to improve your understanding of the material.

Practice Proof 1

Now read this short theorem and proof and self-explain each line, either in your head or by making notes on a piece of paper, using the advice from the preceding pages.

Theorem
There is no smallest positive real number.

Proof
Assume, to the contrary, that there exists a smallest positive real number.
Therefore, by assumption, there exists a real number r such that for every positive number s, $0 < r < s$.
Consider $m = \frac{r}{2}$.
Clearly, $0 < m < r$.
This is a contradiction because m is a positive real number that is smaller than r.
Thus there is no smallest positive real number.   □

Practice Proof 2

Here's another more complicated proof for practice. This time, a definition is provided too. Remember: use the self-explanation training after *every* line you read, either in your head or by writing on paper.

Definition
An *abundant* number is a positive integer n whose divisors add up to more than $2n$. For example, 12 is abundant because $1 + 2 + 3 + 4 + 6 + 12 > 24$.

Theorem
The product of two distinct primes is not abundant.

Proof
Let $n = p_1 p_2$, where p_1 and p_2 are distinct primes.
Assume that $2 \leq p_1$ and $3 \leq p_2$.
The divisors of n are $1, p_1, p_2$, and $p_1 p_2$.
Note that $\dfrac{p_1 + 1}{p_1 - 1}$ is a decreasing function of p_1.
So $\max\left\{\dfrac{p_1 + 1}{p_1 - 1}\right\} = \dfrac{2 + 1}{2 - 1} = 3$.
Hence $\dfrac{p_1 + 1}{p_1 - 1} \leq p_2$.
So $p_1 + 1 \leq p_1 p_2 - p_2$.
So $p_1 + 1 + p_2 \leq p_1 p_2$.
So $1 + p_1 + p_2 + p_1 p_2 \leq 2p_1 p_2$.   □

Remember...

Using the self-explanation strategy had been shown to substantially improve students' comprehension of mathematical proofs. Try to use it every time you read a proof in lectures, in your notes or in a book.

Bibliography

[1] M. Anderson and T. Feil. *A First Course in Abstract Algebra.* Chapman and Hall / CRC, 2005.

[2] J.H. Conway and R.K. Guy. *The Book of Numbers.* Copernicus, 1996.

[3] R. Courant and H. Robbins. *What is Mathematics?* Oxford University Press, 1978.

[4] V. Deaconu and D.C. Pfaff. *A Bridge to Higher Mathematics.* CRC Press, 2017.

[5] H.-D. Ebbinghaus, H. Hermes, F. Hirzebruch, M. Koecher, K. Mainzer, J. Neukirch, A. Prestel, and R. Remmert. *Numbers.* Springer, 1991.

[6] W.J. Gilbert and S.A. Vanstone. *An Introduction to Mathematical Thinking: Algebra and Number Systems.* Pearson Prentice Hall, 2005.

[7] G.H. Hardy and E.M. Wright. *An Introduction to the Theory of Numbers.* Oxford University Press, 1979.

[8] O.A. Ivanov. *Making Mathematics Come to Life.* American Mathematical Society, 2009.

[9] B.L. Johnston and F. Richman. *Numbers and Symmetry: An Introduction to Algebra.* CRC Press, 1997.

[10] T.W. Körner. *Where do Numbers Come From?* Cambridge University Press, 2019.

[11] E. Landau. *Foundations of Analysis.* Chelsea Publishing Company, 1960.

[12] C.H.C. Little, K.L. Teo, and B. van Brunt. *The Number Systems of Analysis.* World Scientific, 2003.

[13] I.K. Rana. *From Numbers to Analysis.* World Scientific, 1998.

[14] G.A. Spooner and R.L. Mentzer. *Introduction to Number Systems.* Prentice Hall, 1968.

[15] F.W. Stevenson. *Exploring the Real Numbers.* Prentice Hall, 2000.

[16] I. Stewart and D. Tall. *The Foundations of Mathematics.* Oxford University Press, 2015.

[17] J. Stillwell. *Elements of Number Theory.* Springer, 2002.

Index

Printed in the United States
by Baker & Taylor Publisher Services